Excel® for Chemists

Second Edition

Excel® for Chemists
A Comprehensive Guide

Second Edition

E. Joseph Billo
Department of Chemistry
Boston College
Chestnut Hill, Massachusetts

WILEY-VCH

New York • Chichester • Weinheim • Brisbane • Singapore • Toronto

For ordering and customer service, call 1-800-CALL-WILEY.

Library of Congress Cataloging-in-Publication Data:

Library of Congress Cataloging-in-Publication Data is available.
0-471-39462-9

Printed in the United States of America.

10 9 8 7 6 5 4 3 2 1

SUMMARY OF CONTENTS

CONTENTS

PREFACE

Since the publication of the first edition of this book in 1997, two new versions of Excel for the PC have appeared: Excel 97 and Excel 2000 (the corresponding Macintosh versions are Excel 98 and Excel 2001). This second edition of *Excel for Chemists* has been revised and updated, not only to take into account the changes that were made in Excel 97 and Excel 2000, but also to incorporate much new material.

The material concerning charts has been changed extensively to reflect the changes that were made to the ChartWizard. The chapters on programming with VBA have been revised, and the chapters on creating command macros and custom functions using VBA have been completely re-written.

There are three completely new chapters in this edition:

- Array formulas are now covered in depth in a separate chapter, rather than being discussed in the chapter on Excel formulas.

- Creating a worksheet with controls, such as option buttons, check boxes or a list box, is now covered in depth in a separate chapter.

- Using arrays in VBA is now covered in depth in a separate chapter.

In addition, an extensive list of shortcut keys — over 250 shortcut keys for PC or Macintosh — has been provided in the appendix.

Much of the material in this book has been incorporated in a course titled "Excel for Scientists and Engineers" that has been presented to over 1300 scientists in the past four years — not only chemists, but also scientists in many other disciplines. Many changes in this edition were made in light of the experience gained in teaching these courses.

January 2001

E. Joseph Billo
Department of Chemistry
Boston College
Chestnut Hill, Massachusetts

PREFACE TO THE FIRST EDITION

Most chemists deal with numbers on a daily basis. They record, calculate, summarize, graph, and report numerical data. Much of this work is done with the aid of a spreadsheet program on a personal computer. Many chemists use spreadsheet programs to *record* data in tabular form, but few have learned to take advantage of the tremendous *scientific calculating power* that is contained within the current versions of these programs. The aim of this book is to show you, a professional chemist, how to use the premier spreadsheet program, Microsoft Excel, to handle chemical calculations, from the relatively simple to the highly complex.

For example, you may need to

• calculate the percentages of carbon, hydrogen, nitrogen, oxygen, and other elements in a newly synthesized compound in order to compare the results of an elemental analysis with the theoretical values

• test various rate laws for a chemical reaction to see which equation best fits the observed data

• create a chart of the concentration of the acid-base forms of a new radiopharmaceutical as a function of pH, to illustrate the species distribution near pH 7

• resolve a UV spectrum into its individual Gaussian components in order to obtain the absorbance contribution of a shoulder peak

• apply linear regression to tensile strength data of polymer samples, to determine the effect of composition and molding conditions

• calculate a binding constant for a host-guest complex from the shift of NMR line position with changes in concentration of the guest molecule

• perform non-linear least-squares curve fitting to obtain the pK_a values of a polyprotic acid from a titration curve

Microsoft Excel can perform all these calculations, and more. You may have access to commercial software programs designed for some of these situations, but often you'll find that these programs don't handle the data you want to treat, or the model you want to fit, in exactly the right way. My purpose in writing this book is to demonstrate that it's relatively easy to "program" Excel to perform the calculations or other data manipulation needed for your specific application. Furthermore, if you use a range of commercial programs to perform data analysis, you'll have to learn (and remember) the commands and idiosyncrasies of each program.

This book is divided into four parts. Part I covers the basics of spreadsheet operations — entering data, cutting and pasting, formatting, creating charts, and so on. Part II shows how to use Excel's wide range of worksheet functions to perform sophisticated chemical calculations, how to create macros to automate spreadsheet tasks or to carry out repetitive calculations, and how to customize menus or toolbars to suit your own particular needs. Part III covers mathematical techniques that are particularly useful in a spreadsheet environment — matrix mathematics, numerical differentiation and integration, basic statistics, graphical and numerical methods of analysis — and shows how you can apply them easily using Excel. Part IV applies the techniques introduced in Parts I, II and III to a wide range of chemical problems.

The intent of this book is not simply to provide a series of templates that can be applied to particular situations (although there are lots of useful spreadsheet templates, macros and other tools on the disk that accompanies this book), but to show how you can create your own spreadsheets or macros to solve completely different chemical problems.

ACKNOWLEDGMENTS

Lev Zompa, University of Massachusetts-Boston, for spectrophotometric data used in Chapter 19.

Ross Kelly, Boston College, and Steve Bell, ICI Australia, for NMR data used in Chapter 20.

Allan D. Waren, Cleveland State University, for discussion about the Solver algorithms, and Edwin Straver, Frontline Systems Inc., for information about the inner workings of the Solver.

Dick Stein, University of Massachusetts-Amherst, and Stan Israel, University of Massachusetts-Lowell, for guidance on polymer databases.

Kavitha Srinivas, Boston College, for guidance about statistics.

Kenneth Kustin, Brandeis University, and Richard Haack, G. D. Searle Inc., Skokie IL, for reading the manuscript and offering helpful comments.

Barbara Goldman, executive editor, Camille Pecoul Carter, managing editor, Brenda Griffing, copy editor and Perry King, associate editor, electronic services, for their assistance and guidance during the publishing process.

My wife, Joanne, for encouragement and patience during the two years it took to write this book.

E. Joseph Billo
Chestnut Hill, Massachusetts

BEFORE YOU BEGIN

MACINTOSH AND WINDOWS VERSIONS OF EXCEL

This book is intended both for users of Excel for the Macintosh and for users of Excel for Windows. There are very few differences between the Mac and PC versions of Excel. I've tried to provide even-handed treatment to users of either type of computer. As you read through this book you'll see illustrations taken from Excel for the Macintosh and from Excel for Windows.

The small differences that *do* exist between Mac and Windows versions of Excel are mostly in the keystrokes that are used to perform some Excel operations. I've "piggybacked" these different instructions within a particular section. For example, in the sections on array formulas, you'll read "to enter an array formula, press COMMAND+ENTER (Macintosh) or CONTROL+SHIFT+ENTER (Windows)". These keystroke differences are also listed in Appendices E and F.

In the rare cases of instructions that are markedly different depending on whether you are using Excel for Windows or Excel for the Macintosh, I've placed those instructions in separate sections.

WHICH VERSION OF EXCEL ARE YOU USING?

This book is for users of Excel 2000 for Windows or Excel 2001 for Macintosh, as well as for those using Excel 97 (the previous version for PC users) or Excel 98 (the previous Macintosh version). The majority of worksheet functions, menu commands, toolbuttons and dialog boxes are identical or near-identical in all four versions. For the most part, you can follow the instructions no matter which version you're using. In a very few cases, you'll find instructions specifically for Excel 97/98 only.

TYPOGRAPHIC CONVENTIONS

As you read through this book, you'll see several different fonts and capitalization styles within the text. Here are the conventions that I've used.

- Names of keyboard keys are in ALL CAPS: TAB, SHIFT, CONTROL, OPTION, SHIFT, COMMAND, RETURN. (In Windows, the key is CTRL, but in this book CONTROL is used for both Windows and Macintosh.)

- Menu headings and menu commands are in boldface type: **File**, **Format**, **Delete**....

- Dialog box titles and options are in Title Case: "The Rename Sheet dialog box...", "... press Cancel".

- Occasionally, menu commands and dialog box options are combined for clarity and conciseness: "... use **Paste Special** (Values)...".

- Cell references are in Geneva font: "In cell A9 ...".

- Worksheet functions and macro functions are in Geneva: SUM, ACTIVATE.

- General (i.e., placeholder) arguments in functions or in text are in Geneva italic; required arguments are in bold italic: LINEST(*known_y's*, *known_x's*, *const*, *stats*).

- Specific arguments in functions or in text are in Geneva, not italic: ACTIVATE(SourceSheet), "... to copy the SourceSheet, you must ....".

- Visual Basic statements are in Geneva; VBA reserved words are bold: **For** Counter = Start **To** End **Step** Increment.

SPECIAL FEATURES IN THIS BOOK

This book has a number of features that you should find useful and helpful. There are over 50 **Excel Tips t**o simplify and improve the way you use Excel. For example:

*Excel Tip. To **Fill Down** a value or formula to the same row as an adjacent column of values, select the source cell and double-click on the Fill Handle.*

Throughout the book you'll see **"How-To" Boxes** that outline, in a clear and systematic manner, how to accomplish certain complex tasks. For example:

To Create a Chart with a Secondary Y Axis
(two different Y Axis scales and the same X Axis)

1. Select all data series to be plotted (the X Axis data series, two Y Axis data series).
2. Create an XY chart.
3. Click on the data series whose axis you want to change.
4. Choose **Selected Data Series...** from the **Format** menu and choose the Axis tab (see Figure 5-19).
5. Press the Secondary Axis button. A preview of the combination chart will be displayed. If the chart is suitable, press the OK button.

THE CD-ROM

The CD-ROM that accompanies this book contains most of the worksheets that are discussed in the book. The files are in Excel 97 format, so that they can be opened using either Excel 97/98 or Excel 2000/2001. The document names have .xls file extensions, so that they are compatible with Excel for Windows. Macintosh users can delete these file extensions if they wish.

The files on the CD-ROM are contained in the Excel for Chemists folder and are read-only. To work with a document and save the changes, you must first copy the files to your hard drive. If you are using a PC, you can run the INSTALL.EXE file on the CD and unzip the files to your hard drive. If you are a Macintosh user, copy the Excel for Chemists folder to your system.

If you have trouble, please contact John Wiley's tech support system at (212) 850-6753.

A complete list of all files on the CD-ROM, with short descriptions, is in Appendix G.

PART I

THE BASICS

1

WORKING WITH EXCEL

This chapter covers the basics of working with Excel: navigating around the worksheet, entering values and formulas, and formatting and editing a worksheet. If you are an experienced Excel user, you can probably skip this chapter; however, even experienced users may find a few useful tips in this chapter.

THE EXCEL DOCUMENT WINDOW

An Excel workbook is a *document* that appears in its own *document window*. Although you can have several workbooks open at the same time, and can see

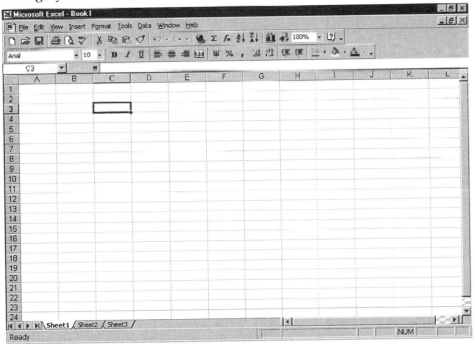

Figure 1-1. The Excel 2000 for Windows document window.

3

them all displayed on the screen simultaneously, only one workbook can be the *active workbook*. The default Excel 2000 workbook contains three worksheets; only one worksheet in the workbook can be the *active worksheet*.

An Excel worksheet consists of 256 columns (labeled A, B, C, ... IV) and 65,536 rows (labeled 1, 2, 3, ...). The rows and columns define *cells* (A1, H27, etc.), which constitute the worksheet. Information can be entered into a cell from the keyboard after the cell has been selected, usually with the mouse pointer. Data can also be entered into a cell, or many cells, by calculation. The Excel 2000 for Windows document window is shown in Figure 1-1. Depending on your monitor, your screen may show a different number of rows or columns.

Reading from the top down you'll see the *application title bar*, the *menu bar* (with **File, Edit, View,** etc. menus), the *Standard toolbar* (with New, Open and Save toolbuttons), the *Formatting toolbar* (with Bold, Italic and Alignment toolbuttons, for example), the *formula bar*, the rows and columns of cells, the *sheet tabs* and the *horizontal scroll bar* and, at the bottom, the *status bar*. The formula bar contains the Name Box or *cell reference area* (displaying the cell reference of the currently selected cell) and the editing area. As you enter values at the keyboard, they appear in the editing area of the formula bar. When you begin to type an entry, the Enter ☑ and Cancel ☒ buttons appear.

The Excel 2001 for the Macintosh document window is almost identical (although the menu bar and toolbars are somewhat different).

CHANGING WHAT EXCEL DISPLAYS

You can display or not display most components of the Excel window, such as menubars, scrollbars, the formula bar, gridlines, row and column headers.

To turn off the display of the scrollbars, the formula bar, gridlines, row and column headers, or to restore them if they are missing, choose **Options...** (Excel for Windows) or **Preferences...** (Excel for the Macintosh) from the **Tools** menu and choose the View tab; then check or uncheck the appropriate box.

Choose **Toolbars** from the **View** menu to display the submenu of available toolbars. The Standard and Formatting toolbars are the default toolbars, but you can display other toolbars by choosing them from the submenu. To learn more about customizing toolbars, see Chapter 19.

If you choose **Options...** from the **Tools** menu and choose the General tab, you can change several default settings that will apply to all future workbooks, such as the number of sheets in a new workbook. You can also switch from using A1-style references in formulas to R1C1-style references; the labels in the column header row of each worksheet change from A, B, C, ... to 1, 2, 3, ... Formulas using R1C1-style references will not be discussed in this book.

If you place the tip of the mouse pointer on one of the toolbuttons, a yellow ScreenTip box appears, describing the button's function. You can deactivate ScreenTips by choosing **Toolbars** from the **View** menu, then **Customize...** from the submenu to display the Customize dialog box. Choose the Options tab and de-select the Show ScreenTips On Toolbars check box.

MOVING OR RE-SIZING DOCUMENTS (EXCEL FOR WINDOWS)

To change the size of a workbook or worksheet, click and drag any of its borders or corners; the mouse pointer changes shape when you click on a border or corner. You can adjust the document to any size you desire. If you click on the Minimize button (the "underline" symbol in the upper right corner of the document) the document will be minimized so that only the title bar is visible. To restore it to its full size, click the Maximize button (the open square in the upper right corner of the title bar) .

To change the position of a document within the Excel window, click on the title bar and drag the document. It can even extend off-screen.

MOVING OR RE-SIZING DOCUMENTS (EXCEL FOR THE MACINTOSH)

To change the size of a workbook or worksheet, click and drag the lower right corner of the document. You can adjust the document to any size you desire. To restore it to its full size, click the Maximize button, the open square in the upper right corner of the title bar, or anywhere in the title bar.

To change the position of a document within the Excel window, click on the title bar and drag the document.

NAVIGATING AROUND THE WORKBOOK

The default Excel 2000 workbook contains three worksheets. If you want a workbook with more than three sheets, you can insert additional worksheets, or choose **Options...** from the **Tools** menu and choose the General tab, change the Sheets In New Workbook default, then create a new workbook.

To select a worksheet, simply click on the sheet tab. If the workbook contains a large number of worksheets, the tab for the sheet that you want to select may not be visible. Use the tab scroll buttons to the left of the sheet tabs to scroll through the sheet tabs. From left to right, these four buttons allow you to jump to the first sheet tab, scroll toward the first sheet tab, scroll toward the last sheet tab, or jump to the last sheet tab. When the desired sheet tab is visible, click on it.

Excel Tip.   *To display a shortcut menu that lists all sheets in the workbook, right-click on any of the tab scroll buttons (Excel for Windows) or hold down the CONTROL key and click on any of the tab scroll buttons (Excel for the Macintosh). You can then select the desired sheet.*

SELECTING MULTIPLE WORKSHEETS

To select multiple worksheets, select the sheet at one end of the range, move to other end of the range, hold down the SHIFT key and select the sheet at the other end of range. All the sheet tabs in the range will be selected.

To select all the sheets in a workbook, you can right-click on any sheet tab (Excel for Windows) or hold down the CONTROL key and click on any sheet tab (Excel for the Macintosh); this displays a shortcut menu that will allow you to Select All Sheets.

To make a non-adjacent selection (e.g., Sheet1 and Sheet3), hold down the CONTROL key (Windows) or the COMMAND key (Macintosh) while selecting.

CHANGING WORKSHEET NAMES

When you create a new workbook, the sheet tabs have the names Sheet1, Sheet2, etc. To rename a sheet, double-click on the sheet tab. The sheetname will be highlighted and you can enter a more descriptive name, as, for example, in Figure 1-2. Click outside the sheet tab to exit from edit mode.

| 26 | 90 | 139.6 | 292.3 | |
| 27 | 95 | 150.5 | 312.7 | |

◄ ◄ ► ►◄ \ UC Heat Transfer Data / Sheet2 /

Ready

Figure 1-2. Descriptive sheet names are helpful.

REARRANGING THE ORDER OF SHEETS IN A WORKBOOK

To move a sheet, just click and drag the sheet tab. The mousepointer shape becomes an icon showing a sheet at the end of the arrow pointer (Figure 1-3). An arrow above the sheet tab indicates where the copy will be inserted.

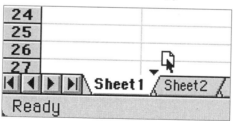

Figure 1-3. Moving a sheet tab.

To make a copy of a worksheet, hold down the CONTROL key (Excel for Windows) or the OPTION key (Excel for the Macintosh) while dragging the sheet tab. A small + sign appears in the icon.

Excel Tip. *To make copies of multiple sheets, select the sheets, begin dragging, then hold down the CONTROL key.*

You can also choose **Move or Copy Sheet...** from the **Edit** menu to move a worksheet. In addition to moving sheets within the active workbook, this menu command allows you to move one or more worksheets to another workbook.

NAVIGATING AROUND THE WORKSHEET

You can move around a worksheet either by means of the mouse or by using keystrokes.

Use the arrows in the *vertical* and *horizontal scroll bars* (the gray bars on the right edge and at the bottom of the window) to scroll through the worksheet. A single click of the mouse on an arrow moves the worksheet one row or column. The position of the *scroll box* (the white square in the gray bar) indicates the position of the window relative to the worksheet. You can also scroll through the worksheet by clicking on an arrow and holding down the mouse button, by dragging the scroll box with the mouse, or by clicking in the gray space on either side of the scroll box. Table 1-1 lists keystroke commands for cursor movement.

Table 1-1. Keys for Cursor Movement

Arrow keys	Move left, right, up, down one cell
RETURN	Move down one cell
TAB	Move right one cell
HOME	Move to the beginning of a row
END	Move to the end of a row
PAGE UP	Move to the top of the window
PAGE DOWN	Move to the bottom of the window
CONTROL+(arrow key)*	Move, in the direction of the arrow, to the end of a block of cells (a range of cells containing values and bounded by empty cells).

*On the Macintosh, use either CONTROL+(key) or COMMAND+(key).

SELECTING A RANGE OF CELLS ON THE WORKSHEET

You can select a range of cells on the worksheet in several ways:

- Click on the cell in one corner of the range, hold down the mouse button and drag to the cell in the opposite corner of the range. The range of cells will be highlighted. The size of the selection (e.g., 10R × 3C) is displayed in the Reference Area of the formula bar.

- Select the cell in one corner of the range, move to the cell in the other corner of the range, hold down the SHIFT key and select the cell in the opposite corner of the range. The range of cells will be highlighted.

- Select a complete row or column of cells by clicking on the row or column heading. The row or column will be highlighted.

SELECTING NON-ADJACENT RANGES

To select non-adjacent ranges, select the first range, then hold down the CONTROL key (Windows) or the COMMAND key (Macintosh) while selecting the second range. Both cell ranges will be highlighted (Figure 1-4).

	A	B	C	D
3				
4	t	[A]	[B]	[C]
5	0	1.0000	0.0000	0.0000
6	5	0.7788	0.2101	0.0111
7	10	0.6065	0.3537	0.0398
8	15	0.4724	0.4474	0.0802
9	20	0.3679	0.5041	0.1281
10	25	0.2865	0.5334	0.1801
11	30	0.2231	0.5428	0.2341
12	35	0.1738	0.5380	0.2882
13	40	0.1353	0.5233	0.3413
14	45	0.1054	0.5020	0.3927
15	50	0.0821	0.4763	0.4416
16				

Figure 1-4. Selecting non-adjacent ranges.

To extend the range of a cell selection you just made, hold down the SHIFT key, select the last cell in the selection and drag to include the additional cells. Alternatively, hold down the SHIFT key and use any of the arrow keys to extend the selection.

SELECTING A BLOCK OF CELLS

A *block of cells* is a range of cells containing values and bounded by empty cells. There are several ways to select cells within a block:

- Use CONTROL+SHIFT+(arrow key) to select in the appropriate direction.

- Select a cell at a boundary of the block (at the top, bottom or side of the block). Move the mouse pointer over the edge of the selected cell until the pointer changes to the arrow pointer (Figure 1-5 Left). Hold down the SHIFT key and double-click on the bottom edge of the selected cell to select all cells in the column from the top to the bottom of the block, as shown in Figure 1-5 Right. You can select cells from top to bottom, from bottom to top, from left to right or from right to left within a block. You can also select multiple columns or rows in the same way.

pKa (317 nm)	pKa (407 nm)
7.01	7.12
7.0	6.99
7.00	7.04
6.98	7.01
7.00	7.02
6.96	6.97
6.96	6.98
6.95	6.96

pKa (317 nm)	pKa (407 nm)
7.01	7.12
7.05	6.99
7.00	7.04
6.98	7.01
7.00	7.02
6.96	6.97
6.96	6.98
6.95	6.96

Figure 1-5. Using the Drag-and-Drop pointer to select a block of data. Left: selecting a cell edge. Right: selecting the block of data by double-clicking while holding down the SHIFT key.

ENTERING DATA IN A WORKSHEET

To enter a value in a worksheet cell, select the cell with the mouse pointer, which appears as a large open cross when it passes over cells. Clicking on the desired cell highlights it, indicating that this is the *active cell*, the cell in which you can now enter a value. As you type in a value, the characters appear in the formula bar and the active cell. You can complete the entry in several ways.

- Press the Enter button ☑ in the formula bar. The cell remains selected.
- Press the RETURN key (Macintosh) or the ENTER key (Windows). This moves the selection to the cell below (although you can change the default option so that the selection is not moved).

To cancel the entry and revert to the original contents of the cell, press the Cancel button ☒ or the ESC key.

Excel Tip.   *To enter the same value in a range of cells, select the range of cells, type the value, then press COMMAND+RETURN or CONTROL+ RETURN (Macintosh) or CONTROL+ ENTER (Windows).*

	A	B	C	D
1	Type	Entered	Displayed in Cell	Displayed in Formula Bar
2	percent	15%	15%	15%
3	scientific	2e-3	2.00E-03	0.002
4	currency	$50	$50	50
5	currency	$20000	$20,000	20000
6	debit	(5000)	-5000	-5000
7	fraction	2 5/8	2 5/8	2.625
8	date	7/4	4-Jul	7/4/2001*
9	date	8-3-38	8/3/38	8/3/1938
10	time	4:30	4:30	4:30:00 AM
11	time	16:00	16:00	4:00:00 PM
12	time	4 p	4:00 PM	4:00:00 PM
13			* enters current year	

Figure 1-6. Number formats recognized by Excel.

ENTERING NUMBERS

Excel has a remarkable ability to recognize the format of the value that you have entered: as a number, a percent, a debit value, as currency, in scientific notation, as a date or time, or even as a fraction. The number will be displayed in the cell in the proper format, but the number equivalent of the value will appear in the formula bar. Figure 1-6 illustrates number formats recognized by Excel.

If you enter a fraction less than 1, such as 1/3, it will be interpreted as a date ("3-Jan"). To prevent Excel from converting the fraction to a date, enter a zero and a space before the fraction (0 1/3). The zero indicates that the entry is a number, and the value will appear in the formula bar as 0.333333333333333.

HOW EXCEL STORES AND DISPLAYS NUMBERS

Excel can accept numbers in the range from $\pm$1E-307 to $\pm$9.99999999999999E+307.

Excel stores numbers with 15-significant-figure accuracy. These are displayed in the formula bar and used in all calculations, no matter what number formatting has been applied. Thus the fraction 1/3 appears in the formula bar as 0.333333333333333, and π as 3.14159265358979.

Excel switches between floating-point and scientific notation for best display of values. The formula bar can display numbers up to 21 characters, including the decimal point. Thus 1E-19 entered on the keyboard will appear as 0.0000000000000000001 (21 characters) in the formula bar, while 1E-20 will appear as 1E-20. Similarly, 1E20 appears as 100000000000000000000, while 1E21 appears as 1E21. Since a total of 21 characters can be displayed, the number of significant figures determines the magnitude of a number less than 1 that can be displayed in non-E format in the formula bar. Thus 1.2345E-15 appears as 0.0000000000000012345, while 1.23456E-15 is displayed as 1.23456E-15.

ENTERING TEXT

If you enter text characters (any character other than numbers, the decimal point, or the characters +, -, *, /, ^, $, %) in a cell, Excel will recognize the entry as text. For example, Chestnut Hill MA 02167-3860 is a text entry. A cell can hold up to 255 characters of text. You can distinguish text entries from number entries in the following way: in a cell that has not been alignment-formatted (e.g., left, centered, right, etc.), text entries are left-aligned, and numbers are right-aligned. Of course, if you format the alignment of a cell to be right-aligned, its value will be right-aligned whether the value is a number or text.

Sometimes it is necessary to enter a number or a date as a text value. To do this, begin the entry with a single quote.

ENTERING FORMULAS

Instead of entering a number in a cell, you can enter an equation (called a *formula* in Microsoft Excel) that will calculate and display a result. Usually formulas refer to the contents of other cells by using *cell references*, such as A2, a reference to a cell, or B5:B12, a reference to a range of cells. The value displayed in a cell containing a formula will be automatically updated if values elsewhere in the worksheet are changed. Formulas can contain values, arithmetic operators and other operators, cell references, the wide range of Excel's worksheet functions, and parentheses.

The rules for writing formulas (the *syntax*) are as follows:

- A formula must begin with the equal sign (=).
- The *arithmetic operators* are addition (+), subtraction (-), multiplication (*), division (/) and exponentiation (^). Other types of operator are described in Chapter 3.
- Parentheses are used in the usual algebraic fashion to prevent errors caused by the *hierarchy of arithmetic operations* (multiplication or division is performed before addition or subtraction, for example).

Some examples of simple formulas:

=A1+273.15	Adds 273.15 to the value in cell A1
=A2^2+13*A2-5	Evaluates the function $x^2 + 13x - 5$, where the value of x is stored in cell A2.
=SUM(B3:B47)	Sums the values contained in cells B3 through B47

=(-C3+SQRT(C3^2-4*C2*C4))/(2*C2)   Finds one of the roots of the quadratic equation whose co-efficients a, b and c are stored in cells C2, C3 and C4 respectively.

Excel formulas are discussed in much greater detail in Chapter 3.

Excel Tip. *Formulas that return the wrong result because of errors in the hierarchy of calculation are common. When in doubt, use parentheses.*

ADDING A TEXT BOX

You can add visible comments or other information to a worksheet by typing them into one or more worksheet cells. Another way to add comments, in a much more flexible form, is by using a *text box*.

To create a text box, press the Text Box toolbutton ⊞. The mouse pointer will change to a crosshair. Position the crosshair pointer where you want to place the text box, and click and drag to outline it (the text box can be moved and sized later). An empty text box will be displayed with a blinking text cursor. Type the desired text within the box.

Text box input has many features of a simple word processor: you can **Cut**, **Copy** or **Paste** text, make individual portions of text bold, italic or underlined, use different font styles, etc., as shown in Figure 1-7. The text within the box can be formatted with the Alignment toolbuttons or with the **Alignment** command.

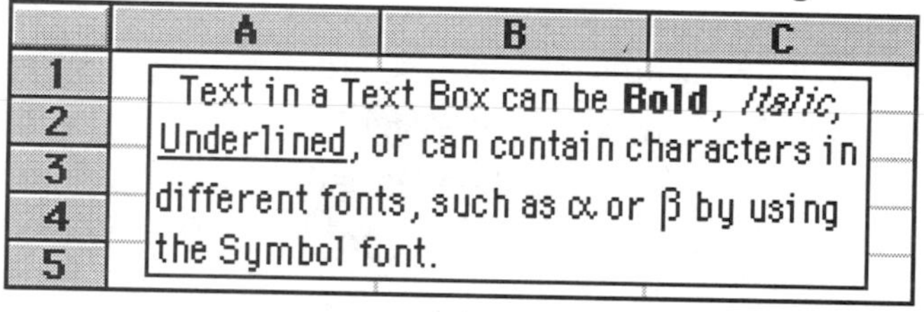

Figure 1-7. A text box.

To move a text box, click the mouse pointer anywhere within the text box and drag it to its new position. To re-size a text box, select it (black handles will appear), then place the mouse pointer over one of the black handles and click and drag to move the border of the box. If you hold down the CONTROL key while dragging, the text box will align with the cell gridlines.

ENTERING A CELL COMMENT

You can attach comments to a cell, for documentation purposes, in the form of a *comment*. A comment appears on the worksheet in a small box similar to a ToolTip. A small red triangle in the upper left corner of the cell indicates that the cell contains a comment. When the mouse pointer is moved over a cell that contains a cell comment, the cell comment appears.

To add a comment to a cell, choose **Comment...** from the **Insert** menu. Enter the text of the comment in the box (Figure 1-8), then press the Enter key (Windows) or Return key (Macintosh). To edit a comment, select the cell containing the comment, then choose **Edit Comment...** from the **Edit** menu. To delete a comment, select the cell containing the comment, then choose **Clear** from the **Edit** menu, and choose **Comments** from the submenu.

Comment indicators are not printed when you print a worksheet. You can turn screen display of comments and/or comment indicators on or off by choosing **Options** from the **Tools** menu, choosing the View tab and pressing the appropriate button in the Comments category.

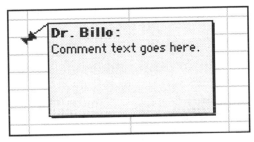

Figure 1-8. A cell comment.

EDITING CELL ENTRIES

You can edit cell entries in one of two ways — either in the formula bar or by using the Edit Directly In Cell feature. When you select a cell that contains an entry, the contents of the cell appear in the formula bar. As soon as you begin to enter a new value, the old value disappears. To make minor editing changes in the old entry, place the mouse pointer in the text at the point where you want to edit the entry. The mouse pointer becomes the vertical insertion-point cursor. You can now edit the text in the formula bar using the **Copy, Cut, Paste** or **Delete** commands or keys. Complete the entry using the Enter button in the formula bar, or by pressing the Enter key on the keyboard.

To use the Edit Directly In Cell feature, double-click on the cell. The text can now be edited in the cell in the same way as in the formula bar.

You can format individual characters in a cell using Bold, Italic, Underlined, etc., or with different fonts, by highlighting the character(s) in the formula bar, then applying the formatting.

EXCEL'S MENUS: AN OVERVIEW

In Excel 2000 for Windows, the Worksheet Menu Bar has the following pull-down menus: **File, Edit, View, Insert, Format, Tools, Data, Window** and **Help**. The **File, Edit, Format** and **Window** menus are discussed in this chapter. Commands in other menus will be discussed in later chapters.

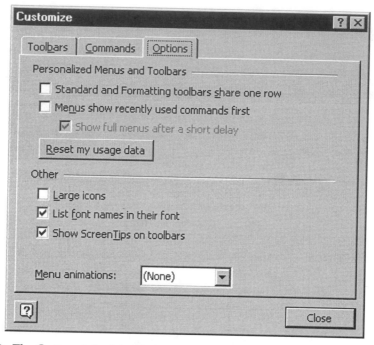

Figure 1-9. The Options tab of the Customize dialog box allows you to specify how toolbars and menus are displayed.

A significant change was made in the appearance of Excel's menus in Excel 2000. The default for menus is that they display recently used commands first, with a drop-down button at the bottom of the menu to display the remaining commands. If you prefer to work with "old style" menus, choose **Customize...** from the **Tools** menu, choose the Options tab and uncheck the Menus Show Recently Used Commands First box (Figure 1-9).

The default for toolbars is that the Standard and Formatting toolbars share one row. A button allows you to expand the toolbar to display the remaining toolbuttons. If you prefer to have "old style" toolbars, uncheck the Standard And Formatting Toolbars Share One Row box as well.

The way in which a command appears in a menu provides information about its form or availability:

- A menu command with an ellipsis (...), such as **Save As...**, indicates that the command opens a dialog box to obtain user input.

- Many Excel menus contain submenus, indicated by the ▶ symbol at the right edge of the menu.

- Some menu commands are dimmed (i.e., appear as gray characters) when the menu command is unavailable. Others appear on the menu only when they are available.

- Some menu commands change the text of their command depending on circumstances. For example, if you use **Comment** to add a comment to a cell, the command changes to **Edit Comment** so that you can edit the text of the comment.
- Some menu commands are preceded by a check mark if the choice has been selected previously. To remove the selection, depending on the command you either click on the check mark or select the command again.

SHORTCUT MENUS

Excel also provides "context-sensitive" *shortcut menus.* If you press the right mouse button (Windows) or press COMMAND+OPTION (Macintosh) while you select a worksheet element with the mouse pointer, a menu is displayed containing commands that apply to the selection. For example, if you select a column while holding down the right mouse button, a shortcut menu containing editing and formatting commands appropriate for a column appears.

MENU COMMANDS OR TOOLBUTTONS?

Many menu commands can be carried out by using toolbuttons. Toolbuttons are more convenient; they often combine a whole series of actions — menu selection plus dialog box options — into a single click of the mouse button.

Some buttons mentioned in this chapter don't appear on either the Standard or Formatting toolbar. To make them available for use, you can display other toolbars, or you can customize a toolbar (see Chapter 19).

OPENING, CLOSING AND SAVING DOCUMENTS

Most menu commands for managing documents are in the **File** menu. For the most part, the menu is similar to the **File** menu in other Windows or Macintosh applications, with **New...**, **Open...**, **Close**, **Save**, **Save As...**, **Page Setup...**, **Print Preview...**, **Print...** and **Exit** (Windows) or **Quit** (Macintosh) commands. The **Save Workspace...** command is specific to Excel.

OPENING OR CREATING WORKBOOKS

Use the **Open...** command to locate and open an existing document; use **New...** to create a new document. **New...** displays a dialog box in which you have a choice of opening either a new worksheet or any of the built-in or user-created template sheets.

To open an existing workbook or worksheet from the desktop, simply double-click on it. This will open the document (and will start Excel as well if it wasn't already running). If you start Excel first, it will open a new blank workbook.

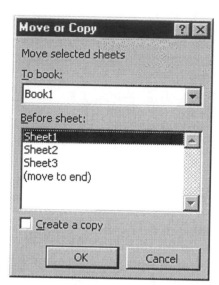

Figure 1-10. The Move or Copy Sheet dialog box.

USING MOVE OR COPY OR DELETE SHEET

The default Excel 2000 workbook contains three worksheets, but you can add or remove sheets. Three Excel commands permit you to add or remove sheets from a workbook. The **Delete Sheet** command in the **Edit** menu permanently removes the active sheet from the workbook. To add a worksheet to a workbook, use the **Worksheet** command in the **Insert** menu (you can also insert a chart sheet). To move or copy sheets within a workbook, or from one workbook to another, use the **Move or Copy Sheet...** command (Figure 1-10) in the **Edit** menu.

USING CLOSE OR EXIT/QUIT

You can **Close** a document either with the **Close** command from the **File** menu or, more conveniently, by using the Close button on the document title bar. You will be asked if you want to save changes.

If you hold down the SHIFT key while you pull down the **File** menu, the **Close** command becomes **Close All**. That way you can close all open Excel documents at once.

When you use the **Exit** command (Windows) or the **Quit** command (Macintosh), you close all open documents (you will be asked if you want to save changes) and then exit from Excel.

USING SAVE OR SAVE AS...

When you **Save** a newly created workbook, the **Save** dialog box will prompt you to assign a name to the document. Excel for Windows automatically appends a three-letter filename extension (e.g., .xls) to identify the file format type.

Earlier versions of Excel for Windows limited document names to a maximum of eight characters, and no spaces were allowed. In Excel 2000, document names can be much longer: the complete path to the file, including drive letter, server name, folder path, file name and the three-character file name extension, can contain up to 218 characters. File names can include spaces but not any of the following characters: slash (/), backslash (\), greater-than sign (>), less-than sign (<), asterisk (*), question mark (?), quotation mark ("), pipe symbol (|), colon (:), semicolon (;). Sheetnames can include most of the preceding characters, but can't include question mark (?), colon (:), backslash (\) or asterisk (*) (wildcard characters or file delimiters).

In Excel 2001 for the Macintosh, names can have up to 31 characters; upper- or lowercase, and spaces are allowed. Any keyboard character except the colon (:) can be used in a filename.

You can use **Save As...** to create a backup copy of a worksheet by giving the copy a different name.

THE TYPES OF EXCEL DOCUMENT

Excel uses a variety of file formats for its documents. Table 1-2 lists the filename extensions used by Excel for Windows. Excel for the Macintosh assigns a specific name or appends a descriptor to some, but not all, of the file types.

Table 1-2. Excel for Windows Filename Extensions and Macintosh Equivalents

Filename extension	File type	Macintosh equivalent
.XLA	Add-in macro	Appends "Add-In" to filename
.XLB	Toolbar configuration	Filename is "Excel Toolbars (n)"*
.XLC	Excel 4.0 chart	No identifier
.XLM	Excel 4.0 macro	No identifier
.XLS	Excel 4.0 worksheet	No identifier
.XLS	Excel workbook	No identifier
.XLT	Template	No identifier
.XLW	Workspace configuration	Proposes "RESUME" as filename

* n indicates Excel version.

USING SAVE WORKSPACE...

You can use **Save Workspace** if you have to interrupt your work, when you have several workbooks open at once. The **Save Workspace** command saves the current configuration of the workspace. Excel for the Macintosh proposes the filename RESUME. When you open the RESUME file, the workbooks will be restored to their former size and position on screen.

PRINTING DOCUMENTS

Menu commands for printing documents are located in the **File** menu.

USING PAGE SETUP...

The **Page Setup** dialog box contains four tabs: Page, Margins, Header/Footer, Sheet. Use the Page tab to choose Portrait or Landscape orientation (Figure 1-11). Use the Margins tab (Figure 1-12) to change margins; use the Header/Footer tab to select header or footer text from a list of built-in options or create a custom header or footer. Use the Sheet tab of the Page Setup dialog box to enter a Print Area (range of cells that will be printed) or Rows to repeat at top (row or rows that will be printed at the top of each printed page), or to turn on or turn off the printing of gridlines, row and column headings, etc.

In Excel 2000, the default header and footer, which will be printed on every page unless you choose otherwise, are *Sheetname* and *Page #*, respectively. To enter a different header or footer, choose the Header/Footer tab in the Page Setup dialog box, which will display list boxes with a wide range of built-in formats for header and footer. You can also create custom headers or footers by pressing the Custom Header... or Custom Footer... button. The Header or Footer dialog boxes are identical; each enables you to enter filename, sheetname, page number, date, time or other information.

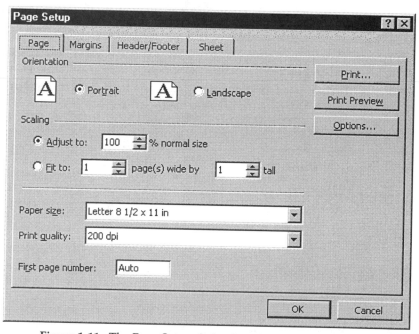

Figure 1-11. The Page Setup dialog box showing the Page tab.

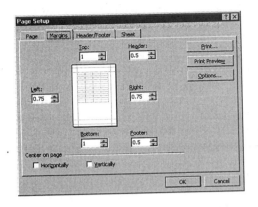 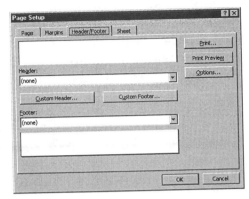

Figure 1-12. The Page Setup dialog box, showing (Left) the Margins tab and (Right) the Header/Footer tab.

To squeeze the maximum amount of worksheet information on a single page, you may want to decrease the margin widths. The default margin values are 0.75 inch left and right and 1 inch top and bottom. If you set the margins to zero, the header and footer information will still be printed, usually right on top of data in your worksheet, so delete the header and/or footer information by choosing "(none)" from the list box.

You can choose Print Row And Column Headings and/or Print Cell Gridlines by choosing the Sheet tab. If you de-select Cell Gridlines, they will still be displayed on the screen but they will not be printed. If you turn off screen display of gridlines by choosing the Display command from the Options menu (described later), the Cell Gridlines check box will also be cleared in the Page Setup dialog box and gridlines will not be printed.

You may need to use Print Black and White if your worksheet uses color. Colors may be printed as various patterns by your printer; to remove the patterns and produce text in cells in black and white, check the Print Black and White Cells box.

USING PRINT PREVIEW

Print Preview is useful in other ways besides showing what your finished worksheet will look like when printed. If you preview your worksheet and then return to the document window, page breaks will be displayed on the worksheet as dashed lines, to assist you in adjusting column widths, for example, before printing.

USING PRINT...

If you choose the **Print** command and simply press the OK button, Excel will print the rectangular array of sheets that includes all filled cells. It's a good idea to use **Print Preview** before printing; the total number of pages to be printed will be displayed in the status bar. This will tell you whether you can print the whole worksheet, or whether you need to specify a range of pages to be printed.

If you choose **Print Preview** or **Page Setup**, Excel displays the automatic page breaks as dashed lines in the worksheet. You can also insert a forced page break if you want to print a portion of a worksheet page. To insert a horizontal page break, select an entire row as if you were going to insert a row. Then choose **Page Break** from the **Insert** menu. The page break will be inserted immediately above the row selected. Excel displays forced page breaks as dashed lines that are heavier than the dashed lines used to indicate automatic page breaks. A forced vertical page break is inserted in a similar fashion; the page break is inserted immediately to the left of the selected column. If you want to insert both a vertical and a horizontal page break, select a single cell within the worksheet; the page breaks will be immediately above and to the left of the cell.

PRINTING A SELECTED RANGE OF CELLS IN A WORKSHEET

To print a selected range of cells within a worksheet, first select (highlight) the range to be printed, then choose **Print...** from the **File** menu. Then press the Selection button in the Print What category box in the lower left corner of the dialog box. The selected range will be printed, but the selection will not be "remembered" the next time you print.

To set a range of cells to be printed each time you choose Print, you must use **Set Print Area**. You can do this in at least three different ways:

- Choose **Page Setup** from the **File** menu, and choose the Sheet tab. Click in the Print Area text box to select it. Now select the range of cells that you want to print (you can move the dialog box out of the way if necessary), and press the OK button. To cancel a Print Area selection, delete the text within the Print Area text box.

- First, select the range of cells to be printed, then choose **Print Area** from the **File** menu, and choose **Set Print Area** from the submenu. The range to be printed will be indicated by Page Break lines. Choose **Remove Print Area** from the submenu to cancel the Print Area.

- Use the Set Print Area toolbutton 📧 . In Chapter 19 you'll learn how to install the Set Print Area Toolbutton on a toolbar.

If the Print Area you selected requires more than one page, you can choose **Page Setup** and change the value in the Reduce/Enlarge box to less than 100%. Sheets printed with values less than about 60% are difficult to read, though. To obtain the appropriate reduction value automatically, after you've selected the area to be printed, choose the Page tab and press the Fit To 1 Pages Wide By 1 Tall button.

PRINTING ROW OR COLUMN HEADINGS
FOR A MULTI-PAGE WORKSHEET

If you are printing a multi-page worksheet, you can duplicate row or column headings automatically on each printed page. Choose **Page Setup...** from the **File** menu, and choose the Sheet tab, as in Figure 1-12 (Right). Select the Rows To Repeat At Top or the Columns To Repeat At Left text box by clicking the cursor in it. Now select the range of cells that you want to print on every page as a title (you can move the dialog box out of the way if necessary). Then click the OK button. The headings will appear at the top or left of each printed page.

EDITING A WORKSHEET

You can use commands in the **Edit** menu to modify the arrangement of values in a worksheet.

INSERTING OR DELETING ROWS OR COLUMNS

To insert an entire row of blank cells, click on a row heading, which selects an entire row. Then choose **Rows** from the **Insert** menu. A new row will be inserted above the row you selected. (Insertion occurs above or to the left of the selected row or column, respectively.) Multiple rows or columns can be inserted in a similar fashion, by selecting as many rows or columns as you want to insert.

To insert an extra cell or cells within a row or column, select the cell range above or to the left of which you want to insert cells, then choose **Insert...**. (Note that **Insert** has become **Insert...** because the Insert dialog box will be displayed.) Excel usually makes a pretty good guess whether the cells should be shifted to the right or down to make the proper insertion, but always check to make sure. Then click OK in the dialog box (Figure 1-13).

When inserting partial rows or columns, take care that other parts of the worksheet do not become mis-aligned.

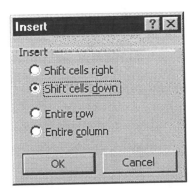

Figure 1-13. The Insert Cells... dialog box.

Complete rows or columns are deleted in a similar manner. Click on the row or column header to select the rows or columns to be deleted, then choose **Delete** from the **Edit** menu. To delete cells *within* a row or column, select them with the mouse, then choose **Delete...**. You will be asked whether the cells should be moved up or to the left. When deleting partial rows or columns, take care that other parts of the worksheet do not become mis-aligned.

USING CUT, COPY AND PASTE

Single cells, ranges of cells, or whole rows or columns can be copied or cut from the worksheet and inserted or pasted into other locations. In general the destination must be the same size as the copied or cut cells. First, select the cell or range of cells that you wish to copy or cut. Then choose **Copy** or **Cut** from the **Edit** menu, or press the Copy 📄 or Cut ✂ toolbutton, or press CONTROL+C (Copy) or CONTROL+X (Cut). A *marquee* (a dashed line) will appear around the selected cells and a copy of the cells is placed on the Clipboard. Next, select the destination range. You can now transfer the copy to the destination by choosing **Paste** from the **Edit** menu, by pressing the Paste toolbutton 📋 or by pressing CONTROL +V.

Excel Tip. *It's much better to select a single cell, then **Paste**, instead of selecting a destination range that is the same size as the copied or cut range. The selected cell will be the upper left corner of the pasted range of cells.*

You can also **Copy** or **Cut** text in the formula bar and **Paste** it in a worksheet cell. Select the text to be copied or cut, then press the Copy or Cut toolbutton or choose the appropriate command from the **Edit** menu. Complete the operation by clicking the Enter button in the formula bar. Then **Paste** in the desired cell.

Excel Tip. *Use the ESC key to cancel a **Copy** or **Cut** operation.*

USING PASTE SPECIAL...

When you **Copy** a cell and **Paste** it, Excel transfers the cell's contents, format and any comment. You can choose to transfer only some of these cell attributes by using **Paste Special...**. The Paste Special dialog box (Figure 1-14) permits you to paste only Formulas, Formats or Comments. In addition, you can convert formulas to constants by choosing Values.

Excel Tip. *You can use the Paste Values 🔢 toolbutton instead of the **Paste Special...** menu command. See Chapter 19 for instructions on how to make this button, and many others, available.*

If you press one of the Operation buttons in the Paste Special dialog box, the value in the destination cell will be added to, subtracted from, multiplied by or divided by the value in the copied cell.

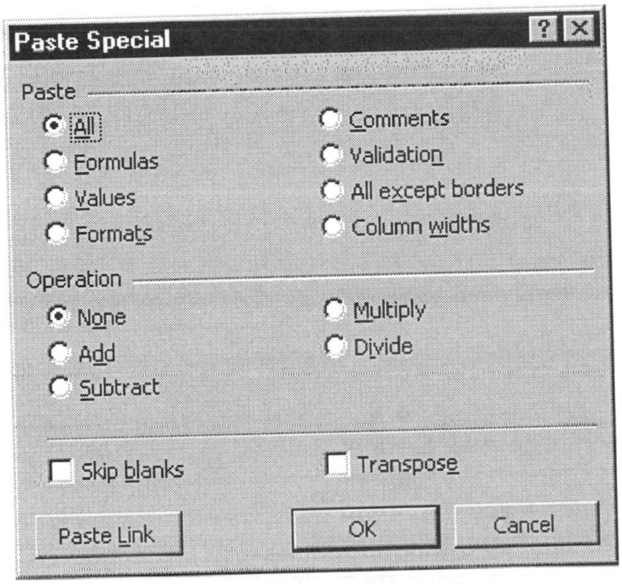

Figure 1-14. The Paste Special dialog box.

If the cell in either the source or the destination contains a formula, then the formula will be enclosed in parentheses and joined to the contents of the destination cell by the arithmetic operator. Relative references in the source will be changed in the same way as in a normal **Paste** operation. You can also **Copy** cells that contain formulas and press both the Values button and one of the Operation buttons to either Add, Subtract, Multiply or Divide.

If you check the Skip Blanks check box, only non-blank cells in the source will be pasted.

USING PASTE SPECIAL TO TRANSPOSE ROWS AND COLUMNS

If values in the source range are arranged in rows, you can convert the data to column format, or vice versa, as shown in Figure 1-15.

First **Copy** the cells, then select a cell or range in which you want the transposed values to be placed. Choose **Paste Special** and check the Transpose box, then press OK.

	A	B	C	D
1	pH	6.00	6.20	etc.
2	Abs	0.903	0.861	etc.

	F	G
1	pH	Abs
2	6.00	0.903
3	6.20	0.861
4	etc.	etc.

Figure 1-15. Rows and columns transposed. (Left) Before using and (Right) after using **Paste Special** (Transpose).

USING CLEAR

When you choose **Clear** from the **Edit** menu, a submenu is displayed with All, Formats, Contents, Comments. If you choose Formats from the submenu, for example, you can remove only formats from selected cells. Choosing Contents will delete the cell value but not the format.

Excel Tip. *The easiest way to clear a range of cells is to use the Erase tool* ⬧.

To remove only formatting from a cell, use the Clear Formats button ⬧. *See Chapter 19 for instructions on how to make these buttons available.*

USING INSERT

If you **Copy** a cell or range, then select a cell, range, row or column to specifiy the destination, then click on the **Insert** menu, the **Cells...** menu command becomes **Copied Cells....** A dialog box will ask whether you want to shift cells down or to the right. If you decide that it's easier to insert new rows or columns first, then **Copy** and **Paste**, you'll have to clear the clipboard before **Copied Cells...** becomes **Cells....** You can do this either (i) by selecting an empty cell and pressing DELETE, (ii) by choosing **Clear...** from the **Edit** menu, then exiting from the dialog box by pressing the Cancel button or, most conveniently, (iii) by pressing ESC.

TO COPY, CUT OR PASTE USING DRAG-AND-DROP EDITING

You can also **Copy**, **Cut** or **Paste** using Excel's "Drag-and-Drop" method. With this method you **Cut** and **Paste** a selection by using only the mouse pointer.

To Cut and Paste by Using Drag-and-Drop

1. Select the range of cells to be moved.
2. Position the mouse pointer over a border of the selection (top, bottom or side). The mouse pointer will change to an arrow.
3. Drag the selection toward the desired position. The border of the selection will be indicated as you drag it (Figure 1-16).
4. Position the selection as desired and release the mouse button.

To **Copy** the selection instead of cutting, hold down CONTROL (Windows) or OPTION (Macintosh) while dragging. A small plus sign will appear near the arrow pointer.

To **Insert** the selection, hold down the SHIFT key while dragging. The insertion point of the selection will be indicated by a horizontal or vertical bar as you drag (Figure 1-17).

	A	B	C	D
1	t, sec	A(obsd)		
2	0.2	0.0047		
3	0.6	0.0129		
4	1.0	0.0163		
5	1.4	0.0188		
6	1.8	0.0201		
7				
8				
9				
10				
11				
12				
13				
14				
15				

Figure 1-16. Cutting and Pasting cells using Drag-and-Drop editing.

	A	B	C	D
1	t, sec	A(obsd)		
2	0.2	0.0047		
3	0.6	0.0129		
4	1.0	0.0163		
5	1.4	0.0188		
6	1.8	0.0201		
7				
8				
9				
10				
11				

Figure 1-17. Inserting cells using SHIFT+ Drag-and-Drop.

Excel Tip. To use this method, Drag and Drop must be "turned on". Choose **Options** *from the* **Tools** *menu, choose the Edit tab and check the Allow Cell Drag And Drop box.*

DUPLICATING VALUES OR FORMULAS IN A RANGE OF CELLS

To duplicate a value or formula in one cell into a range of cells, highlight the cell whose value you want to duplicate, plus cells below or to the right of where you want the value duplicated. Then choose **Fill** from the **Edit** menu and choose **Down, Right, Up** or **Left** from the submenu.

If the cell contains a number or a text label, the value will be duplicated in the rest of the cells. If the cell contains a formula, the formula will be copied into the selected cells, except that Microsoft Excel uses *relative referencing* when formulas are copied. For example, if cell A2 contains the formula =A1+1, and **Fill Down** is used to copy the formula into a range of cells below cell A2, the formula copied into cell A3 will be =A2+1, and so on.

Cell references are adjusted when you **Insert** or **Delete** rows or columns, too. If you **Insert** a new column to the left of column A in the preceding example, the formula in cell B2, which used to be cell A2, will read =B1+1.

To use the **Across Worksheets** option in the **Fill** submenu, you must have selected multiple sheets beforehand. Select adjacent or non-adjacent sheets in the same way you select cells. To select an adjacent range of worksheets click on the tab of the first sheet, then hold down the SHIFT key and click on the tab of the last sheet in the range. To select non-adjacent sheets, hold down the CONTROL key (Windows) or the COMMAND key (Macintosh) while selecting. When you choose **Across Worksheets** from the submenu, the Fill Across Worksheets dialog box (Figure 1-18) will appear; you can choose to fill Contents, Formats or both.

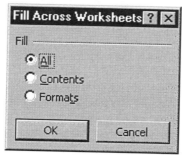

Figure 1-18. The Fill Across Worksheets dialog box.

ABSOLUTE, RELATIVE AND MIXED REFERENCES

To keep the address of a cell fixed when you use the **Fill** commands, precede both its letter and number designation by a dollar sign (e.g., $B$1). Thus the formula =A1+$B$1 in cell A2, when filled down, yields =A9+$B$1 in cell A10 (Figure 1-19). You will find this *absolute cell addressing* useful if you wish to use numerical constants in formulas. Occasionally it is useful to use *mixed references*. A *relative reference* in a formula, such as A1, becomes A2, A3, etc., as you **Fill Down** a formula into cells below the original formula. An *absolute reference* such as $A$1 remains $A$1 as you **Fill Down**. A *mixed reference* is a reference such as A$1 or $A1; the row or the column designation, respectively, will remain constant when you **Fill Down** or **Fill Right**.

	A	B
1	1	0.001
2	=A1+$B$1	
3	=A2+$B$1	
4	=A3+$B$1	
5	=A4+$B$1	
6	=A5+$B$1	
7	=A6+$B$1	
8	=A7+$B$1	
9	=A8+$B$1	
10	=A9+$B$1	

	A	B
1	1	0.001
2	1.001	
3	1.002	
4	1.003	
5	1.004	
6	1.005	
7	1.006	
8	1.007	
9	1.008	
10	1.009	

Figure 1-19. Two views of the same worksheet, showing formulas (left) and values (right). The formula in cell A2 has been filled down into A3:A10.

RELATIVE REFERENCES WHEN USING COPY AND CUT

If you **Copy** and **Paste** a formula, its references will be transferred using relative referencing. Thus, if you **Copy** the formula =A1+1 from cell A2 and **Paste** it in cell A10, the formula in cell A10 will be =A9+1. If you **Copy** the formula from cell A2 and **Paste** it in cell C2, the formula in cell C2 will be =C1+1. (This is probably not the formula you want.)

On the other hand, if you **Cut** the formula in cell A2 and **Paste** it anywhere in the worksheet, it will still be the formula =A1+1.

Thus the difference (with respect to cell references) between **Copy** and **Paste** and **Cut** and **Paste** is that **Cut** adjusts relative references so that they still refer to the original cells, while **Copy** does not adjust relative references, with the result that they refer to different cells.

The best way to copy a formula to a different row and column without altering relative references is to **Copy** it from the formula bar, click the Enter box to complete the **Copy** operation, then **Paste** in the destination cell.

USING AUTOFILL TO FILL DOWN OR FILL RIGHT

Excel's AutoFill feature lets you **Fill Down** or **Fill Right** simply by using the mouse pointer. To use AutoFill, select a cell by clicking on it. You will see a small black square on the lower right corner of the selected cell. Position the mouse pointer exactly over the small black square (the *fill handle*, or AutoFill handle). The mouse pointer becomes a small black cross. Click and drag in the usual way to select a range of cells. If the cell contains a formula, it will be duplicated to the rest of the range just as if you had used **Fill Down** or **Fill Right**. If the cell contains a number or a text label, the value will be duplicated in the rest of the cells. With AutoFill you can also **Fill Up** and **Fill Left**.

Excel Tip. To **Fill Down** *a value or formula to the same row as an adjacent column of values, select the source cell and double-click on the fill handle.*

USING AUTOFILL TO CREATE A SERIES

AutoFill provides an additional feature: you can use it to create a series. There are three ways to create a series. For example, to create the series 1, 2, 3, ... in column A, you can either:

- Enter the value 1 in cell A1, enter the formula =A1+1 in cell A2 and then use **Fill Down** to create the series. You can then use **Copy** and **Paste Special** (Values) to convert the formulas to values.

- Use **Series**... from the **Fill** submenu of the **Edit** menu. With **Series** (Figure 1-20) you enter the start value, the end value and the increment.

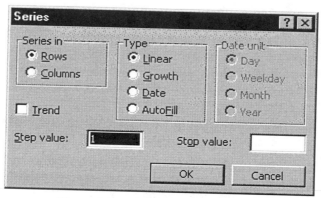

Figure 1-20. The Series dialog box.

- Use AutoFill. This is by far the simplest, most convenient method. If you select a cell containing a number formatted as a date or time, or a text label containing a number, and use AutoFill to fill a range of cells, AutoFill creates a series using the selected cell as the starting value. For example, if you select a cell displaying 30-Jan and use AutoFill to **Fill Right**, the series 30-Jan 31-Jan 1-Feb 2-Feb will be created. The value of the series being entered in the active cell is displayed in the Reference box of the formula bar as you move the AutoFill handle. If you select two or more cells, AutoFill will create a series based on the cells you select, as shown in the second and third examples of Figure 1-21.

	A
1	January

	B
1	1
2	3

	C
1	Compound 1
2	Experimental
3	Calculated
4	

	A
1	January
2	February
3	March
4	April
5	May
6	June
7	July
8	August
9	September
10	October
11	November
12	December

	B
1	1
2	3
3	5
4	7
5	9
6	11
7	13
8	15
9	17
10	19
11	21
12	23

	C
1	Compound 1
2	Experimental
3	Calculated
4	
5	Compound 2
6	Experimental
7	Calculated
8	
9	Compound 3
10	Experimental
11	Calculated
12	

Figure 1-21. Some examples of the use of AutoFill to produce a series. (Above) cells selected for, and (Below) series produced by AutoFill.

Excel Tip.   To use AutoFill to fill a cell range by copying the same cell contents to all the cells in the range (in other words, to prevent AutoFill from creating a series), hold down the CONTROL key (Windows) or COMMAND key (Macintosh) as you position the black cross pointer over the fill handle. A small plus sign will appear to the right of the black cross pointer. Click and drag in the usual way to fill rather than create a series.

FORMATTING WORKSHEETS

You can use commands from the **Format** menu to change and improve the appearance of the worksheet and to modify the way number values are displayed.

USING COLUMN WIDTH... AND ROW HEIGHT...

When Excel creates a blank worksheet, all rows are the same default height, all columns the same width. You can change the width of columns, or the height of rows, to improve the appearance of a worksheet or to eliminate wasted space so that you can get more information on a single page. You can also hide rows or columns by reducing their height or width to zero. The data they contain will still be still there, but it will be hidden.

To change the width of a cell or column, choose **Column** from the **Format** menu, then choose **Width** from the submenu. You can enter the new width of the column; one unit corresponds to the width of one character of the current font. Column widths can also be changed by choosing **Autofit Selection** from the submenu. You can adjust the column width to fit a single selected cell or to be the best fit to the widest entry in a whole column.

You can also change the column width by using the mouse pointer. Place the cursor on the separator bar between column headings, on the right of the column whose width you want to change. The cursor changes to a double-headed arrow: ✛. Click the mouse button and drag to the right or left to change column width. The column width is displayed in an "InfoBox" as you drag the separator bar.

> *Excel Tip.* To adjust several columns at a time to the same width, select the columns, thenperform the column width adjustment with the mouse pointer on any of the selected columns. When you release the mouse button, all the columns will have the adjusted width. You can also get a "best fit" simply by double-clicking on the row or column separator. To adjust several rows or columns at once, select the columns, then double-click on any row or column separator.

USING ALIGNMENT...

The **Alignment** command provides a number of formatting options for the alignment of values in cells. Choose **Cells...** from the **Format** menu and choose the Alignment tab. There are option buttons for both horizontal and vertical alignment (Figure 1-22). The Vertical orientation options are useful if you want

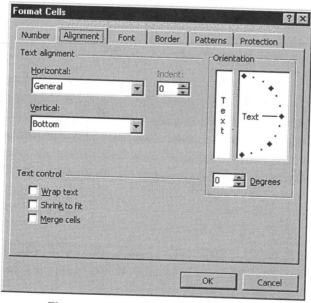

Figure 1-22 The Alignment dialog box.

to add a text label to a narrow data column. The Orientation "inclinometer" allows you to display text on an angle.

To use the Merge Cells option, select the columns across which you want the text to be centered, as indicated in Figure 1-23. (The text can be in any of the selected cells.)

	A	B	C	D	E
1	Centered across selected columns				
2					

Figure 1-23. Using Merge and Center .

Alternatively, for the most common horizontal alignment options you can use the alignment toolbuttons ▤ ▤ ▤ ▦ to format cells left, centered, right or centered across selected cells.

You can also format text entries in cells so that the text wraps and is displayed in more than one line (Figure 1-24). Excel breaks the text at a space character. Text can be both aligned vertically and wrapped.

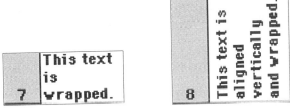

Figure 1-24. Examples of using Wrap Text

When you select a text box, the Alignment dialog box has a slightly different form. The Automatic Size check box, when selected, fits the borders to the text.

USING FONT...

The Font tab in the Format Cells dialog box allows you to format individual cells in any of the installed fonts (Figure 1-25). In addition, you can format individual characters in various font styles or sizes, or as strikethrough, superscript or subscript characters. You can also insert Greek letters in text labels, as shown in Figure 1-7, by using the Symbol font.

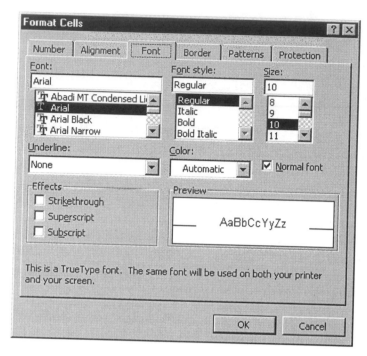

Figure 1-25. The Font dialog box.

THE ALTERNATE CHARACTER SET

There is another way to enter some useful characters. From the keyboard, you can enter symbols that are in the so-called *alternate character set*. In Excel for the Macintosh these characters are obtained by pressing OPTION+(key) or OPTION+SHIFT+(key). The characters produced may be different for each different font. The ones shown in Figure 1-26 are obtained with the Geneva font, the default font for Excel for the Macintosh.

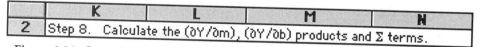

Figure 1-26. Special characters in Excel for the Macintosh by using the OPTION key.

In Excel for Windows, the characters are obtained by holding down the ALT key and typing the four-digit ASCII code for the character, *using the numeric key-pad*. The range of useful characters obtainable in this way is rather limited. Table 1-5 gives some characters for Macintosh; for characters obtained using Arial, one standard font for Excel 2000 for Windows, use the ASCII codes shown in parentheses in Table 1-5.

Table 1-5. Some Useful Alternate Characters*

ß	OPTION+s	(0223)	∂	OPTION+d		
Δ	OPTION+j		μ	OPTION+m	(0181)	
π	OPTION+p		Π	OPTION+P		
Σ	OPTION+w		Ω	OPTION+z		
÷	OPTION+/	(0247)	±	OPTION++ (plus)	(0177)	
≈	OPTION+x		≠	OPTION+=		
«	OPTION+\	(0171)	»	OPTION+SHIFT+\	(0187)	
≤	OPTION+, (comma)		≥	OPTION+. (period)		
√	OPTION+v		∫	OPTION+b		
∞	OPTION+5		Å	OPTION+A	(0197)	
°	OPTION+SHIFT+8	(0176)	·	OPTION+SHIFT+9	(0225)	

*For Macintosh. Four-digit ASCII codes in parentheses are for Windows.

If you use a different font, you'll have to experiment to see what alternate characters are produced.

ENTERING SUBSCRIPTS AND SUPERSCRIPTS

You can enter subscripts and superscripts in text, as in Figure 1-27. First, highlight the characters to be subscripted or superscripted. Choose the Font tab in the Format Cells dialog box, check the Subscript or Superscript check box (see Figure 1-25), press the OK button, then enter the text by pressing the ENTER key or the Enter button.

	A
1	Chemical formulas with subscripts
2	$C_{12}H_{22}O_{11}$
3	$BaCl_2 \cdot 2H_2O$
4	$H_2SO_4 + 2NaOH = Na_2SO_4 + 2H_2O$

Figure 1-27. Subscripts in Excel.

USING BORDER AND PATTERNS

The Border tab in the Format Cells dialog box allows you to place a border around one or more sides of a selected cell or range. This is useful if you want to emphasize comments, instructions or values. The Patterns tab is used to change the background color or pattern of cells.

The Border option is often used to underline headings, or in a sheet in which the gridlines have been removed, to create a custom form. Figure 1-28 illustrates

BOSTON COLLEGE					DATE	6/5/1996		

OFFICE OF THE UNIVERSITY REGISTRAR
CHESTNUT HILL, MASSACHUSETTS 02167

ACADEMIC YEAR 1996

SEMESTER 96S

COURSE   CH 222 01   INTRO/INORGANIC CHEM

INSTRUCTOR   BILLO, E JOSEPH

MEETING PLACE

STUDENT'S NAME	ID NUMBER	SCHOOL	CLASS	MAJOR	CREDIT	GRADE

Figure 1-28. Using Border to create a custom report form.

a portion of a template sheet produced in this way. To remove gridlines, choose **Options** from the **Tools** menu and choose the View tab, then uncheck the Gridlines check box.

The built-in template sheets provided with Excel 2000 ("Spreadsheet Solutions") are good examples of the use of Border and Patterns to create custom forms.

USING THE FORMAT PAINTER TOOLBUTTON

The Format Painter toolbutton ✎ copies and pastes formats from one cell or range to another cell or range. To use it, do the following:

- Select the range with the desired format(s).
- Click the Format Painter toolbutton (this copies the format).
- Click on a cell or drag across a range of cells to paste the format(s).

Excel Tip. To use the the Format Painter button to "paint" a format on a series of non-adjacent cells or ranges, select the cell with the desired format, then double-click on the Format Painter button. This will keep the button in the "pressed" position, allowing you to click on several cells or ranges to paste the format. When you're done, click once on the button to return it to the "unpressed" state.

NUMBER FORMATTING

The formatting described in the preceding sections — bold text, italic text, alignment of text, adjusting column widths, etc. — is sometimes referred to as *stylistic formatting*. In addition, it is possible to change the way number values are displayed in cells. This type of formatting, called *number formatting*, is described in the following sections.

USING EXCEL'S BUILT-IN NUMBER FORMATS

To change the way a number is displayed in a cell, choose **Cells...** from the **Format** menu and choose the Number tab to display the number format categories — Number, Currency, Date, Time, Percentage, Scientific, etc. For example, selecting a cell and then choosing Number from the list of categories will display the value in the cell to two decimal places (the default); you can change the number of displayed decimal places using the Decimal Places box. You can also display values as percentages or in exponential notation, as indicated in Figure 1-29.

If you choose the Custom category you will see the number formatting code that was applied to the cell. If you scroll through the list of codes, you'll see that many of them are quite complex (Figure 1-30). For the meaning of the built-in number format code symbols, see Table 1-3 or go to "Number Format Codes" in Excel's On-Line Help.

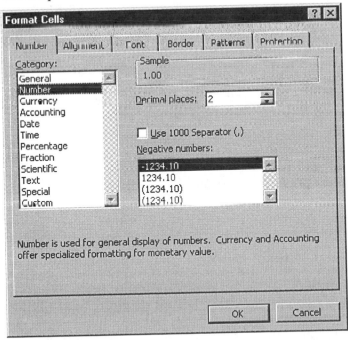

Figure 1-29. The Number Format dialog box.

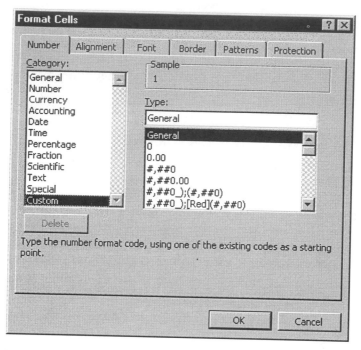

Figure 1-30. The Number Format dialog box showing number formatting codes.

CUSTOM NUMBER FORMATS

You can create custom number formats. First, choose the Custom category from the list of numbers formats. This will display the list of (so far) built-in number formats. To add a new, user-defined number format (it will be added at the bottom of the list), type the format in the Type box. For example, if you want to display numbers to four decimal places, type 0.0000 in the Type box. The new format will be stored in the list of formats so that you can apply it to other cells; the format will be available in all sheets in the workbook.

Table 1-3 lists the formatting symbols you can use to create your own custom formats.

To format a column of social security numbers, enter the format 000-00-0000, which will display the number 12545842 as e.g., 012-54-5842. To format a column of telephone numbers, use the custom format (###)   ###-####. This will format a cell entry such as 6175523619 in the format (617) 552-3619. The format #.??????? was used to format the table of atomic weight values shown in Figure 1-31, so that they are aligned on the decimal. (Note that, since the format contains seven ? symbols and the atomic weight of Na has only six digits to the right of the decimal point, there is an additional space to the right of the number.)

	A	B
1	H	1.00797
2	O	15.9994
3	Na	22.989768
4	S	32.066

Figure 1-31. Values aligned on the decimal point by using the ? formatting symbol.

You can create some fairly sophisticated number formats. For example, the format $#.0,, (dollar sign, number sign, period, zero, comma, comma) formats financial entries rounded to millions, with one decimal; the value 21180000 is displayed as $21.2.

You can use number formatting to add units to a number value. For example, the format #" g" appends the grams unit g to a number value; the value 50 is displayed as 50 g, as shown in Figure 1-32.

	A
1	50 g

Figure 1-32. Units added to a value by means of number formatting.

Table 1-3. Number Formatting Symbols

#	Placeholder for digit.
0	Placeholder for digit. Displays an extra zero if the number has fewer digits than the number of zeros specified in the format.
?	Placeholder for digit. Same as #, except that numbers are aligned on the decimal point. Also used when formatting a number as a fraction.
,	Thousands separator (if used with #, 0 or ?). Used alone, it rounds and truncates to the thousands place (millions place if two commas are used, etc.)
%	Converts to percent .
E	Converts to scientific format. Use E- to include sign with negative exponents only, E+ to include sign with both positive and negative exponents.
/	Converts to a fraction. Usually used in the form # ??/?? or ##/##. The number of ? or # symbols determines the accuracy of the display.
"text"	Text characters can be included in a format by enclosing them in double quotes. Although you can usually get away without the quotes, it's safer to use them.
@	Text placeholder. If the cell contains a text entry, the text is displayed in the format where the @ symbol appears.
[RED]	Displays the characters in the cell in red. You can also use [BLUE], [GREEN], [YELLOW], etc.

As you saw earlier, Excel recognizes the format of data typed into cells. If you enter a value such as 6/23/00 and then choose **Cells...** from the **Format** menu, choose the Number tab and choose the Custom category, you'll see that Excel has recognized the value as a date and applied the m/d/yy number format. Now choose the format d-mmm-yy from the menu and observe the Sample display at the top of the dialog box. The data is displayed as 23-Jun-00.

You can create custom date formats by using the year, month and day formats listed in Table 1-4. Day or month formats can have one-, two-, three- or four-letter formats; year formats can have either two- or four-letter formats. For example, the number format dddd, mmmm d, yyyy applied to a date entered as 8/3/38 will display Wednesday, August 3, 1938.

Table 1-4. Date Formatting Symbols

d	Displays the day as a number without leading zeros (1-31)
m	Displays the month as a number without leading zeros (1-12)
dd	Displays the day as a number with leading zeros (01-31)
mm	Displays the month as a number with leading zeros (01-12)
ddd	Displays the day as an abbreviation (Sun-Sat)
mmm	Displays the month as an abbreviation (Jan-Dec)
dddd	Displays the day as a full name (Sunday-Saturday)
mmmm	Displays the month as a full name (January-December)
yy	Displays the year as a two-digit number, e.g., 97
yyyy	Displays the year as a four-digit number, e.g., 1997

VARIABLE NUMBER FORMATS

Different number formats can be applied to positive, negative, zero and text values entered into a cell. A complete format consists of four sections, separated by semicolons, for positive, negative, zero and text values, respectively. If only one number format is specified, it applies to all values. If two number formats are specified, then the first one applies to positive numbers and zero, the second to negative numbers. For example, the format $#,###;[Red]$#,### formats positive amounts in black, Excel's default color, and negative amounts in red.

CONDITIONAL NUMBER FORMATS

Conditional number formats can be created by using the syntax [*condition, value*] *format statement*. *Condition* is one of the symbols <, >, =, >=, <=, <>; *value* may be any number. Format statement may be any built-in or custom format. For example, the number format [>1] "Number too large" accepts any input less than 1 but otherwise issues an error message.

Several conditions may be combined using semicolons. The number format
$$[>999]\#.\#\#,,\%;\#" \quad ppm"$$
displays the values 110 and 21560 as 110 ppm and 2.16%, respectively.

USING THE NUMBER FORMATTING TOOLBUTTONS

You can also format number values in cells by using the number formatting
toolbuttons shown following.

+.0 .00	Increases the number of decimal places.
.00 +.0	Decreases the number of decimal places.
%	Formats the number in percent style, with no decimal places.
$	Formats the number in currency style, with two decimal places.
,	Formats a number with commas and two decimal places.

> *Excel Tip.*   *There isn't a toolbutton to format number values in exponential*
> *format. Apply exponential format conveniently by using the shortcut key se-*
> *quence CONTROL+SHIFT+^. See Appendix E for a complete list of shortcut*
> *keys.*

FORMATTING NUMBERS USING "PRECISION AS DISPLAYED"

To permanently change *all values* stored on a worksheet to their displayed
values, use the Precision As Displayed option. Once this command has been in-
voked, you can't restore the original values.

To apply Precision As Displayed, choose **Options** from the **Tools** menu and
choose the Calculation tab. Check the Precision as Displayed box (shown in Figure 10-7) then press OK. Because this is an irreversible change, Excel asks you to
confirm the change.

To change only a *selected range of values* to "precision as displayed", use the
FIXED worksheet function (see "Text Functions" in Chapter 3).

> *Excel Tip.*   *You can apply the same formatting to multiple worksheets simply*
> *by grouping the sheets (click on the first sheet tab in the range of sheets to be*
> *formatted, then hold down the SHIFT key and click on the last sheet in the*
> *range). When you apply the desired formatting to the active sheet it will be ap-*
> *plied to all sheets in the group.*

PROTECTING DATA IN WORKSHEETS

Sometimes you'll want to protect data in a worksheet, either from changes by other users, or changes entered accidentally by yourself.

USING PROTECTION

You can protect either a whole workbook or a worksheet within a workbook. Within a worksheet, you can lock cells so that they cannot be changed, and you can hide cells so that the user can't view the cell contents.

The process for doing this is somewhat complicated. First you select cells to be locked and/or hidden, and set their status using the Protection tab of the **Cells...** command in the **Format** menu. Then you put the status into effect by choosing the **Protection** command in the **Tools** menu.

Before you begin, it's important to know that when a new worksheet is opened, the status of *all* cells in the document is Locked. Thus if you select a range of cells to be locked, and set the status using **Protect Sheet...**, you will find that all cells in the document are locked.

To lock only a limited range of cells in a document (as you will most often want to do), you must first set the status of all the cells in the document to Unlocked and then select the range of cells that you want to be locked.

To Lock a Range of Cells in a New Document

1.   Select all cells in the document by clicking on the row/column header button in the upper left corner of the worksheet.

2.   Choose **Cells...** from the **Format** menu and choose the Protection tab, uncheck the Locked option, and press the OK button. This unprotects all cells in the worksheet.

3.   Now select the range of cells that you want to protect. Choose **Cells...** from the **Format** menu, choose the Protection tab, check the Locked option, and press the OK button.

4.   Choose **Protection** from the **Tools** menu and choose **Protect Sheet...** from the submenu. You can enter a password if you wish (Figure 1-33). If you merely want to prevent yourself from making accidental changes, no password is necessary. If you want to protect the document from changes by others, you need a password; make sure that you will be able to retrieve it when you need it.

PROTECTING THE CONTENTS OF A WORKBOOK
BY MAKING IT A READ-ONLY WORKBOOK

If you make a workbook read-only, users can view formulas in cells, and change values and formulas, but the changes cannot be saved.

Figure 1-33. The Protect Sheet with Password dialog box.

To make a workbook read-only, the document should be closed. In the Windows **Start menu**, choose **Programs**, and then **Windows Explorer**. In the **Exploring** window, open the drive or folder that contains the file and select the document name. Choose **Properties** from the **File** menu, choose the General tab, and check the Read-only check box.

To make a Macintosh document read-only, select the name or icon of the document on the desktop. Choose **Get Info** from the **File** menu, and check the Locked check box.

CONTROLLING THE WAY DOCUMENTS ARE DISPLAYED

Use the **Window** menu to switch between one Excel document and another. All open documents are listed at the bottom of the **Window** menu; the active document is indicated with a check mark.

Use **Hide** to hide a workbook. Most commonly you'll use **Hide** with workbooks that contain macros. A macro is still "active" even when it's hidden.

To hide a worksheet in a workbook, choose **Sheet** from the **Format** menu and choose **Hide** from the submenu.

VIEWING SEVERAL WORKSHEETS AT THE SAME TIME

Although only one worksheet at a time can be the active window, Excel provides a number of ways to examine data in several different worksheets, or different areas of the same worksheet, at the same time.

USING NEW WINDOW AND ARRANGE...

If you have more than one document open, you can view them simultaneously in a number of ways. One way is to re-size and move the documents so that the desired part of each can be seen at the same time. Another way is to use the **New Window** and **Arrange...** commands in the **Window** menu. The latter method can be used to view multiple documents, or multiple sheets in the same workbook, as described in the following paragraph.

To view multiple worksheets in the active workbook, choose **New Window**. A second window will be opened for the active workbook. If, for example, the workbook is named Lookup, the windows will be named Lookup:1 and Lookup:2. Activate each window in turn and click on the sheet that you want to display. Now choose **Arrange...** from the **Window** menu, Excel displays the Arrange Windows dialog box (Figure 1-34). You can arrange the windows horizontally (one above the other) or vertically (side by side). (The active document will be on top or on the left, respectively.) If you have created a separate chart sheet from data in a worksheet, **Arrange**... provides a convenient way to work with a sheet and observe changes in the associated chart. With **Arrange...**, chart documents are reduced in size so that the whole chart appears in the window; worksheet documents are not reduced in size. Figure 1-35 illustrates a worksheet/chart combination displayed using the **Arrange** (Vertical) option.

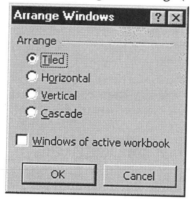

Figure 1-34. The Arrange Windows dialog box.

With three open documents, the **Tiled** option arranges the documents with the active sheet occupying the left half of the screen; the other two sheets each occupy one-quarter of the screen, one above the other. With four documents **Tiled**, each occupies one-quarter of the screen. Click on any document to make it the active sheet. Double-click anywhere on the solid border between the windows to undo the arrangement.

DIFFERENT VIEWS OF THE SAME WORKSHEET

As your worksheets get larger and more complicated, it becomes impossible to view all of a single worksheet at one time, or even all cells in one row or column at one time. Excel provides several convenient ways to display separate portions of a single worksheet on the screen at the same time, so that you can view one part while entering or changing data in another part.

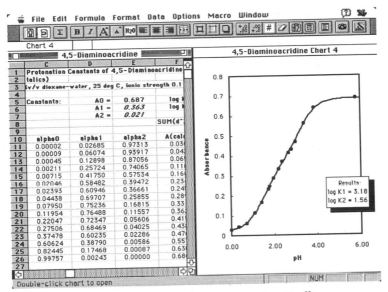

Figure 1-35. Two windows arranged vertically .

USING NEW WINDOW

When you choose **New Window,** a second window of the active document is opened. You can then re-size and move the windows so that the desired parts of the worksheet can be seen at the same time. This is useful if you want to **Cut** or **Copy** several cell ranges and then **Paste** them into another area of a worksheet, but the two areas of the worksheet are far apart.

Excel Tip. You can set different display options for the two windows. Display values in one window and formulas in another to see the effect of changes.

USING SPLIT SCREENS

To use the **Split** feature to split a document window horizontally into two windows, select an entire row as if you were going to insert a row. Then choose **Split** from the **Window** menu. This creates a split in the window, above the selected row, with each part of the window displaying the active document. Each part of the document now has its own scroll bar, and you can scroll one part of the document while the other part remains fixed. A vertical split is accomplished in the same way.

You can also split the document window by placing the mouse pointer on either split button (the small rectangles at the right end of the horizontal scroll bar and at the top of the vertical scroll bar), then click and drag the button.

The document window can be split both horizontally and vertically, by first selecting a single worksheet cell, then choosing **Split,** as illustrated in Figure 1-36.

To remove a split, choose **Remove Split** from the **Window** menu, or slide the split button back to its original position.

Excel Tip.   To remove a split from a window, it's not necessary to slide the split button back to its original position at the top or left-hand side of the scroll bar. Just place the pointer on the split button and double-click.

USING **FREEZE PANES**

Freeze Panes can be used to create a similar split document window, but the upper or left part of the window is fixed and cannot be scrolled. Split panes are useful to display fixed row or column headings (or both) while scrolling through the rest of the worksheet.

To use the **Freeze Panes** feature to split a document window horizontally into two windows, select an entire row as if you were going to insert a row. Then choose **Freeze Panes** from the **Window** menu. To split the window both horizontally and vertically, select the cell whose upper left corner defines the location of the split, then choose **Freeze Panes**.

	A	E	F	G	H	I	J	K
1	Grade Sheet							
2			Hour Exams					
3	Name	#1	#2	#3	Oral report	Paper	Final Exam	Total
4								
7	FERREIRO, Kathy	24	16	32	20	45	52	63.0
8	GANGE, Eric	28	13	43	20	40	51	65.0
9	GREALEY, John	22	14	40	17.5	40	56	63.2
10	HAPPERSBACH, Bill	28	12	30	17.5	45	59	63.8
11	HOGAN, Derek	37	17	37	20	50	60	73.7
12	LAROZI, Patrick	27	12	38	20	45	58	66.7

Figure 1-36. A document with a two-way split.

COPYING FROM EXCEL TO MICROSOFT WORD

Very likely you will want to copy information from Excel and paste it into a Word document. You may want to copy a range of cells, a chart, or even an image of an Excel dialog box.

There are three ways to copy from Excel to Word: by using **Copy** and **Paste**, by using the Camera icon on the toolbar or by making a "screen shot". Each produces a different "product". The first two can be linked dynamically to the source worksheet.

USING COPY AND PASTE

If you **Copy** a portion of a spreadsheet and **Paste** it into Microsoft Word, the copied data will appear in the Word document as a table (Figure 1-37). Dotted lines will indicate the table cells on the screen, but these do not appear in the printed document. Data in individual cells in the table can be selected and edited.

To copy a chart, select the chart by clicking in the chart border. Then choose **Copy** from the **Edit** menu. A marquee will appear around the chart. Activate the Word document, position the cursor where you want the chart to appear in the document, then choose **Paste**. The appearance of the copied chart will be a little different, depending on whether you copied a chart in a chart sheet or an embedded chart. Most noticeably, the copy of an embedded chart will have a border; to remove the border from the chart before copying, choose **Object...** from the **Format** menu, then choose the Patterns tab, then Border = None.

If you hold down the SHIFT key while choosing the **Edit** menu, the **Copy** command will be replaced by **Copy Picture....** This will allow you to copy a simple image of selected cells or a chart.

When pasted, the chart in the Word document will be the same size that it was on the screen. To re-size it, click on it. Black handles will appear. Selecting one of the handles and dragging it crops the image; selecting a handle and dragging while pressing SHIFT re-sizes the image. To re-size the image to a desired size, display the ruler and scroll the page until the image is adjacent to the ruler, then re-size.

Excel Tip   It's generally a good idea to postpone copying until the size and proportions of a chart in an Excel document have been changed to exactly the way you want the chart to appear in the Word document. If a pasted chart has to be reduced significantly in size to fit on a page, the axis labels, text characters and data markers may become unreadably small.

	QUIZ #											
	1	2	3	4	5	6	7	8	9	10	11	12
AGRAWAL, Sunda	5	4	2	5	4	2	5	1	2	3	4	2
ALI, M Saqwat	0	2	0	2	3	2	0					
ALVARADO, Annelise	5	5	5	5	5	4	5	5	4	5	5	5
AMATO, John	2	4	2	2	2	2	4	3	1	1	2	1
ANTOINE, Mark	2	2	3	2	5	0	3	1	1	2	4	3
ATKINSON, Gordon		3	5	3	3	2	5	3		2		3

Figure 1-37. Portion of a worksheet after **Copy** and **Paste**.

MAKING A "SCREEN SHOT" (MACINTOSH)

You can take a picture of the whole Excel screen (a "screen shot"), then use familiar graphic editing tools to select any portion of it. You'll need to have the SimpleText utility available.

First, arrange the window as desired. Then press COMMAND+SHIFT+3 (you'll hear the "shutter" sound). A screen shot of the complete document window will be placed in the hard disk, numbered consecutively Picture1, etc. When you open PictureN, you will start the SimpleText utility; the cursor will be a crosshair. Click and drag the cursor to select the desired portion of the image, or select the whole image using **Select All** from the **Edit** menu, then **Copy**. Finally, activate the Word document and **Paste**. The graphic object (see e.g., Figure 1-38) can be re-sized as described earlier. The data are not dynamically linked to the source worksheet.

If you press COMMAND+SHIFT+4, the mouse pointer will become a crosshair. You can now select a rectangular area of the screen. When you release the mouse button, you'll hear the shutter sound; again, the image will be saved on the hard disk.

With CAPS LOCK on, COMMAND+SHIFT+4 creates a picture file of a window, such as a dialog box.

To copy the image to the Clipboard instead of saving it to the hard disk, hold down the CONTROL key as you select the screen or window.

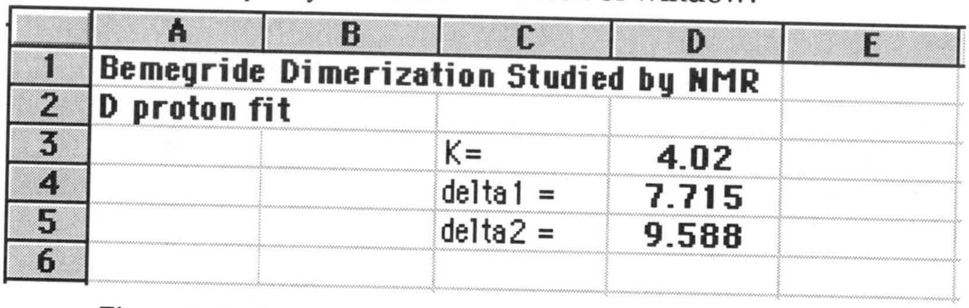

	A	B	C	D	E
1	Bemegride Dimerization Studied by NMR				
2	D proton fit				
3			K =	4.02	
4			delta1 =	7.715	
5			delta2 =	9.588	
6					

Figure 1-38. Portion of a worksheet copied as a "screen shot".

MAKING A "SCREEN SHOT" (WINDOWS)

To make a screen shot of the entire screen, press the Print Screen key, or, to make a screen shot of the active window, press ALT+Print Screen. Then activate Microsoft Word and **Paste** the graphic (which was captured on the Clipboard) into your Word document. You can move or re-size the graphic in Microsoft Word. To crop a portion of the graphic, you'll have to import the Clipboard contents into a graphics program such as Paintbrush or Photoshop.

USEFUL REFERENCES

John Walkenbach, *Microsoft® Excel 2000 Bible*, IDG Books Worldwide, San Mateo, CA, 2000.

John Walkenbach, *Excel 2000 Formulas*, IDG Books Worldwide, San Mateo, CA, 2000.

2

CREATING CHARTS:
AN INTRODUCTION

Nothing can be as helpful as displaying data in graphical form. With Excel you can quickly and easily create a chart, simply by selecting the data to be plotted and choosing the way you want the data to be displayed; Excel does the rest. In this chapter you'll learn the basics of creating Excel charts.

ONLY ONE CHART TYPE IS USEFUL FOR CHEMISTS

Excel 2000 provides a gallery of 14 standard chart types — bar charts, column charts, line charts and pie charts, among others. Since Excel originated as a financial tool, most of the chart types are those that are useful for displaying financial and related information — a bar chart to show sales figures for each business quarter, a line chart to show stock values each day over a one-month period, etc. Only one kind of chart, the XY or scatter plot, is of general usefulness for displaying scientific data. It is the only one in which numeric values are used along both axes. All other charts plot the numeric y values vs. equal increments on the X Axis and use the x values only for labels (called *categories* by Excel). The line chart, which looks like an XY chart, is actually only a bar chart of y values with the y values (the tops of the bars) shown as marker points connected by straight lines.

CREATING A CHART

There are two ways to create a chart: either as a separate chart sheet in a workbook, or as a chart *embedded* in a worksheet, so that you can see both the data and the chart at the same time. An embedded chart is useful if you want to see how a curve changes as you change its parameters. As you change the values in worksheet cells, the chart will update automatically.

CREATING A CHART USING THE CHART WIZARD

You can use the Chart Wizard toolbutton ▐▐▌ to create a chart. To use the Chart Wizard, first select the data to be plotted, e.g., a column of x values and a column of y values. The data can be in rows or columns. If the rows or columns are not adjacent, hold down the CONTROL key (Windows) or COMMAND key (Macintosh) while selecting the separate rows or columns of data. Then press the Chart Wizard toolbutton (alternatively, you can press the Chart Wizard button first and select the data range later). The first of a series of four dialog boxes will appear. The first Chart Wizard dialog box (Figure 2-1) lets you select the desired chart format. There are tabs for two categories: Standard Types and Custom Types (Custom Types are discussed in Chapter 5). When you select a chart type from the Chart Type category box, the chart subtypes will be displayed on the right. There are five subtypes of XY Scatter chart: marker points with no connecting lines, marker points connected by straight lines, straight lines with no marker points, points connected by a smooth curve and a smooth curve with no marker points.

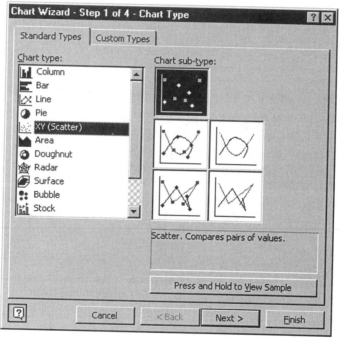

Figure 2-1. The first Chart Wizard dialog box: choosing chart type.

In general, the Smooth Curve options are not useful, especially if you are plotting data that obeys a particular mathematical relationship, or data with some experimental scatter. To produce a chart with a smooth calculated curve through experimental data points, see "Plotting Experimental Data Points and a Calculated Curve" in Chapter 5.

The second dialog box (Figure 2-2) displays a preview of the chart and allows you to enter or change the range of data to be plotted. The second and subsequent dialog boxes also provide the Cancel and Back buttons, in case you want to change what you've already selected.

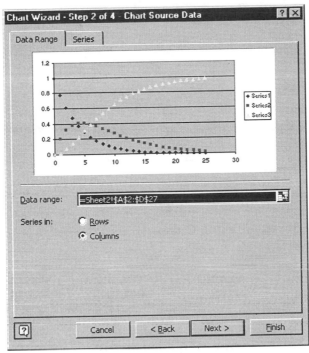

Figure 2-2. The second Chart Wizard dialog box: a preview of the chart.

The third dialog box (Figure 2-3) allows you to add or change titles, chart axis labels, gridlines, legend or data labels. (All of these can be added or removed after the chart has been completed.)

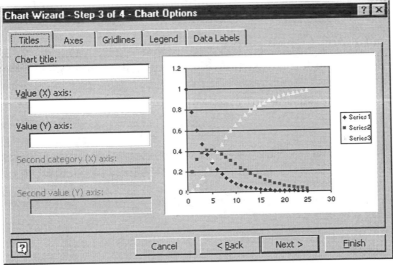

Figure 2-3. The third Chart Wizard dialog box: adding or removing chart elements.

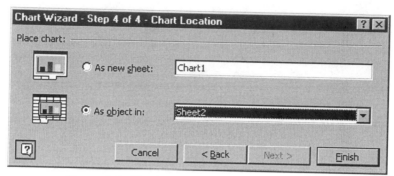

Figure 2-4. The fourth Chart Wizard dialog box:
determining the location of the completed chart.

The fourth dialog box allows you to create the chart either as a new chart sheet, which will be inserted in the workbook, or as an embedded chart on a worksheet (Figure 2-4).

ACTIVATING, RE-SIZING AND MOVING AN EMBEDDED CHART

To activate (select) an embedded chart for moving, re-sizing or formatting, you simply click once on the chart. When selected, an embedded chart has black "handles" at the sides and corners. In earlier versions of Excel you double-clicked on a chart to activate it but, beginning with Excel 97, only a single click is necessary to activate an embedded chart.

To move an embedded chart, place the mouse pointer anywhere inside the chart and click and drag the chart to its new location. To change the size of the chart, click on one of the handles, then drag to move that edge or corner inward or outward.

FORMATTING CHARTS: AN INTRODUCTION

Excel scales and formats each newly created chart automatically. It does a good job, but usually there is plenty of room for improvement. Excel provides a wide range of tools for modifying a chart. A few of these are discussed in this chapter, and further details are provided in Chapter 5, "Advanced Charting Techniques".

Chart formatting is accomplished by using commands in the **Chart** menu in the Chart Menu Bar, or by using the **Format** command in the **Edit** menu in the Chart Menu Bar.

USING THE CHART MENU

When you activate an embedded chart (by clicking on it) or switch to a chart sheet, the Worksheet Menu Bar is replaced by the Chart Menu Bar. The first four commands in the **Chart** menu — **Chart Type...**, **Source Data...**, **Chart**

Options..., Location... — correspond to the four dialog boxes of the Chart Wizard. You can use these commands to modify a chart that you've already created.

USING CHART TYPE... IN THE CHART MENU
TO SWITCH FROM ONE CHART TYPE TO ANOTHER

If you created a chart as an XY Scatter plot, for example, you can change it to another chart format by choosing **Chart Type...** from the **Chart** menu, then choosing any of the other chart types. Fifteen standard types, as well as a number of more specialized custom types, such as a combination Line-Column chart, are available.

USING CHART OPTIONS... IN THE CHART MENU
TO ADD TITLES, GRIDLINES OR A LEGEND

To add text to the chart, choose **Chart Options...** from the **Chart** menu and choose the Titles tab. The resulting dialog box, identical to Figure 2-3 but now called Chart Options, lets you add text for the chart title or the labels for the X- or Y axes. When you enter text in the Chart Title input box, for example, the title text will appear in the sample chart. Excel wraps the text if it is too long to fit on one line. In the same way you can add or remove gridlines or a legend.

USING LOCATION... IN THE CHART MENU
TO MOVE OR COPY AN EMBEDDED CHART
TO A SEPARATE CHART SHEET OR VICE VERSA

If you created an embedded chart and would like to convert it to a separate chart sheet, simply activate the chart, choose **Location...** from the **Chart** menu, press the As New Sheet option button, then press OK. The embedded chart will be converted into a chart on a separate chart sheet. To put a copy of an embedded chart on a separate chart sheet, first make a copy of the chart, then follow the same procedure.

FORMATTING THE ELEMENTS OF A CHART

The various components that make up a chart —Plot Area, Axes, Data Series, Arrows, Text, Gridlines, etc. — are the *chart elements*. The **Format** menu allows you to customize any of these chart elements. You can change the weight and style of lines, the color and shape of plotting symbols, the numerical ranges of axes, etc.

Excel Tip   As you move the mouse pointer around the chart, yellow Chart Tip boxes appear, with the name of the chart element that the mousepointer is currently passing over. If you find these Chart Tips distracting, you can deactivate them by choosing Options from the Tools menu. Choose the Chart tab and un-check the Chart Tips check box.

SELECTING CHART ELEMENTS

Select a particular chart element by clicking once on it with the mouse pointer. The selected element will be indicated by the appearance of "handles"; in addition, the name of the chart element will appear in the Name Box (the cell reference area of the formula bar). It's a good idea to look at the text in the Name Box when you choose a chart element, just to make sure that you've selected the correct chart element.

Excel Tip. *Sometimes it's difficult to select a chart element by clicking on it (for example, if two chart elements are almost superimposed). Instead of selecting with the mouse pointer, you can use the up and down arrows on the keyboard to select chart elements. This allows you to select each chart element in turn (Chart, Plot, Axis, Series 1, etc.); the name of the selected chart element is displayed in the reference area of the formula bar. By using the left and right arrows, you can select related chart elements within a group (e.g., Series 1 Point 1, Series 1 Point 2, etc.).*

FORMATTING CHART ELEMENTS

Once the desired chart element has been selected, choose **Format** from the Chart Menu Bar. In Excel 2000 there is a single context-sensitive menu command in the **Format** menu, which appears as **Selected Axis...**, **Selected Data Series...**, etc., depending on which chart element is selected.

When you choose the command, a dialog box will appear with the formatting possibilities for the selected chart element. There are different **Patterns** dialog boxes or tabs (for Chart, Plot, Series, Axis, Arrow, Gridline, Text, etc.), different **Font** dialog boxes or tabs (for Chart, Axis, Text, etc.) and so on. For example, with a Series selected, Patterns permits you to change or remove the plotting symbol or the line. With an Axis selected, you can use the Patterns tab (Figure 2-5) to change the style or weight of the axis, change or remove or add major and minor Tick Marks, and change or remove the Tick Labels.

Excel Tip. *If you double-click on a chart element you will go directly to the Format dialog box for that element.*

With an axis selected, the Scale tab (Figure 2-6) enables you to change the scale range and where the X Axis crosses the Y Axis, for example. The Number tab (not shown) permits you to use the same number formats available for worksheet cells in the **Format Cells...** command to change the number formatting of the axis labels.

Excel Tip. *You can also number-format chart axes by using number-formatting toolbuttons, such as Increase Decimal* ⊞ *or Decrease Decimal* ⊞.

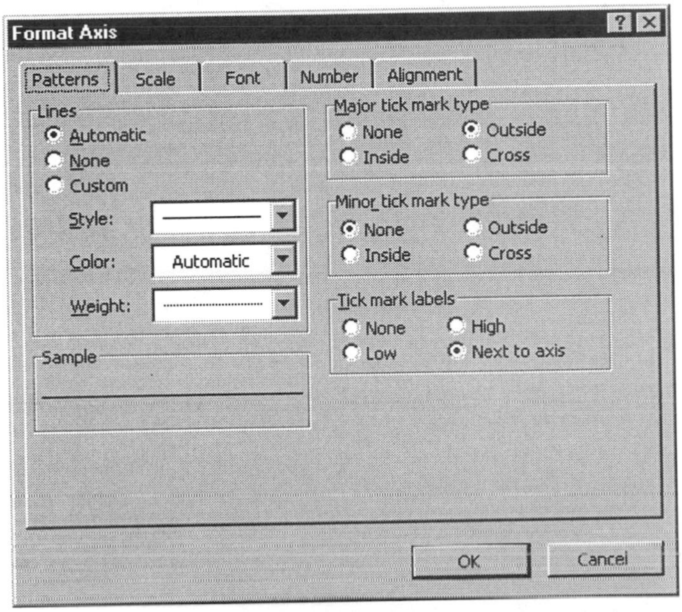

Figure 2-5. The **Patterns** dialog box for axes.

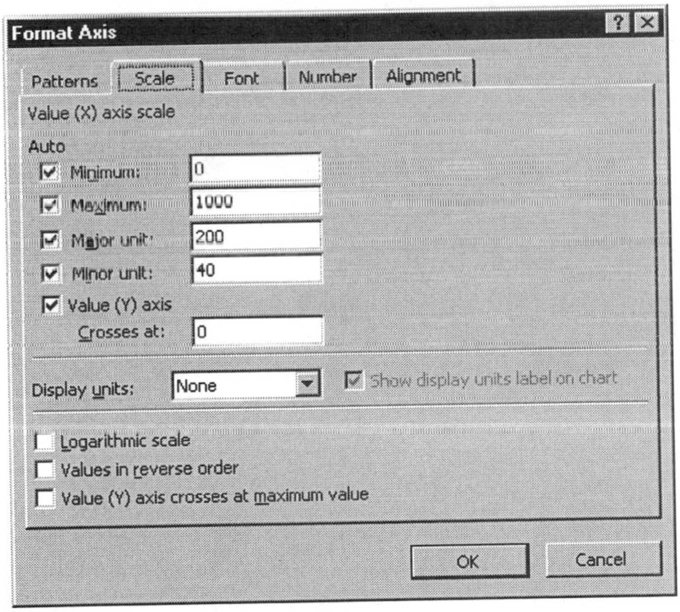

Figure 2-6. The **Scale** dialog box for axes.

Occasionally you will need to change the axis scales. Excel always creates axis scales that include zero, but this is not suitable for every case, as in Figure 2-7, where the wavelength data ranged from 350 nm to 850 nm.

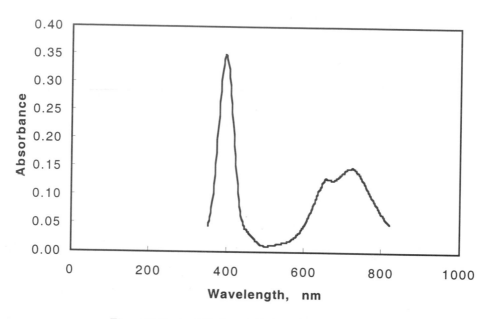

Figure 2-7. An XY chart as originally created by Excel.

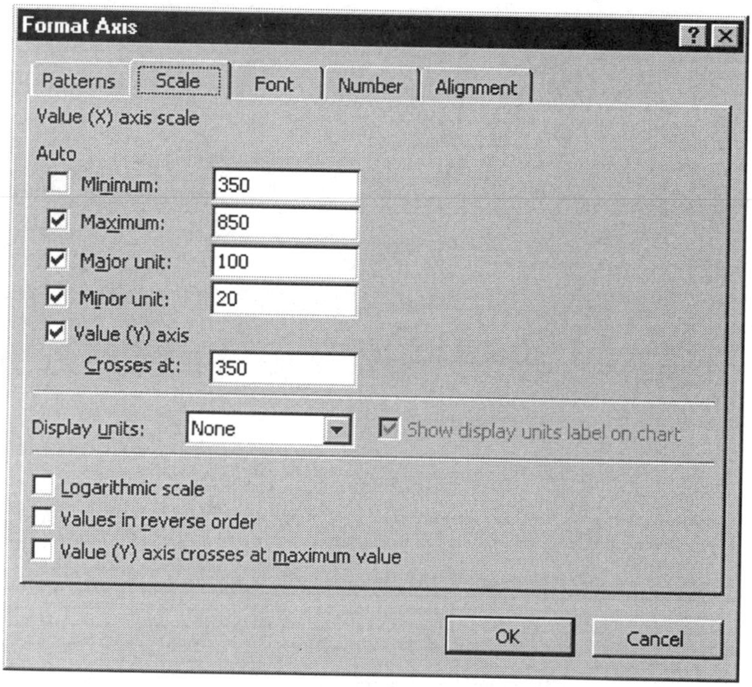

Figure 2-8. The Format Axis dialog box.

To modify the X Axis scale, first select the axis by clicking on it. Black "handles" will appear at the ends of the axis. Then choose **Selected Axis...** from the **Format** menu and choose the Scale tab. Enter 350 (the lowest x value) for the new value of Minimum, and 850 (the largest x value) for the new value of Maximum, as in Figure 2-8, and press OK.

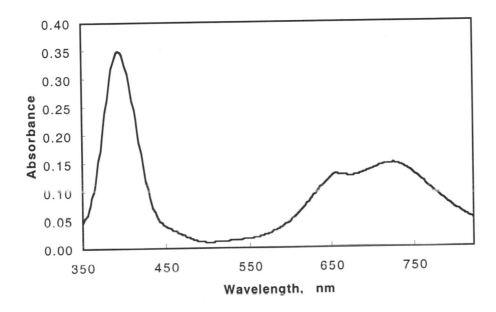

Figure 2-9. The chart in Figure 2-7 after adjustment of the scale of the X Axis.

Figure 2-9 shows the same chart after the horizontal axis scale has been adjusted

When you change a value in any of the five input boxes in the Scale tab, the Auto check box next to the entry becomes unchecked. When Excel creates a chart, it automatically scales the axes to conform with the data but does not adjust any of the axis parameters for which Auto is unchecked.

PART II

ADVANCED
SPREADSHEET
TOPICS

3

CREATING
ADVANCED WORKSHEET FORMULAS

This chapter shows you how to use Excel's wide range of worksheet functions to construct sophisticated worksheet formulas. In addition, you'll learn how to use *named references* in formulas, which makes constructing complicated formulas an easier and more error-free task. You'll also learn techniques of formula editing and worksheet troubleshooting.

THE ELEMENTS OF A WORKSHEET FORMULA

A worksheet formula consists of some or all of the following: values, cell references, operators and worksheet functions.

OPERATORS

Worksheet operators are either arithmetic, text, logical or reference operators.

The *arithmetic operators* for addition, subtraction, multiplication, division and exponentiation are familiar ones and have already been mentioned in Chapter 1.

There is only one *text operator*: the & (ampersand) symbol. It is used to concatenate text, or text and variables. For example, if cell G256 contains the value 2001, the formula ="Chemical Inventory for "&G256 displays Chemical Inventory for 2001. You'll encounter many other uses of the & operator in the chapters that follow.

The *logical operators* compare two values and produce a logical result, either TRUE or FALSE. The logical operators are the following: = (equal to), > (greater than), < less than), >= (greater than or equal to), <= (less than or equal to) and <> (not equal to). Note that >=, <= and <> must be typed as shown, or Excel will not recognize them as operators and will produce an "Error in formula" message. In worksheet formulas, FALSE can be represented by zero and TRUE by any non-zero value.

There are three *reference operators*: the range operator (colon), the union operator (comma) and the intersection operator (space). The *range operator* produces a reference that includes all the cells between and including the two references, e.g., G3:L3. The *union operator* produces a reference that includes the

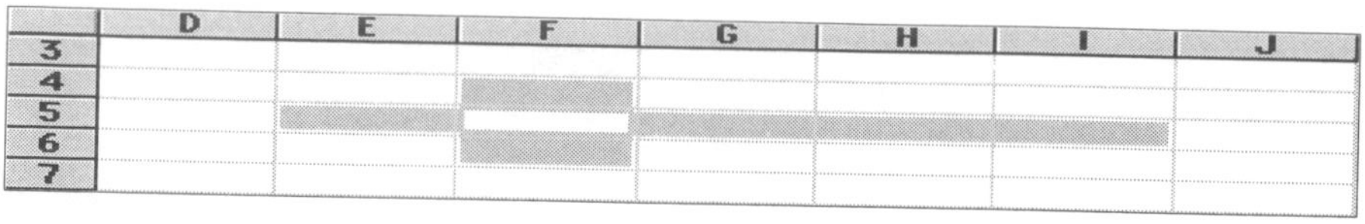

Figure 3-1. A single cell reference (F5) produced by the intersection operator.

two references, e.g., G3,L3. More than one union operator can be used in a single reference. The *intersection operator* produces a reference to the cells common to two references. For example, the reference F4:F6 E5:I5 refers to cell F5, as illustrated in Figure 3-1. The intersection operator is particularly useful when used with *named references* (see "Using Create Names" later in this chapter).

ABSOLUTE, RELATIVE AND MIXED REFERENCES

Cell references can be absolute, relative or mixed. A *relative reference*, such as A1, becomes B1, C1, and so on, as you **Fill Down** a formula into cells below the original formula. An *absolute reference* such as $A$1 remains $A$1 as you **Fill Down**. A mixed reference is a reference such as A$1 or $A1; use *mixed reference* style if you want to control how the reference is changed as a formula is duplicated using **Fill Down** or **Fill Right**. In the first case the row designation will remain constant when you **Fill Down** or **Fill Right**; in the second case the column designation will remain constant. For example, the reference $D2 becomes $D3 when duplicated to the cell in the next row down;, it remains $D2 when duplicated to the cell in the next column to the right, however.

> *Excel Tip. You can use F4 (Windows) or COMMAND+T (Macintosh) to toggle between cell reference types. Select the cell reference by double-clicking on it in the formula bar (or just put the insertion point cursor anywhere in the reference), then press F4 to cycle through the formats, in the sequence: relative (e.g., A1), absolute (e.g., $A$1), mixed (e.g., A$1), mixed (e.g., $A1). If you are typing a formula, you can use F4 after typing the cell reference; Excel converts the reference to the immediate left of the insertion point.*

CREATING AND USING 3-D REFERENCES

You can create references that extend across several worksheets in a workbook. If you visualize a workbook as consisting of many worksheets stacked one above the other, a 3-D reference refers to a range of cells extending vertically. For example, you could calculate a total in the same cell, e.g., cell D1, on each of 12 sheets, and then calculate a grand total, the sum of the values in cell D1 of all 12 sheets, on a summary sheet, e.g., Sheet13, by using the 3-D formula =SUM(Sheet1:Sheet12!D1). You can enter a 3-D reference by typing it or by selecting.

There are two ways you can enter a 3-D reference by selecting. Follow either of the procedures in the following box.

To Enter a 3-D Reference by Selecting

1. Select the sheet tab of the first sheet in the 3-D reference, e.g., Sheet1.

2. Hold down the SHIFT key and select the last sheet in the range, e.g., Sheet12.

3. Select the cell or range, e.g., D1.

or...

1. Select the sheet tab of the first sheet in the 3-D reference, e.g., Sheet1.

2. Select the desired cell or range therein.

3. Hold down the SHIFT key and select the last sheet in the range, e.g., Sheet12.

Many worksheet functions can be used in a 3-D formula, for example AND, AVEDEV, AVERAGE, COUNT, COUNTA, DEVSQ, LARGE, MAX, MEDIAN, MIN, OR, PRODUCT, SMALL, STDEV, SUMSQ, VAR.

Excel Tip   If you want to insert rows or columns in a 3-D range such as Sheet1:Sheet12!D1:E10, be sure to group all of the sheets in that range before inserting.

CREATING AND USING EXTERNAL REFERENCES

External references are used to establish links between documents. You can create a link between worksheets within the same workbook, or between worksheets in different workbooks. By linking worksheets, you can utilize the data from one worksheet in a formula in another worksheet, merge data from several worksheets in a summary sheet, or simplify a complicated model by breaking it up into manageable portions.

Linked worksheets are linked dynamically; that is, when you make changes in a *source worksheet*, the changes automatically occur in the *dependent sheet*.

Links between worksheets are established by means of *external references*. An external reference is simply a reference that includes the filename of the source worksheet, separated from the reference or name by an exclamation point, e.g., =SheetName!CellReference if the link is between sheets in the same workbook or [WorkbookName]SheetName!CellReference if the link is between sheets in different workbooks.

If either the workbook name or the sheet name contain spaces, the [WorkbookName]SheetName must be enclosed in single quotes (e.g., ='[logTplot.xls]Phase diagram data'!$A$1).

Since these external references are generally long and complicated, you should never attempt to type an external reference in a formula. Instead, enter the reference by selecting.

There are two ways to enter an external reference in a formula. With the first method, you begin typing the formula in the destination cell, then enter the external reference by selecting the appropriate range in the source document. With the second method, you copy the range in the source document, then use **Paste Special...** to paste it into the formula in the destination cell. These two methods are described in the following sections.

CREATING AN EXTERNAL REFERENCE BY SELECTING

To enter a reference from a source document into a formula in a dependent document, first select the cell where the value is to be entered and begin typing the formula. At the point in the formula where you want to enter an external reference, click on the sheet tab of the source (use the **Window** menu to switch to the source document if the source sheet is in a different workbook) and select the cell containing the value to be entered (let's say it's in cell H12 of the worksheet Expt #XVIII-32). When you press the Enter button, Excel returns you to the dependent document, and the formula ='Expt #XVIII-32'!$H$12 is entered in the cell.

CREATING AN EXTERNAL REFERENCE BY USING PASTE LINK

To use this method, **Copy** the cell or range in the source document, then switch to the destination. Select the cell or range where the contents are to be pasted, choose **Paste Special...** from the **Edit** menu and press the Paste Link button.

This method can be used only to create a link from a cell or range in the source to a cell or range in the destination; it cannot be used to insert an external reference in a formula.

THE EXTERNAL REFERENCE CONTAINS
THE COMPLETE DIRECTORY PATH

When the source sheet is not open, the external reference displays the complete path of the directory or folder containing the sheet (e.g., 'Macintosh HD:Lecture and Laboratory:CH351 Lecture:All 95F: Homework Grades 95F'!$A$3:$Y$57).

UPDATING REFERENCES AND RE-ESTABLISHING LINKS

If you open a dependent worksheet before opening its source document or documents, you'll get an "Update references to unopened documents?" message. If you press the Yes button, the data must be updated and there can be a delay, depending on the size and complexity of the documents. If you press No, you'll get either whole columns of #VALUE! error values, or even worse, values that are not updated. So it's advisable to open all source documents before beginning to open dependent documents. For the same reason, always save the source worksheet before saving the dependent sheet linked to it.

If you edit a source worksheet, the cell references in the dependent worksheet will be updated, provided the dependent worksheet is open. The same is true if you change the name of a source document or move it to a different directory. If you rename or move a source document while the dependent document is closed, references will not be updated and you'll get a "Can't find document" message, and will have to re-establish the links between the documents.

ENTERING WORKSHEET FORMULAS

There are a number of useful techniques that you can use for entering worksheet formulas or editing formulas that you have entered in worksheet cells.

- To edit a formula in a cell, you can simply select the cell and then edit the formula in the formula bar.

- Alternatively, you can use Excel's Edit Directly In Cell feature: use function key F2 (Windows) or COMMAND+U (Macintosh) or double-click on the cell to enter edit mode. Excel uses colors to show range references in formulas and the corresponding ranges on the worksheet. You can use the right and left arrow keys to move through the formula, or CONTROL+(arrow key) to jump to the next element of the formula, or CONTROL+SHIFT+(arrow key) to select the next element of the formula.

- Type formulas in lowercase to facilitate detection of typographical errors. When you enter a formula, Excel converts functions and cell references to uppercase. If you type the formula =offset(d1,5,1), Excel will convert it to =OFFSET($D$1,5,1) when you enter the formula; but if you type "ofsett" instead of "offset", Excel won't recognize it and will display the error message #NAME?. When you examine the formula, you'll easily see that the incorrect function name remained in lowercase letters.

- Enter cell or range references in formulas by selecting, not by typing. This makes it less likely that you will enter an incorrect reference (e.g., C25 instead of B25) and also makes entering complicated references (such as external references) much easier.

- If complicated formulas contain terms identical to those used in other cells, you can **Copy** that part of the formula and **Paste** it into the new formula. Here's one method: before beginning to type the new formula, select the cell containing the formula you want to copy. In the formula bar, highlight the part of the formula you want to copy, **Copy** it, then click the Check Box in the formula bar. Now select the cell into which you want to type the new formula, type the new formula until you reach the part that you've copied, then **Paste** in the formula fragment.

Excel Tip.   *To select (highlight) a word or reference for editing in the formula bar, double-click on it.*

USING NAMES INSTEAD OF REFERENCES

A name can be applied to a cell, a range of cells, a value or a formula. Most often you'll use names for cell references. Using names makes it easier to create and to decipher complex formulas. For example, the formula

=pKa+LOG(base/acid)

is easier to understand than the formula

=$E$1+LOG(B2/C2)

The **Name** submenu of the **Insert** menu contains several commands for working with names: **Define...**, **Paste...**, **Create...**, **Apply....** and **Label....** You will probably find **Define...** and **Create...** most useful. Use **Define...** to assign a single name to a cell or range; use **Create...** to create names for several cells or ranges, based on row and/or column titles.

USING DEFINE NAME

To assign a name to a cell reference, first select the cell or range. Then choose **Define...** from the **Name** submenu to display the Define Name dialog box (Figure 3-2). Excel will propose a name in the Name box, using text from the cell immediately above or to the left of the selected cell. The absolute reference of the selected cell will appear in the Refers To box. Edit the name if desired, then press OK.

The first character of a name must be a letter. Subsequent characters can be letters or numbers or the period or underline character. Spaces are not allowed (the space character is the intersection operator); Excel will substitute an underline character for a space in any name that it proposes based on text in worksheet cells.

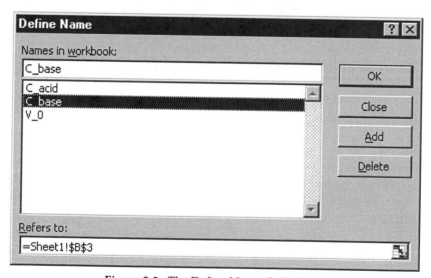

Figure 3-2. The Define Name dialog box.

To indicate breaks between words in a name, use a period or underline character, or use capitalization.

Names that look like references (e.g., A1), will not be accepted by **Define Name.** Since Excel can also use the R1C1 reference style, the letters R and C cannot be used as names.

Instead of a reference, you can type a numeric value or a formula in the Refers To box. You can use named formulas to simplify long worksheet formulas by assigning names to parts of the formula. See "Returning an Array of Unique Entries in a List" in Chapter 4 or "A Drop-down List Box on a Worksheet" in Chapter 8 for examples of this technique.

USING CREATE NAMES

You'll find **Create Names** very useful if you have worksheets with constants or other values arranged in a table format, as in Figure 3-3. This command allows you to assign names to ranges *en masse.*

	A	B
1	V_0	50.00
2	C_acid	0.002056
3	C_base	0.1381

Figure 3-3. A table of constants and names.

To assign names to cells, select the cells to be named and the adjacent cells containing the names. Choose **Name** from the **Insert** menu and choose **Create...** from the submenu. Excel displays the Create Names box; in Figure 3-4 the Left Column check box is checked, indicating that Excel proposes to use text in the cells in the left column for names. Press the OK button. The names will be assigned to the appropriate cells.

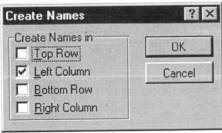

Figure 3-4. The Create Names dialog box.

Names can be enclosed in parentheses for clarity in the worksheet, as in Figure 3-5; the parentheses will be ignored in creating the names. Here the cells F4:G9 are selected and Excel proposes to use names in the Right Column. Equal signs and colons are also ignored, as in Figure 3-6.

	B	C	D	E	F	G
4	Initial volume, mL				50.00	(V_0)
5	Initial Concentration of aedach.3HBr, M				0.002056	(CL)
6	pH correction factor C				-0.11	(C_)
7	Concentration of titrant NaOH, M				0.1381	(CB)
8	Buret calibration factor				0.990	(calib)
9	pcKw				13.78	(pcKw)

Figure 3-5. An example of selecting cells for Create Names.

	F	G	H	I
4		Protonation constants:		
5		logK1H =	10.42	
6		logK2H =	9.74	
7		logK3H =	8.21	
8		logK4H =	5.44	
9				

Figure 3-6. Another example of selecting cells for Create Names.

If the data table is a two-dimensional one, as in Figure 3-7, cells are referenced both by row and by column. Excel proposes the row and column titles as names, as shown in the Create Names dialog box of Figure 3-8. Excel will apply the name max to the range F5:I5, the name band1 to the range F5:F7, etc. The intersection operator (the space character) can then be used to identify the named variables. For example, band3 A_0 refers to cell H6.

	D	E	F	G	H	I	J
2							
3			band1	band2	band3	band4	
4		max	29.25	22.65	18.56	10.11	
5		A_0	1.12	0.15	0.87	1.57	
6		s	1.60	1.58	1.36	1.99	
7							

Figure 3-7. A two-dimensional data table for Create Names.

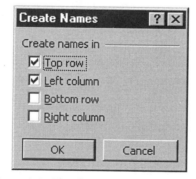

Figure 3-8. The Create Names dialog box.

USING THE DROP-DOWN NAME LIST BOX

You can also define a name by using the Name box (the cell reference area in the formula bar). Simply select the cell or range on the sheet (the range will be displayed in the Name box), click the cursor in the Name box (the typing area will be highlighted), type the name, then press ENTER. (Excel does not propose a name based on a text label above or to the left of the selected range, as it does when you use Define Name or Create Names; you have to type in the name yourself.)

You can use the Name box to jump to a cell or range that has previously been assigned a name. Use the ⏷ button to actuate the drop-down Name list (Figure 3-9) and select the desired name from the list; when you release the mouse button the cell or range will be highlighted. (You can also type into the Name box the name of the cell or range to which you want to move.)

ENTERING A NAME IN A FORMULA BY SELECTING

Earlier in this chapter it was recommended that you enter cell or range references in formulas by selecting, not by typing. If a name has been assigned to a cell or range, Excel will enter the name in the formula, rather than the reference, when you click on the cell or range.

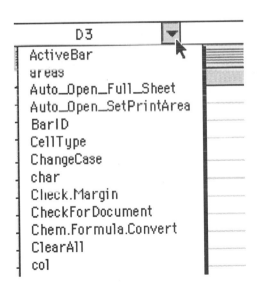

Figure 3-9. The drop-down Name list.

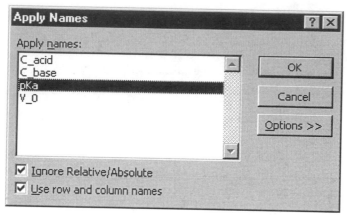

Figure 3-10. The Apply Names dialog box.

USING APPLY NAMES

Use **Apply Names** (Figure 3-10) if you have created a spreadsheet with formulas using cell references and now want to replace the cell references with names. First, use **Define Name** or **Create Names** to assign names to the references. Then choose **Apply Names**. The names that you have assigned will be shown in the list box. The Ignore Relative/Absolute box should usually be checked. Select a name from the list and press OK. All cells containing the reference will be replaced by the name.

Excel Tip. *You can select more than one name at a time from the list box. In Excel for Windows, press the CONTROL key and then click on the names you want to select. In Excel for the Macintosh, press the COMMAND key and click on the names you want to select.*

USING PASTE NAME

The **Paste Name** command allows you to select a variable name from the list of names in the worksheet and **Paste** it into a cell or formula. There doesn't seem to be any advantage to this command over simply typing in the name, or selecting the named cell or range.

DELETING NAMES

The Create Names dialog box lists all names that have been assigned in the workbook, even if they are no longer used or valid. If you have removed an unwanted name by deleting the cell, row or column in which it was located, the reference to that name in the Refers To box will be #REF!. Use the Delete button to delete unwanted or invalid names from the list.

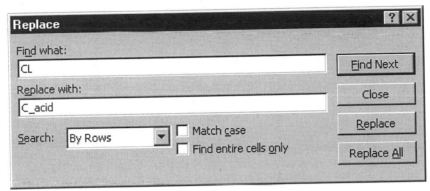

Figure 3-11. The Replace dialog box.

CHANGING A NAME

You can easily change all occurrences of a name in a spreadsheet by using **Replace...** from the **Edit** menu. The Replace dialog box (Figure 3-11) asks you to specify the search text and the replacement text. Select Look At Part; otherwise the only cells modified will be those that contain the search text and nothing else.

The Match Case box is not checked when the dialog box is initially displayed, but because this default option can often cause problems, it's usually a good idea to check it. In the example of Figure 3-11, where CL is changed to C_acid, unless the Match Case box is checked, all occurrences of the letters "cl" will be replaced, including in text entries such as $NaClO_4$.

Although all occurrences of the name are changed in the worksheet, the name definition in the Define Name box is not changed, so you'll have to change it there also. Until you do so, all formulas containing the changed name will display the #NAME? error value. Note that when you change CL to C_acid in the Define Name dialog box, the old name CL will remain in the list until you delete it.

NAMES CAN BE LOCAL OR GLOBAL

When you assign a name to a cell or range in a workbook, the name is a global or workbook-level name. The name can be used in all sheets in the workbook.

If you want to create a name that is a local or worksheet-level name, that is, one that is available only in a particular sheet, you can do so as described in the following box.

To Assign a Local or Worksheet-Level Name to a Cell or Range

1. First select the cell or range. (In this example Sheet1 is the active sheet.)
2. Choose **Define...** from the **Name** submenu to display the Define Name dialog box. Excel will propose a name in the Name box.
3. Edit the name by typing e.g., **Sheet1!** before the name.
4. Press OK.

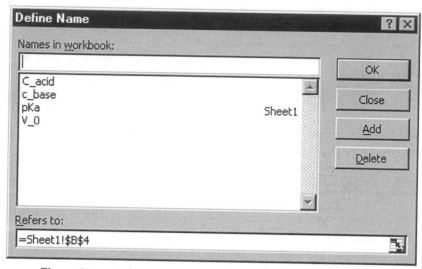

Figure 3-12. A sheet-level name in the Define Name dialog box.

If, for example, the name that you had assigned to a cell in Sheet1 was Total, the name must be edited in the Names In Workbook box to appear as Sheet1!Total. However, you can still use the name Total in your formulas in Sheet1.

A sheet-level name appears in the Define Name dialog box only when that sheet is the active sheet, with the sheetname shown on the right side of the list box, as in Figure 3-12. The name appears in the drop-down Name list box (in the formula bar) for that sheet, but does not appear when the drop-down list is selected in other sheets.

You can define the same worksheet-level name in a range of sheets, each sheet having its own value for the named reference. You can define the name by following the procedure in the preceding box, once for each sheet, or you can use the method described in the box on the next page to assign the name in all sheets at once.

If you use **Move or Copy Sheet...** in the **Edit** menu to create a copy of a sheet that has defined names, the copy will have the same names but they will be local.

THE **LABEL...** COMMAND

The **Label...** command was introduced in Excel 97. It also allows you to use names (except that they're called labels) in formulas. It is intended primarily to label rows and column headings of tables. The **Label...** command doesn't provide any features that are not available by using Define Name or Create Name, and in fact is much less versatile than using names.

To use the **Label...** command, you select the range that contains the text labels, not the cells containing values, then choose **Label...** from the submenu. Excel displays the Label Ranges dialog box, where you specify either Row Labels or Column Labels.

Labels are worksheet-level, not global.

To Assign the Same Worksheet-Level Name to the Same Cell or Range in Several Sheets

1. Group the worksheets (select the first sheet in the range, then hold down the SHIFT key and select the last sheet in the range).

2. Select the cell that you want to contain the name, and type the name. The text will be entered in that cell in all of the grouped sheets.

3. Select the cell or range that you want to name and the adjacent cell containing the name.

4. Use Create Names to assign the name to the cell or range in the range of selected sheets. This procedure will assign a global name in the workbook but will "overwrite" this global name with a local name in all of the selected sheets *except the active sheet*.

5. Use Define Name to change the global name to a local name in the active sheet, as described in the preceding box.

6. Choose Define Name and delete the global name, if you wish.

EXCEL WILL CREATE LABELS AUTOMATICALLY

You can use range labels in formulas without ever assigning names or labels; Excel will search through the active sheet to find suitable text labels and will use the labeled cells in your formulas. For example, in the worksheet containing the table shown in Figure 3-7, you could use the expression band1 max in a formula, even though you had not assigned range names or range labels.

If you use this automatic feature of Excel, make sure that you haven't used the same label text (e.g., max) elsewhere in the worksheet.

WORKSHEET FUNCTIONS: AN OVERVIEW

Even though Excel is primarily a business tool, it provides a wide range of functions that are useful for scientific calculations. There are over 300 worksheet functions, organized in ten categories: Database, Date & Time, Engineering, Financial, Information, Logical, Lookup & Reference, Math & Trig, Statistical and Text. This chapter provides examples using selected worksheet functions from the Date & Time, Logical, Lookup & Reference, Math & Trig, Statistical and Text function categories. Database functions are described in Chapter 6. Excel's Engineering functions are available only when you load the Analysis ToolPak Add-In; they include functions to perform conversions from one number system to another (e.g., decimal to hexadecimal) and functions to operate on complex numbers.

Appendix A lists selected worksheet functions in the Database, Date & Time, Information, Logical, Lookup & Reference, Math & Trig, Statistical, Lookup & Reference, and Text categories. Appendix B provides an alphabetical list of these worksheet functions along with the required syntax, some comments on the required and optional arguments, one or more examples and a list of related functions.

FUNCTION ARGUMENTS

Most worksheet functions require one or more *arguments*: the values that the function uses to calculate a return value. The arguments are enclosed in parentheses following the function name, e.g., SQRT(125) or SUM($F$3:$F$28) or SUBSTITUTE(PartNumber, "-1995", "-1996"). A few functions, such as PI() or NOW(), do not require arguments, but the opening and closing parentheses must still be provided.

Function arguments are either required or optional. In the following sections of this chapter and in the appendices, required arguments are shown in bold and optional arguments in non-bold text, in accordance with Microsoft's convention.

When an optional argument is omitted, a default option for that argument is used, so you need to know what is assumed when the argument is omitted. For example, the worksheet function LOG(**number**, base) returns the logarithm of a number to a particular base. If the optional argument *base* is omitted, the function returns the base-10 logarithm.

Most arguments must be of a particular data type (number, text, reference, array, logical or error). Most argument names indicate the data type that is required, by using the words *number*, *text*, *reference*, etc., or by appending _num or _number, etc. to the argument name. For example, the syntax of the SUBSTITUTE function is SUBSTITUTE(**text**, **old_text**, **new_text**, instance_num) The first three arguments must be text, the fourth must be a number. Some functions can operate on arguments of any data type, indicated by the use of *value* as an argument name.

In every case, a function can use a cell reference as an argument, but the cell must contain a value of the correct data type.

MATH AND TRIG FUNCTIONS

Excel's mathematical and trigonometric functions (50 of them in Excel 2000) include functions that correspond to the following "scientific calculator" functions: $\sqrt{x}$, $\log x$, $\ln x$, Σ, e^x, π, $\sin x$, $\cos x$ and $\tan x$. See Appendix A for a complete listing.

FUNCTIONS FOR WORKING WITH MATRICES

Excel provides functions for the manipulation of arrays or matrices: TRANSPOSE(*array*) returns the transpose of an array, MDETERM(*array*) returns the matrix determinant of an array, MINVERSE(*array*) returns the matrix inverse of an array, MMULT(*array1*, *array2*) returns the matrix product of two arrays and SUMPRODUCT(*array1*, *array2*, ...) returns the sum of the products of corresponding array elements. These functions are discussed more fully in Chapter 9.

STATISTICAL FUNCTIONS

Excel 2000 provides 78 statistical functions, including functions that return the mean, median, maximum and minimum values, average deviation, standard deviation, variance, n^{th} quartile and rank. Many of these functions are described in Appendices A and B.

A statistical function of considerable use for chemists is LINEST (*linear estimation*). It returns the least-squares regression parameters of the linear function that best describes a data set. LINEST is discussed in detail in Chapter 11.

LOGICAL FUNCTIONS

Logical functions allow you to use different formulas in a cell, depending on the values in other cells. The logical functions provided with Excel are IF, AND, OR and NOT. The latter three are almost always used in combination with the IF function, as are the comparison operators.

THE IF FUNCTION

The syntax of the IF function is IF(*logical_test*, *value_if_true*, *value_if_false*). *Logical_test* is an expression that evaluates to either TRUE or FALSE, e.g., B3<C3, SUM(F3:F28)<>0, etc. Since FALSE can be represented by 0 and TRUE by any non-zero value, the formula =IF(G27, *value_if_true*, *value_if_false*) tests whether G27 is non-zero or zero.

If *value_if_true* or *value_if_false* is omitted, the IF function returns TRUE or FALSE in place of the missing expression. To avoid this, use a null string ("") instead of omitting the expression.

A common use of an IF function is to prevent the display of error values when values are missing or inappropriate. In the table shown in Figure 3-13, the following formula is used in cell C4 to calculate the percentage change in freshman chemistry enrollment from one year to the next:

=100*(B4-B3)/B3

but if the formula is filled down to cells in which both operands are missing, as in row 17 of Figure 3-13, the formula returns the #DIV/0! error message.

	A	B	C
1		**Enrollment**	
2	**Year**	**Chem I**	**% Change**
3	1989	363	
4	1990	328	-10
5	1991	358	9
6	1992	285	-20
7	1993	257	-10
8	1994	255	-1
9	1995	255	0
10	1996	329	29
11	1997	414	26
12	1998	481	16
13	1999	495	3
14	2000	536	8
15	2001		-100
16	2002		#DIV/0!
17	2003		#DIV/0!
18	2004		#DIV/0!

Figure 3-13. A worksheet displaying error values.

If the formula is replaced by

=IF(B4<>0,100*(B4-B3)/B3,"")

the calculation is not performed for cells in which the operand in column B is missing, as shown in Figure 3-14. Note that the formula =IF(B4,100*(B4-B3)/B3,"") is equivalent.

NESTED IF FUNCTIONS

IF functions can be nested. Most commonly a second IF function is used for *value_if_false*. Up to seven IF functions can be nested. The following formula performs different operations depending on whether C3-B3 is positive, zero or negative.

=IF(C3-B3>0, C3-B3, IF(C3-B3=0, C3, "unable to calculate"))

Nested IF functions allow you to calculate, in one cell, values that require different formulas depending on the value in one or more different cells. Otherwise you'd have to enter a different formula for each separate case and manually select the cells in which it was to be entered.

	A	B	C
		Enrollment	
1			
2	Year	Chem I	% Change
3	1989	363	
4	1990	328	-10
5	1991	358	9
6	1992	285	-20
7	1993	257	-10
8	1994	255	-1
9	1995	255	0
10	1996	329	29
11	1997	414	26
12	1998	481	16
13	1999	495	3
14	2000	536	8
15	2001		
16	2002		
17	2003		
18	2004		

Figure 3-14. Display of error values suppressed by using an IF function.

EXAMPLE. In the titration of a weak acid with a strong base, four different equations for the pH of the solution are required: at $V = 0$, pH $= \sqrt{C_{HA}K_a}$; before the equivalence point, pH $= pK_a + \log(C_A/C_{HA})$; at the equivalence point, pH $= 14 - \sqrt{C_{HA}K_a}$ and after the equivalence point, pH $= 14 - pOH$. You can calculate the pH of the solution by entering different formulas into cells in a column labeled "pH", as follows: for the initial point of the titration, =-LOG(SQRT(C_acid*10^-pKa); for the remaining points before the equivalence point, =pKa+LOG(A/HA); at the equivalence point, =14-(-LOG(SQRT(A*10^-(14-pKa)))); beyond the equivalence point, =14-(-LOG((C_base*V_base-V_0*C_acid)/V_tot)).

The following formula, using nested IF statements, calculates the pH using a single equation (the formula is entered all in one cell, of course):

=IF(V_base=0,-LOG(SQRT(C_acid*10^-pKa)), IF(HA>0,pKa+LOG(A/HA),

IF(C_base*V_base>V_0*C_acid,14-(-LOG((C_base*V_base-
V_0*C_acid)/V_tot)),   14-(-LOG(SQRT(A*10^-(14-pKa)))))))

Here, combining the calculations into a single expression makes for a more compact spreadsheet, and doesn't require you to decide which cells require which formula. The downside is that it's a pretty complicated formula. If you are relatively new to Excel, you'll probably find it easier to break such calculations up into parts, each in a different row or column of your worksheet.

AND, OR AND NOT

The AND and OR functions are similar to the comparison operators — they produce a logical result, either TRUE or FALSE, and are almost always used in conjunction with IF. AND and OR can take up to 30 arguments. AND(*logical1*, *logical2*,...) returns TRUE if all of its logical arguments are TRUE; OR(*logical1*, *logical2*,...) returns TRUE if at least one of its logical arguments is TRUE.

EXAMPLE. The following formula (all in one cell, of course) calculates the pK_a values of a diprotic weak acid from the pH and the parameter n-bar (symbolized by $\bar{n}$ in printed equations; see, e.g., Chapter 22, using one of two different formulas, one if n-bar is between 1.2 and 1.8, the other if n-bar is between 0.2 and 0.8; otherwise the formula returns "".

=IF(AND(n_bar>1.2,n_bar<1.8),pH+LOG((n_bar-1)/(2-n_bar)),
IF(AND(n_bar>0.2,n_bar<0.8),pH+LOG((n_bar)/(1-n_bar)),""))

The NOT function reverses the logical value of its argument. For example, the formula

=IF(NOT(ISERROR(logical_expression)),value_if_true,value_if_false)

returns value_if_true if logical_expression does not return an error value.

You can use OR to test whether a value is equal to one of the values in an array. For example, the formula

=OR(month={"Jan","Feb","Mar","Apr","May","Jun","Jul","Aug","Sep",
"Oct","Nov","Dec"})

returns TRUE if month contains the value Oct.

It's better to restrict use of this approach to arrays entered within formulas. If you define an array of months of the year elsewhere and use a formula such as =OR(month=array), you must remember to enter the formula as an array formula, that is, by pressing COMMAND+ENTER (Macintosh) or CONTROL+SHIFT+ENTER (Windows). Otherwise the formula returns FALSE unless month=Jan. You can find out much more about array formulas in the following chapter.

DATE AND TIME FUNCTIONS

Excel records dates and times by means of a serial value. There are two different serial value systems — the 1900 Date System, used by Excel for Windows, and the 1904 Date System, used by Excel for the Macintosh. In Excel for Windows, a date is calculated as the number of days elapsed since the base date, January 1, 1900; in Excel for the Macintosh, the base date is January 1, 1904. Thus in Excel for Windows, July 1, 2001 is represented as the serial value 37073, the number of days elapsed since the base date. Dates can extend to the year 9999 but not prior to 1900 (or prior to 1904 in Excel for the Macintosh).

Times are represented by the decimal part of the serial number. You can use either *number formatting* or *worksheet functions* to convert these date serial values into comprehensible dates.

As you've already seen, dates and times can be entered into worksheet cells using any one of several convenient formats: July 1 can be entered as 7-1, 7/1, July 1, Jul 1 or 1 July, among others. All these date entries produce the date 7/1/xxxx in the formula bar (xxxx is the four-digit year) and the displayed date 1-Jul. Excel enters the current year unless a different year is specified. If you enter the year, either as a two-digit or four-digit number, Excel displays the date in a different format, as e.g., 7/1/xx.

If you enter a two-digit year between 00 and 29, Excel assumes the 21st century; thus, 7/4/18 appears in the formula bar as 7/4/2018.

Times are also recognized by Excel. If you enter 10:00 in a cell, it will be recognized as a time, and 10:00:00 AM will appear in the formula bar. Excel assumes a 24-hour clock (military time) unless you indicate differently. You can use AM/PM or am/pm designations with times. Even "2 p" can be used to enter 2:00 PM in a cell.

> **Excel Tip.**   *Enter the current date in a worksheet by using CONTROL+ (semicolon) (Windows) or COMMAND+(hyphen) (Macintosh); to enter the time use CONTROL+(colon) (Windows) or COMMAND+(semicolon) (Macintosh). The date appears in the format mm/dd/yy but can be formatted otherwise.*

Dates can also be entered using the worksheet functions DATE(*yy,mm,dd*), TODAY() or NOW(). DATE is used to enter the serial value of any date. The function TIME(*hh, mm, ss*) performs the same function for a particular time.

TODAY()and NOW() return the serial value of today's date, or date and time, respectively. TODAY() returns an integer number (date only) while NOW() produces a number with decimal (date and time). Both functions return the serial value at the time the function was entered; this value is not updated continuously but *is* updated each time the worksheet is recalculated.

There are two additional functions for entering dates and times. DATEVALUE(*text*) and TIMEVALUE(*text*) convert text arguments into date serial values. For example, TIMEVALUE("8:30 PM") returns the value 0.85416667.

The following worksheet functions operate on the date serial value to return a date or time: YEAR(*value*), MONTH(*value*) and DAY(*value*). *Value* can be a serial value, a cell reference or a date as text. MONTH returns a number between 1 and 12, DAY a number between 1 and 31. The text function TEXT can also be used to format a date serial value; any custom date format can be applied (see "The FIXED and TEXT Functions" later in this chapter).

The HOUR(*value*), MINUTE(*value*) and SECOND(*value*) functions are similar to the DAY, MONTH and YEAR functions.

The WEEKDAY(*value*) function returns a number from 1 to 7 representing the day of the week (Sunday = 1, Monday = 2, etc.).

DATE AND TIME ARITHMETIC

If you keep in mind that Excel stores dates and times as date serial numbers, performing date or time arithmetic is simple. For example, in a kinetics experiment you may have a table of times at which data points were recorded at irregular intervals (Figure 3-15). To analyze the data you need the elapsed time from $t_{initial}$. Subtracting the time values from the initial value yields numbers that are decimal fractions of a day and are converted into minutes by multiplying by 24 × 60. The formula in cell B2 is =(A2-$A$2)*1440.

	A	B
1	Time	Minutes
2	10:01 AM	0
3	10:15 AM	14
4	10:33 AM	32
5	11:00 AM	59
6	12:01 PM	120
7	1:07 PM	186
8	2:15 PM	254
9	3:30 PM	329
10	4:59 PM	418

Figure 3-15. Calculating elapsed times

If you enter 10 AM in cell A1 of a worksheet, and =A1+3 in cell B1, you may at first be confused when cell B1 displays 10:00 AM. But remember that you've added three days, not three hours; if you apply the date format m/d/yy to the cell, you'll see that you've calculated a date three days from the current date. If you change the formula to =A1+3:00, you get an "Error in formula" message. That's because 3:00 is not a numerical value. To obtain the desired result, 1:00 PM, use the formula =A1+"03:00". Excel recognizes that 03:00 is text and evaluates it just as it would if you'd typed it into a cell.

TEXT FUNCTIONS

Excel provides a wide range of worksheet functions that operate on text. You are already familiar with the & operator, to concatenate text or text and values. Most of Excel's text functions select or modify one or more characters within a text string.

THE LEN, LEFT, RIGHT AND MID FUNCTIONS

The LEN(*text*) function returns the number of characters in a text string.

The LEFT(*text*, *num_characters*) function returns the leftmost character or characters in a text string. For example, LEFT("02167-3860",5) returns 02167. If *num_characters* is omitted, the value 1 is assumed. The RIGHT(*text*, *num_characters*) function is similar. If cell B7 contains a nine-digit number, then RIGHT(B7,4) returns the last four digits of the number.

The syntax of the MID function is MID(*text*, *start_num*, *num_characters*); it returns a specific number of characters from a specified position in a text string. For example, if cell A1 contains H_2SO_4, the expression MID(A1,3,1) returns S.

THE UPPER, LOWER AND PROPER FUNCTIONS

Three functions change the case of a text string: the UPPER(*text*) and LOWER(*text*) functions do what their names suggest; the PROPER(*text*) function capitalizes the first letter in each word of a text string, as illustrated in column B of Figure 3-16.

THE FIND, SEARCH, REPLACE, SUBSTITUTE AND EXACT FUNCTIONS

FIND(*find_text*, *within_text*, *start_at_num*) and SEARCH(*find_text*, *within_text*, *start_at_num*) are similar. Each returns the position number of *find_text* within the text string *within_text*. FIND is case-sensitive, SEARCH is not. For example, if cell A4 contains toluene, 2-chloro-, the expression FIND(",",A4,1) returns the value 8. Unless the optional starting position is specified, the functions begin at position 1.

The following two functions are complementary: REPLACE(*text*, *start_num*, *num_characters*, *new_text*) and SUBSTITUTE(*text*, *old_text*, *new_text*, *instance_num*).

REPLACE replaces unspecified characters at a specified position within a text string. Note that, except for the inclusion of a fourth argument, *new_text*, its syntax is similar to that of the MID function.
EXAMPLE: if cell A1 contains the text 2001, REPLACE(A1, 3, 2, "02") returns 2002.

SUBSTITUTE replaces specific characters within a string. For example, if cell A1 contains Et and cell B1 contains (C2H5) then SUBSTITUTE("Et3N", A1, B1) returns the text (C2H5)3N. If the optional argument *instance_num* is specified, only that instance of *old_text* will be replaced. If *instance_num* is omitted, all instances of *old_text* will be replaced.

EXACT(*text1*, *text2*) returns TRUE if the two strings are identical, FALSE otherwise. EXACT is case-sensitive. Simple comparison of strings is not case-sensitive. For example, the formula =("Name"="NAME") returns TRUE. Use EXACT if you want to make a case-sensitive comparison of two strings.

EXAMPLE. The following formula reformats a list of names in which the original names, in column A of Figure 3-16, are in the form LAST_NAME,FIRST_NAME; the reformatted names are in column B.

=PROPER(RIGHT(A1,LEN(A1)-FIND(",",A1))&" "& LEFT(A1,FIND(",",A1)-1))

The function FIND(",",A1) returns the position of the comma in the string. The first names and/or initials are obtained using the RIGHT function; the number of characters to be returned is equal to the length of the string minus the position number of the comma. A space is concatenated, then the last name is obtained using the LEFT function. The PROPER function is used to change the case of the string.

	A	B
47	ANTONIUS,STEPHEN J	Stephen J Antonius
48	BRUNEL,STEVEN D	Steven D Brunel
49	CARRESTO,KATHY E	Kathy E Carresto
50	LAKLIS,CLAIR L	Clair L Laklis
51	PEDROZO,BENITO A	Benito A Pedrozo
52	SOUSSANE,WALID	Walid Soussane
53	WOODSTOCK,PAUL	Paul Woodstock
54	WYNDLAKE,KEVIN D	Kevin D Wyndlake
55	ZILARIO,J PATRICK	J Patrick Zilario

Figure 3-16. A list of names reformatted using text functions.

THE FIXED AND TEXT FUNCTIONS

There are two functions that permit you to apply number formatting to values in formulas.

FIXED(*number, decimals, no_comma_logical*) formats a number as text with a specified number of decimal places, with or without commas.

TEXT(*value, format_text*) converts a number to text and formats it, using the same number formatting symbols used the **Format** menu.

EXAMPLE. When text and a number value are concatenated, the number can no longer be formatted by using menu commands or tool buttons. For example, if cell A1 contains today's date (e.g., July 1 2001), then the formula

="Today is "&A1

displays the result "Today is 37073", probably not the result that you intended. The formula

="Today is "&TEXT(A1,"mm/dd/yyyy")

displays the result "Today is 07/01/2001".

THE VALUE FUNCTION

Occasionally, number values will be entered in cells as text. When these text values are used in formulas, Excel normally evaluates them as numbers, and calculates the desired result. In the rare instance where Excel does not perform this conversion, VALUE(*text*) can be used to convert a text argument to a number.

THE CODE AND CHAR FUNCTIONS

CODE(*text*) and CHAR(*number*) perform opposite functions. CODE returns the numeric code (either Macintosh character codes or ANSI character codes for Windows) for a single character or the first character in a text string. CHAR returns the character corresponding to the character code. For example, CODE("a") returns 97, CHAR(36) returns $.

CHAR(10) is the line feed character, CHAR(13) is the carriage return (the ¶ symbol you see when you choose **Show ¶** from the **View** menu in Microsoft Word). Use one of these characters to insert a line break in text within a formula: CHAR(10) in Excel for Windows, CHAR(13) in Excel for the Macintosh.

EXAMPLE. The formula (all in one cell, of course)

="Missing Reports as of "&TEXT(NOW(),"h AM/PM mmm dd, yyyy")

&CHAR(10)&"(X = a report that has not been received"&CHAR(10)&" or was returned for recalculation)"

in an Excel for Windows worksheet produces the text displayed in Figure 3-17. To enable the line breaks, you must choose **Cells...** in the **Format** menu, choose the Alignment tab and check the Wrap Text box.

If you merely use Wrap Text in **Alignment** from the **Format** menu, Excel will decide where to break the text. By using CHAR(10) or CHAR(13), you get to decide where to break the text.

> **Missing Reports as of 9 AM Dec. 15, 1994.**
> **(X = a report that has not been received**
> **or was returned for recalculation)**

Figure 3-17. Text with line breaks inserted.

Excel Tip.   The following formula will produce the correct line break character when used in either Excel for Windows or Excel for the Macintosh:

="first text line"&CHAR(10+3*(INFO("system")="mac"))&"second line"

LOOKUP AND REFERENCE FUNCTIONS

There are several functions for obtaining values from a table, based on position or value.

THE VLOOKUP AND HLOOKUP FUNCTIONS

The function VLOOKUP(*lookup_value, array, column_index_num, match_type_logical*) looks for a match in the first column of a two-dimensional array and returns a value offset by *index_num* across the row in which the match was found. *Lookup_value* is the value to be found in the first column of array. A *column_index_num* of 2 returns a value from column 2 of array. If *match_type_logical* is TRUE or omitted, VLOOKUP returns the largest array value that is less than or equal to *lookup_value*. The array must be in ascending order. If *match_type_ logical* is FALSE, VLOOKUP returns an exact match or, if one is not found, the #N/A! error value. The array can be in any order.

When you use VLOOKUP, you must always "look up" in the first column of the table, and retrieve associated information from columns to the right in the same row; you cannot use VLOOKUP to "look up" to the left.

The function HLOOKUP(*lookup_value, array, row_index_num, match_type_ logical*) is similar to VLOOKUP, except that it "looks up" in the first row of the array and returns a value from a specified row in the same column.

THE LOOKUP FUNCTION

The function LOOKUP(*lookup_value,lookup_vector,result_vector*) has two syntax forms: vector and array. The vector form of LOOKUP looks in a one-row or one-column range (known as a vector) for a value and returns a value from the same position in another one-row or one-column range. The values in *lookup_vector* must be sorted in ascending order. If LOOKUP can't find *lookup_value*, it returns the largest value in *lookup_vector* that is less than or equal to *lookup_value*.

The array form of LOOKUP is similar to VLOOKUP or HLOOKUP in that it automatically looks in the first column or row of an array, but is limited to returning a value from the same position in the last column or row. Go to Excel's On-line Help for more details.

Excel Tip. Use the vector form of LOOKUP (instead of, e.g., VLOOKUP) when you need to return a value from a column to the left of the column containing the lookup values, or when your lookup values are in a column and your return values are in a row.

THE INDEX AND MATCH FUNCTIONS

The INDEX and MATCH functions are, in a sense, mirror images. The function INDEX(*array, row_num, column_num, area_num*) returns a single value from within a one- or two-dimensional range of cells, based on a specified position in the array. Non-adjacent selections are permitted; they are handled by *area_num*. See Appendix B for details.

The function MATCH(*lookup_value*, *array*, *match_type_num*) returns the relative position of a value in a one-dimensional array. If *match_type_num* = 1, MATCH returns the position of the largest array value that is less than or equal to *lookup_value*. The array must be in ascending order. If *match_type_num* = -1, MATCH returns the position of the smallest value that is greater than or equal to *lookup_value*. The array must be in descending order. If *match_type_num* = 0, MATCH returns the position of the first value that is equal to *lookup_value*. The array can be in any order. If no match is found, #N/A! is returned.

For an example of using VLOOKUP and MATCH, see "Looking Up Values in Tables" in Chapter 9.

USING WILDCARD CHARACTERS WITH MATCH, VLOOKUP OR HLOOKUP

You can use wildcard characters with the MATCH function. If your lookup values are text and *match_type* is set to 0, *lookup_value* can contain the asterisk (*) and question mark (?) wildcard characters. A question mark matches any single character, an asterisk matches any sequence of characters.

These wildcard characters can also be used with HLOOKUP and VLOOKUP, although this capability does not seem to be mentioned anywhere in Microsoft's documentation.

THE OFFSET FUNCTION

The OFFSET(*reference*, *rows*, *columns*, *height*, *width*) function returns a reference offset from a given reference in a one- or two-dimensional range of cells. Thus, although INDEX and OFFSET are similar, INDEX returns only a single value from a one- or two-dimensional range of cells, while OFFSET can return a reference to a range of cells.

The reference argument can be a reference to a single cell or a range. If reference is a range of cells and the optional arguments *height* and *width* are omitted, then OFFSET returns a reference of the same dimensions as reference. To select a single cell, the formula =OFFSET(reference, rows, columns, 1, 1) must be used.

For an example of using OFFSET, see "A Drop-down List Box on a Worksheet" in Chapter 8.

USING INSERT FUNCTION

Because Excel provides such a wide range of functions, it is sometime difficult to remember them, or to enter their arguments correctly. You can use Excel's **Insert Function** to paste a function in a cell, or within a formula that you're typing in the formula bar. To access Paste Function, press the *f_x* button or choose **Function...** from the **Insert** menu to display the Paste Function dialog box (Figure 3-18).

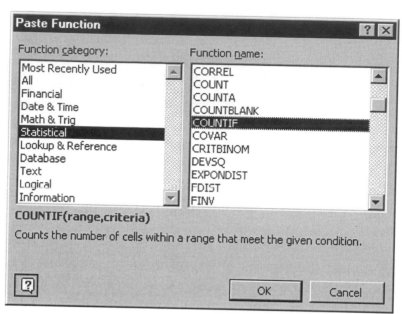

Figure 3-18. The Paste Function Step 1 dialog box.

In the dialog box you must first select a function category in the box on the left; Excel will display all the functions in that category in the Function Name box on the right. When you select a function, its name and syntax appear at the bottom of the dialog box. When you press the OK button, the Step 2 dialog box (Figure 3-19) will be displayed.

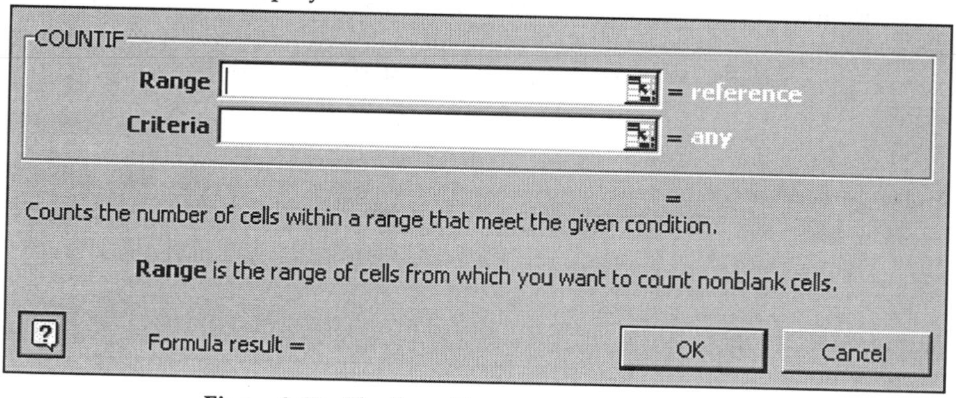

Figure 3-19. The Paste Function Step 2 dialog box.

The Step 2 dialog box presents information about each argument as you enter it: a description of the argument, and whether it is required or optional (the names of required arguments are in bold; optional arguments are non-bold).

As you enter each argument, its value is displayed in the text box on the right. Use TAB to move to the next argument. If you need information about a particular argument (the effect of entering either TRUE or FALSE for a logical argument, for example), press the Help button. When you press Finish, or RETURN or ENTER, the function is entered into the worksheet cell.

> *Excel Tip.* *When selecting functions in the Paste Function dialog box, as long as the All category is selected, if you type a letter, the first function beginning with that letter is selected from the list of functions. For example, if you type the letter D, the DATE function is selected. You can type several letters in succession to zero in on the function you want. If you type R, the RAND function is selected, but if you type R-O-W (rapidly), you will select the ROW function. If you type a string of letters that doesn't correspond to any function, you'll get a beep.*

As you become familiar with the range of functions provided by Excel, you will probably type most of them directly, rather than using the Function Wizard.

A SHORTCUT TO A FUNCTION

Most often you'll know what function you want to enter, in which case it's much faster to type in the function and its arguments instead of using Paste Function.

Occasionally you will not be sure of the arguments and their proper order. If you type the function name in a worksheet cell and then press CONTROL+A, you will go directly to the Step 2 dialog box. You can then enter values for the function's arguments.

If you type the function name in a worksheet cell and then press CONTROL+SHIFT+A, Excel will paste dummy arguments, called *placeholder arguments,* and add the closing parenthesis. For example, after entering =LINEST, press CONTROL+SHIFT+A. The function will be completed and will appear as follows:

    =LINEST(known_y's,known_x's,const,stats)

The first placeholder argument is selected (highlighted) so that you can enter a value. After entering it, double-click on the next argument to select it.

CREATING "MEGAFORMULAS"

As you become more experienced in constructing worksheet formulas, you will probably create more and more complicated ones. You can simplify a complicated calculation by performing it in steps, with intermediate calculations in separate cells of the worksheet. You can even hide the rows or columns containing the intermediate stages of the calculation. But there are advantages to constructing a single "megaformula" in which all the intermediate calculations are combined in a single formula. You'll use less memory and, more importantly, recalculation of the worksheet will take less time.

You saw some examples of megaformulas earlier in this chapter. A good way to begin to construct a megaformula is to break the calculation into steps and store the results in separate cells. When the formulas are working correctly, you can combine them all in a single cell by copying and pasting. Here's an example: a list of names was imported into Excel from a word processing document. The first few entries are shown in Figure 3-20. You want to create a column in Excel containing the last names, and a second column containing the first names and initial, if any.

The formulas to accomplish this are the following (the values returned by the formulas are shown in Figure 3-21): in cell B4, use the formula =LEN(A4)-LEN(SUBSTITUTE(A4," ","")) to determine the number of spaces in the text. In cell C4, use =SUBSTITUTE(A4," ","*",B4) to substitute a marker character for the last space in the name. (SUBSTITUTE accepts the optional argument *instance_number*, which specifies the instance of *find_text* that is to be substituted.) In cell D4, use =FIND("*",C4) to find the location of the marker character, which immediately precedes the last name portion of the string. In cell E4, use =RIGHT(A4,LEN(A4)-D4) to return the last name.

	A
4	Jeffrey N. Adams
5	Nicholas Bartlett
6	Cindy A. Bronstein
7	Ming-Hwang Chung

Figure 3-20. Portion of text imported from Microsoft Word.

	A	B	C	D	E
4	Jeffrey N. Adams	2	Jeffrey N.*Adams	11	Adams
5	Nicholas Bartlett	1	Nicholas*Bartlett	9	Bartlett
6	Cindy A. Bronstein	2	Cindy A.*Bronstein	9	Bronstein
7	Ming-Hwang Chung	1	Ming-Hwang*Chung	11	Chung

Figure 3-21. Portion of worksheet to parse test into separate columns.

Finally, combine the formulas: first, **Copy** the formula in cell D4 from the formula bar (don't include the equal sign) and press the Enter button; then select cell E4; in the formula bar, select D4 in the formula and **Paste** the formula fragment. You can now delete the formula in cell D4. Repeat the process for the formulas in cells C4 and B4, pasting them into the formula in E4. The final megaformula is

=RIGHT(A4,LEN(A4)-FIND("*",SUBSTITUTE(A4," ","*",LEN(A4)-LEN(SUBSTITUTE(A4," ","")))))

A similar formula is used to return the first name plus initial. Finally, **Delete** columns B–D.

A formula can contain up to 1024 characters, so your megaformulas can be quite complicated.

TROUBLESHOOTING THE WORKSHEET

Inevitably, your worksheet formulas will at times contain errors. Earlier in this chapter you learned some of the ways to prevent or discover errors while entering formulas. But even if a formula is syntactically correct, it may produce an error when it is evaluated. This section provides some techniques and tips for tracking down errors in your spreadsheets.

ERROR VALUES AND THEIR MEANINGS

Excel displays an error value in a cell if the formula can't be evaluated. The error values, which are #DIV/0!, #N/A, #NAME?, #NULL!, #NUM!, #REF! and #VALUE!, can give you a good idea of what caused the error. They are listed here in the approximate order of frequency with which they are encountered.

- #DIV/0! is displayed if a formula uses a cell containing zero as divisor, or if the cell is blank.

- #NAME? is displayed when you use a name that Excel doesn't recognize. Most often this occurs when you misspell a name or function, or when you enter a name without having defined it. It will also happen when you enter a text argument in a formula and forget to enclose the text in quotes, or when you use a worksheet function in a formula and forget to include the parentheses and/or arguments.

- #REF! is displayed when a formula refers to a cell that has been deleted or is beyond the range of a reference used as an argument in a function.

- #VALUE! is displayed when the wrong type of argument is used in a function.

- #N/A is displayed when certain built-in functions (especially HLOOKUP, VLOOKUP or MATCH) contain incorrect arguments.

- #NUM! is displayed when a number supplied to a function is not a valid argument, e.g., SQRT(-1).

- #NULL! is displayed if you have used the intersection operator to specify an intersection of two references that do not intersect.

EXAMINING FORMULAS

When you see an error value displayed in a cell, you'll need to examine the formula. There are several things you can do to track down the error. The error value displayed can suggest a good place to begin. For example, if the error value is #NAME?, you most likely have misspelled a variable name or function, or entered a variable name that has not yet been defined.

To track down the source of an error in a lengthy, complicated formula, examine the value of individual references, names, functions or function arguments.

Excel Tip. *To view the current value of a variable or function in a statement in the formula bar, highlight it (e.g., by double-clicking) and press F9 (Windows) or COMMAND+= (Macintosh). The value of the selected portion of the formula will be displayed. Click on the Cancel box in the formula bar or press Undo to restore the statement; otherwise the selected portion of the formula will be permanently replaced by the numerical value.*

FINDING DEPENDENT AND PRECEDENT CELLS

To audit the logic of a complicated worksheet, you may want to find all cells that contain formulas that refer to a given cell (*dependent* cells) or all cells that are referred to by the formula in a given cell (*precedents*). You can search backward or forward, finding the direct precedents, then the cells that are referred to by those cells, and so on.

First, choose **Auditing** from the **Tools** menu. The **Auditing** submenu (Figure 3-22) allows you to trace precedents, dependents or errors. Figure 3-23 illustrates a typical display when a cell is selected and **Trace Precedents** is chosen.

Trace Precedents
Trace Dependents
Trace Error
Remove All Arrows
Show Auditing Toolbar

Figure 3-22. Excel 2000 Auditing submenu.

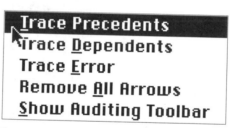

28	t, min	A(observed)	A(calculated)	
29	0	0.855	0.859	
30	1	0.521	0.522	
31	2	0.317	0.317	
32	3	0.195	0.193	
33	4	0.118	0.117	
34	5	0.070	0.071	
35				
36			Σ(difference)2 =	2.49E-05

Figure 3-23. Trace Precedents display.

USING PASTE LIST

The **Paste List** command is useful for worksheet auditing and error tracing. It produces a list of all names used in the worksheet, with their references.

Select the upper-left corner of the range in which you want the list to appear; make sure that nothing will be over-written by the list. Choose **Name...** from the **Insert** menu and choose **Paste...** from the submenu, then press the Paste List button.

Figure 3-24 shows an example of a list of names produced by **Paste List**. Inspection of the list shows that the names log_pcKw and pcKw are duplicates. Of course, it's permissible to assign more than one name to the same reference, but in general it's not good programming practice.

Sorting the list according to reference makes it easier to find duplications.

	0	P
12	calib	='Sheet 1'!$G$8
13	CB	='Sheet 1'!$G$7
14	CL	='Sheet 1'!$G$5
15	C_	='Sheet 1'!$G$6
16	log_pcKw	='Sheet 1'!$G$9
17	n_bar	='Sheet 1'!$E$13:$E$43
18	pcKw	='Sheet 1'!$G$9
19	pH	='Sheet 1'!$B$13:$B$43
20	pH_corr	='Sheet 1'!$D$13:$D$43
21	v	='Sheet 1'!$A$13:$A$43
22	V_0	='Sheet 1'!$G$4

Figure 3-24. Using **Paste List** to audit a worksheet.

USEFUL REFERENCE

Microsoft Excel 97 Worksheet Function Reference, Microsoft Press, Redmond WA, 1997.

4

CREATING ARRAY FORMULAS

Array formulas are undoubtedly Excel's most powerful formulas. With array formulas, you can accomplish things in Excel that you can't accomplish otherwise. This chapter illustrates the point by means of some case studies.

USING ARRAY FORMULAS

In Excel, the terms *range*, *array* and *matrix* are essentially identical. They all simply refer to a range of cells in a worksheet. They can either be one- or two-dimensional.

Array formulas can simplify worksheets, as illustrated in the following four examples, where absorbance values from a first-order rate process are fitted to the equation $A_{calc} = A_0 e^{-kt}$. The worksheet calculates the sum of squares of residuals, which was minimized by changing A_0 and k to obtain the least-squares best fit of the calculated absorbance to the experimental values. The values of A_0 and k obtained in this way were 0.85503 and 0.49537, respectively.

To calculate the sum of squares of the residuals, $\Sigma(A_{obsd}-A_{calc})^2$, you could use the (non-array) approach shown in the worksheet of Figure 4-1.

The sum-of-squares calculation can also be done by using an array formula, shown in the second example (Figure 4-2). The formula in cell D10 is an unusual

	A	B	C	D
1				
2	t, min	A(observed)	A(calculated)	(difference)^2
3	0	0.855	0.855	9.00E-10
4	1	0.521	0.521	7.40E-11
5	2	0.317	0.317	2.25E-07
6	3	0.195	0.193	2.40E-06
7	4	0.118	0.118	1.47E-08
8	5	0.070	0.072	3.34E-06
9				
10			Σ(difference)^2 =	5.98E-06
11			equation used in cell D10:	=SUM(D3:D8)

Figure 4-1. Calculating a sum of squares.

one, {=SUM((C5:C10-B5:B10)^2)} in which a range is subtracted from a range. Normally a formula such as this would produce a #VALUE! error, but if it is entered as an array formula, by pressing CONTROL+SHIFT+RETURN or COMMAND+RETURN (Macintosh) or CONTROL+SHIFT+ENTER (Windows), it will be evaluated correctly. Excel indicates that the formula is an array formula by enclosing it in braces. Don't type the braces as part of the formula; they are added automatically by Excel.

The third example shows how using names for the A_{obsd} and A_{calc} ranges simplifies the array formula and makes it much more self-documenting. The formula in cell D10 is

{=SUM((A_calc-A_obsd)^2)}

The cell ranges B3:B8 and C3:C8 were defined as A_obsd and A_calc, respectively.

	A	B	C	D
1				
2	t, min	A(observed)	A(calculated)	
3	0	0.855	0.855	
4	1	0.521	0.521	
5	2	0.317	0.317	
6	3	0.195	0.193	
7	4	0.118	0.118	
8	5	0.070	0.072	
9				
10			Σ(difference)2 =	5.98E-06
11	equation used in cell D10:		{=SUM((C3:C8-B3:B8)^2)}	

Figure 4-2. Using an array formula to calculate a sum of squares.

	A	B	C	D
1				
2	t, min	A(observed)	A(calculated)	
3	0	0.855	0.855	
4	1	0.521	0.521	
5	2	0.317	0.317	
6	3	0.195	0.193	
7	4	0.118	0.118	
8	5	0.070	0.072	
9				
10			Σ(difference)2 =	5.98E-06
11	equation used in cell D10:		{=SUM((A_calc-A_obsd)^2)}	

Figure 4-3 Using named arrays in an array formula to calculate a sum of squares.

	A	B	C	D
1				
2	t, min	A(observed)		
3	0	0.855		
4	1	0.521		
5	2	0.317		
6	3	0.195		
7	4	0.118		
8	5	0.070		
9				
10			Σ(residuals)^2 =	5.98E-06
11	equation used in cell D10:{=SUM((A_0*EXP(-k*t)-A_obsd)^2)}			

Figure 4-4. Using an array formula to calculate a sum of squares.
Intermediate formulas have been eliminated.

In the two array examples shown in Figures 4-2 and 4-3, the array calculations are carried out just as if separate formulas had been produced using the **Fill Down** or **Fill Right** commands. The values of A_calc and A_obsd in the same row are subtracted.

In the final example, shown in Figure 4-4, the column of A_{calc} values has been eliminated and A_calc is obtained from the formula =A_0*EXP(-k*t). All the intermediate calculations have been compressed into a single formula in cell D10, leaving only the raw data.

ARRAY CONSTANTS

In the same way that a worksheet formula can contain a simple constant, e.g., the value 3 in the formula =3*A1+A2, an array formula can contain an *array constant*. An array constant is included in a formula by enclosing the array of values in braces. In this case *you* must type the braces. When you enter the completed array formula by pressing CONTROL+SHIFT+ENTER, Excel automatically provides braces around the whole formula.

Within an array, values in the same row are separated by commas, and rows of values are separated by semicolons. For example, the array constant {2, 3, 4; 3, 2, –1; 4, 3, 7} represents a 3 × 3 array

Array constants can contain number or text values, but they cannot contain references. Individual text values in an array constant must be enclosed in quotes.

An array constant is usually incorporated in a worksheet formula, e.g.,

=MATCH(char,{188,193,170,163,162,176,164,166,165,187},0).

An array constant can be entered as a named variable in the Define Name dialog box, as shown in Figure 4-5.

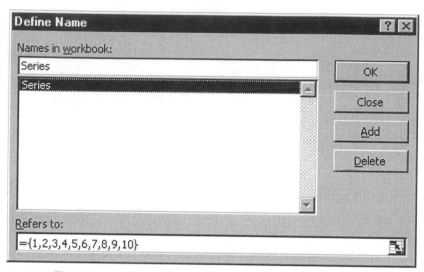

Figure 4-5. An array constant entered as a named variable.

EDITING OR DELETING ARRAYS

Since a single formula is entered in all the cells of an array, you can't change or delete part of an array; if you try, you'll get a "Cannot change part of an array" message.

To edit an array formula, simply select any cell in the array. Then edit the formula in the formula bar. When you begin to edit, the braces surrounding the formula will disappear. To re-enter the edited formula, press CONTROL+SHIFT+ENTER. The formula will be entered into all of the cells originally selected for the array.

Although you can't change part of an array's formulas or values, you can format individual cells. You can also **Copy** values from individual cells of an array and **Paste** them elsewhere.

You can select individual values in an array by using the INDEX function, but this is a "read-only" option. If you need to be able to change individual values in an array, you should enter the values in a range of worksheet cells, rather than as an array.

> *Excel Tip.*   *To select an entire array range (in order, for example, to **Clear** it), select any cell in the array, then press CONTROL+/.*

FORMULAS THAT RETURN AN ARRAY RESULT

You can also create array formulas that *return* an array of different values in a selected range of cells. Formulas that operate on matrices, described in Chapter 9, are examples of formulas that return an array.

To use a worksheet formula that returns an array result, you must first select a suitable range of cells, with dimensions (R × C) large enough to accommodate the returned array, then type the formula in the formula bar, and finally enter the formula by pressing CONTROL+SHIFT+ENTER.   Excel will indicate that the formula is an array formula by enclosing it in braces and will enter the array formula in all the selected cells.

A few of Excel's worksheet functions are array functions, that is, they return arrays as results. One of these, the LINEST statistical function, is described in Chapter 11.

CREATING A THREE-DIMENSIONAL ARRAY ON A SINGLE WORKSHEET

It's possible to create an array with three dimensions by entering an array formula in each cell of a rectangular range of cells. The following example illustrates the use of a three-dimensional array to calculate an "error surface" curve such as the one shown in Figure 5-17. The error-square sum, i.e., the sum of the squares of the residuals, $\Sigma(y_{obsd} - y_{calc})^2$ for a one-dimensional array of data points, was calculated for each cell of a two-dimensional array of trial values. The "best" values of the independent variables are those which produce the minimum error-square sum.

The spreadsheet shown in Figure 4-6 contains kinetic data obtained for the reaction sequence A → B → C. The concentration of the intermediate species B was measured at time intervals from 0 seconds to 100 seconds (only a portion of the data table in rows 4 and 5 is shown); the t and $[B]_{obsd}$ values were defined as named ranges t and B_obs. $[B]_{calc}$ values were obtained from the equation

$$[B] = [A]_0 \frac{k_1}{k_2 - k_1} \{e^{-k_1 t} - e^{-k_2 t}\}$$

which can be found in any standard text on chemical kinetics. The error-square sum was calculated for a range of trial values of k_1 and k_2 using an array formula.

The following array formula is entered in cell C10:

```
{=SUM((B_obs-(C$9/($B10-C$9)*(EXP(-C$9*t)-EXP(-$B10*t))))^2)}
```

(note the use of mixed addressing), and then filled into the range C10:K22 by using AutoFill.

A 3-D chart of the values is shown in Figure 5-17.

Consecutive First Order Reactions (A->B->C)

$$[B]_{calc} = (k_1/(k_2-k_1))*(EXP(-k_1*t)-EXP(-k_2*t))$$

Experimental values

t, sec	0	2	4	6	8	10	12	14	16	18
$[B]_{obsd}$	0.002	0.169	0.292	0.387	0.445	0.476	0.498	0.499	0.493	0.479

Error sum = $\Sigma([B]_{calc}-[B]_{obsd})^2$

	k_1 trial values								
	0.060	0.070	0.080	0.090	0.100	0.110	0.120	0.130	0.140
0.020	0.833	0.829	0.839	0.860	0.889	0.922	0.958	0.995	1.034
0.025	0.505	0.481	0.475	0.483	0.501	0.525	0.553	0.584	0.617
0.030	0.311	0.273	0.257	0.256	0.266	0.284	0.307	0.333	0.361
0.035	0.201	0.154	0.130	0.122	0.127	0.140	0.159	0.181	0.206
0.040	0.144	0.090	0.060	0.048	0.049	0.058	0.074	0.093	0.116
0.045	0.121	0.063	0.029	0.014	0.011	0.017	0.030	0.048	0.069
0.050	0.122	0.060	0.024	0.005	0.000	0.004	0.015	0.031	0.050
0.055	0.138	0.074	0.035	0.015	0.008	0.010	0.020	0.034	0.052
0.060	0.164	0.099	0.058	0.037	0.028	0.029	0.037	0.050	0.067
0.065	0.198	0.131	0.090	0.067	0.057	0.057	0.064	0.076	0.092
0.070	0.236	0.169	0.126	0.102	0.092	0.091	0.097	0.108	0.123
0.075	0.277	0.209	0.166	0.142	0.130	0.129	0.134	0.144	0.159
0.080	0.319	0.252	0.209	0.184	0.172	0.169	0.174	0.184	0.197

(k_2 trial values shown in left column)

At minimum: k_1 = 0.10 k_2 = 0.05

Figure 4-6. Spreadsheet implementation of a three-dimensional array.

EVALUATING POLYNOMIALS OR POWER SERIES USING ARRAY FORMULAS

You can evaluate series functions, for example

$$\ln(1 + x) = x - x^2/2 + x^3/3 - x^4/4 + \ldots$$

or

$$N! = 1 \times 2 \times 3 \times 4 \ldots \times N$$

in a single cell, instead of calculating individual terms in separate cells and then summing, by using an array formula. (Of course, Excel already provides the LN and FACT worksheet functions; the above are only used as examples.)

Let's begin with a simple example, the evaluation of $N!$, and evaluate it for $N = 10$. You could use an array constant as shown in the following example:

=PRODUCT({1,2,3,4,5,6,7,8,9,10})

The formula is an array formula; in order to get the correct answer, you must remember to press CONTROL+SHIFT+ENTER.

USING THE ROW FUNCTION IN ARRAY FORMULAS

Entering an array as an explicit array of constants, as in the preceding example, is not very convenient. Nor does it provide the generality needed for more complicated examples. To generate an array of integers for use in array formulas, use the ROW worksheet function. When used in an array formula, the expression ROW(1:10), evaluates to the array of numbers {1;2;3;4;5;6;7;8;9;10}. Thus the formula to evaluate $N!$ could have been entered as follows:

=PRODUCT(ROW(1:10)).

The problem with this formula is that if you insert a row above the row containing the formula, the formula becomes =PRODUCT(ROW(2:11)). You can prevent this by using the INDIRECT worksheet function, described next.

USING THE INDIRECT FUNCTION IN ARRAY FORMULAS

The INDIRECT function creates a reference specified by a text string. In the example just referred to, the formula

=PRODUCT(ROW(INDIRECT("1:10")))

will not change if rows are inserted.

Often, you'll want to use a variable number of terms in your array formula. You can do this by using INDIRECT. In the following example we want to evaluate $N!$, where cell A1 contains the value for N. In this example let's assume that cell A1 contains 20. We use the formula

=PRODUCT(ROW(INDIRECT("1: "&A1)))

The text argument of the indirect worksheet function evaluates to "1:20", and the expression ROW(INDIRECT("1:"&A1)) evaluates to

{1;2;3;4;5;6;7;8;9;10;11;12;13;14;15;16;17;18;19;20}

USING ARRAY FORMULAS TO WORK WITH LISTS

As well as the use of array formulas themselves, you'll see that the following are invaluable when developing array formulas to operate on lists:

- use names in formulas
- logical operators
- the INDIRECT worksheet function
- the F9 key to display the value returned by a formula: to display the array produced by an expression, highlight the expression in the formula bar and press F9 (Windows) or COMMAND+= (Macintosh).

In the examples that follow, we will build up the array formula step by step, to show how it works.

USING MULTIPLE CRITERIA TO COUNT ENTRIES IN A LIST

Excel's COUNTIF and SUMIF worksheet functions can return a value based only on a single criterion. For example, consider a database, the first portion of which is shown in Figure 4-7. You could use COUNTIF to return the number of students in the Chem 1 class whose major is biology, or the number whose year of graduation (YOG) is 96, but you can't use COUNTIF to find out how many students are biology majors *and* are in the class of 96. You can construct an array formula that will do this.

	A	B	C	D
1	Name	YOG	Major	Grade
2	ALBERTY, Christine	96	THEATER ARTS	77.9
3	ANSARTO, Newman	97	BIOLOGY	78.3
4	BARBIERI, Joseph	94	ENV GEO SCIEN	62.2
5	BASALTA, Patrick	97	BIOLOGY	68.4
6	BLAKELY, Patrick	97	BIOLOGY	73.2
7	BLOOMSBURG, Peter	97	UNCLASSIFIED	67.7
8	BOGACZU, Roberto	97	BIOCHEMISTRY	90.8
9	BONOAN, Elwen	97	BIOLOGY	25.7
10	BOWEN, Armand	97	BIOLOGY	28.6
11	BOYLE, Seamus	97	BIOLOGY	80.3

Figure 4-7. Using an array formula to count using multiple criteria.

A good way to construct a complicated formula such as this one is to proceed step by step, making sure that each part of the formula works properly before proceeding to the next step. The following example illustrates this procedure.

First, use Create Names to assign the names Name, YOG, Major and Grade to the values in columns A, B, C and D, respectively.

We'll begin the development of our formula by creating a formula to count the number of students whose year of graduation is 1996. Enter the formula =YOG in any worksheet cell. Only one value can be displayed in the cell, but if you highlight YOG in the formula bar and press the F9 key (Windows) or COMMAND+= (Macintosh), the array of values will be displayed (only the first portion of the array is shown):

={96;97;94;97;97;97;97;97;97;97;97;97;96;97;97;96;97;97;97;97;96

Don't forget to press the Cancel button in the formula bar or use Undo before continuing, to revert back to the original formula.

Now, change the formula to the logical expression =(YOG=96). Once again, highlight the formula and use function key F9 to display the result in the formula bar (only the first portion of the array is shown):

{TRUE;FALSE;FALSE;FALSE;FALSE;FALSE;FALSE;FALSE;FALSE;FALSE;

Earlier we learned that FALSE can be represented by zero and TRUE by any non-zero value. If a logical expression is included in an arithmetic operation, FALSE becomes zero and TRUE becomes 1. Thus we can convert the array of TRUE and FALSE values to 1's and 0's by the expression =(YOG=96)*1. Again, use function key F9 to display the array of values:

{1;0;0;0;0;0;0;0;0;0;0;0;1;0;0;1;0;0;0;0;1;0;0;0;0;0;0;0;0;0;0;1;
0;1;1;0;1;1;1;0;0;0;0;0;0;1;0;1;1;0;0;1;0;0;0;0;0;0;0;0;0;0;0;0;
0;0;0;1;0;0;0;0;0;0;0;0;0;1;0;0;1;0;1;0;1;0;1;0;1;1;0;0;0;0;0;0;
0;1;0;0;0;0;1;0;0;0;1;0;0;1;1;0;0;0;0;0;0;0;0;0;1;1;0;1;0;0;0;0;
0;0;1;0;1;1;1;1;0;0;0;0;0;0;0;0;1;0;1;0}

To find the number of students whose YOG = 96, we simply have to sum this array of 1's and 0's. The formula =SUM((YOG=96)*1) is an array formula and you must remember to press CONTROL+SHIFT+ENTER.

By now you have probably figured out how to create a formula that handles multiple criteria. Our goal was to find how many students were biology majors in the class of 96. The formula is

=SUM((YOG=96)*(Major="Biology"))

Notice that multiplying the two logical expressions has the effect of converting the result to a number, making it unnecessary to multiply by 1.

COUNTING COMMON ENTRIES IN TWO LISTS

A similar approach can be used to find the number of entries in one list that appear in another list. In the following example, ClassList1 is the class roster of students who took Chemistry for Non-Science Majors Part I; ClassList2 is the class roster for Part II. Part I is not a pre-requisite for Part II, and many students leave after taking only Part I, while other students take only Part II. How many students took both Part I and Part II?

A portion of ClassList1 is shown in Figure 4-8. Each student has a unique ID number. Thus to find the number of students in ClassList1 that appear in ClassList2, we can simply find how many ID numbers in ClassList1 (with N rows) are also found in ClassList2 (with M rows).

Once again, we begin by using the Create Names dialog box to assign the name ID1 to the range of values in column A of ClassList1, and the name ID2 to the range of values in column A of ClassList2.

To find how many ID values in ClassList1 are identical to ID values in ClassList2, we create a matrix of logical values of size $N \times M$, by means of the expression ID1=TRANSPOSE(ID2)), then convert the TRUE and FALSE values to 1's and 0's by multiplying by 1, the 1's occurring when ID1=ID2.

To find the number of values that are common to the two lists we simply need to sum the cells in the matrix. The complete formula is thus

=SUM(1*(ID1=TRANSPOSE(ID2)))

	A	B	C
1	ID Number	LAST NAME	FIRST NAME
2	08086937	Adamcik	Wendy A.
3	46457849	Adams	Daniela
4	83677323	Afayi	Tanya B.
5	16125442	Aleardo	Kathleen S.
6	89105904	Anadorelli	Jacquelyn N.
7	30219115	Apreaza	Katherine M.
8	03068210	Azmealla	Milan S.
9	54029952	Bakaymanno	Berkeley
10	49295168	Barros	Keith
11	48984878	Barros	Lisa N.
12	01624248	Bilodeau	Jonathan C.
13	72881292	Blum	Judith K.

Figure 4-8. Using an array formula to find the number of common entries in two lists.

Again, this formula is an array formula, so you must use CONTROL+SHIFT+ENTER to enter it.

If the lists did not have ID numbers, a similar formula combining the last name (e.g., LN1) and first name (e.g., FN1) into a single string

=SUM(1*(LN1&FN1=TRANSPOSE(LN2&FN2)))

would accomplish the same result.

COUNTING DUPLICATE ENTRIES IN A LIST

You may want to find the number of duplicate entries in a list. Figure 4-9 shows a sample list, Range1, with six entries in column A.

The following expression, used in the formula in cell B3, entered as an array formula, returns the number of duplicate entries in a list:

SUM(1*(Range1=TRANSPOSE(Range1)))-ROWS(Range1)

	A	B
2	Range1	
3	Able	There are 2 duplicate entries
4	Baker	There are 4 unique entries.
5	Charlie	
6	Baker	
7	Dog	
8	Able	

Figure 4-9. Using array formulas to find the number of duplicate or unique entries in a list.

The expression 1*(Range1=TRANSPOSE(Range1)) creates an 2-D array of 1's and 0's, the 1's corresponding to instances of duplicate entries. Summing this expression returns the total number of matches. Since we're comparing a list with itself, all the diagonal elements of the array will be 1's, and so we need to subtract the size of the list, which we do by using ROWS(Range1).

COUNTING UNIQUE ENTRIES IN A LIST

Instead of counting the number of duplicates in a list, you may want to find the number of unique entries. The following expression, used in cell B4 of Figure 4-9:

SUM(1*(ROW(INDIRECT("1:"&COUNTA(Range1)))=MATCH(Range1,Range1,0)))

when entered as an array formula, returns the number of unique entries in the list. Here's how it works. The expression

MATCH(Range1,Range1,0)

compares the individual values in the array Range1 (the lookup_value) to the same array (the lookup_array) and returns the relative position in the array. In this example, the array

{1;2;3;2;5;1}

is returned. The expression

ROW(INDIRECT("1:"&COUNTA(Range1)))

generates an array of integers, in this case

{1;2;3;4;5;6}

Comparing the two arrays returns TRUE for the unique items only, and multiplying the array of TRUE and FALSE values by 1 produces an array of 1's and 0's, in this case   {1;1;1;0;1;0}. Finally, summing the array of 1's and 0's returns the number of unique items in the list. Once again, the formula is an array formula and must be entered using CONTROL+SHIFT+ENTER.

INDICATING DUPLICATE ENTRIES IN A LIST

Instead of simply returning the number of duplicate entries in a list, you probably want to know which entries are duplicates and where they are located in the list. Part of a list of addresses is shown in Figure 4-10; the complete list contains over 100 entries. The list contains many duplicates; an array formula can be used to identify them.

The following formula in cell B2 of Figure 4-10 (List is in column A, beginning in cell A2), when filled down, returns the text "Duplicate in row" N:

=IF(MAX(ROW(List)*(A2=List))=ROW(),"","Duplicate in row
"&MAX(ROW(List)*(A2=List)))

	A	B
1		**Forward-looking**
2	5625 Manzanita Ave Apt 84	
3	Rr 1	
4	722 Briarbend Dr	Duplicate in row 21
5	1639 Olympia Fields St	Duplicate in row 80
6	14333 Diplomat Dr	Duplicate in row 98
7	5000 Fawn Mdws Apt 137	
8	328 N York St	
9	12 Dodge St	Duplicate in row 13
10	15140 El Cameno Real Dr	Duplicate in row 104
11	23407 Western Ave	
12	107 Ledgewood Dr	Duplicate in row 72
13	12 Dodge St	
14	14911 Eleanor Ave	Duplicate in row 86
15	107 Ledgewood Dr	Duplicate in row 72
16	94051 Doyle Point Rd	Duplicate in row 59
17	4903 Evergreen St	Duplicate in row 31
18	3529 S Chase Ave	
19	6147 Fairway Dr	Duplicate in row 33
20	423 N Columbus St	Duplicate in row 45
21	722 Briarbend Dr	
22	17614 N 131st Dr	Duplicate in row 30
23	10023 McCartney Ln	Duplicate in row 76
24	2925 Imperial Ct	Duplicate in row 37
25	111 Ashford Dr	Duplicate in row 120
26	56 Summit Ave	Duplicate in row 95
27		Duplicate in row 84
28	5705 Northfield Rd	Duplicate in row 118
29	17614 N 131st Dr	Duplicate in row 30
30	17614 N 131st Dr	
31	4903 Evergreen St	

Figure 4-10. Using an array formula to identify duplicate entries in a list.

Here's how the formula works. The expression

MAX(ROW(List)*(A2=List))

returns the largest row number of a row where the list entry in cell A2 matches an entry in the list. This formula is "forward-looking"; that is, it lists only entries that are "ahead" of the examined entry, because the MAX worksheet function is used. If desired, a comparable "backward-looking" formula can be constructed.

The IF function was incorporated because if the simple expression is used, a list entry will always "find itself". Including the IF function ignores the case where the row number of the examined entry equals the row number of the element in the array.

RETURNING AN ARRAY OF UNIQUE ENTRIES IN A LIST

Instead of simply returning the number of unique entries in a list, you can create an array formula to return the array of unique values in the example in Figure 4-9.

Let's create the formula in a stepwise fashion. To simplify the formulas, we'll use Define Name to create named formulas (see "Using Create Names" in Chapter 3). These formulas will be located in the Define Name dialog box, not in worksheet cells.

The expression

=ROW(Range1)

returns the array of row numbers of the list Range1, namely {3;4;5;6;7;8}. We'll combine this with an expression we used in an earlier section in this chapter

=ROW(INDIRECT("1:"&COUNTA(Range1)))=MATCH(Range1,Range1,0)

which returns the array {TRUE;TRUE;TRUE;FALSE;TRUE;FALSE} of unique items in the list. We'll name this formula UniqLogicals.

Multiplying the two expressions

=ROW(Range1)*UniqLogicals

returns the array {3;4;5;0;7;0} containing the row numbers of the unique entries. We'll name this formula UniqRows.

The formula

=SUM(1*UniqLogicals)

returns the number of unique items; call this formula CountUniq.

To generate the series of integers {1;2;3;4} we use the formula

=ROW(INDIRECT("1:"&CountUniq))

which we name Series.

Now we'll use the preceding in the LARGE function to obtain the array of row numbers of the unique items. The syntax of LARGE is LARGE(*array,k*) where *k* is a value used to return the k-th largest value in the array. Thus the formula

=LARGE(UniqRows,Series).

returns the array {7;5;4;3}; we'll name this formula UniqArrayRows.

	B
39	Dog
40	Charlie
41	Baker
42	Able

Figure 4-11. Using an array formula to display unique entries in a list.

The array is then used in the INDIRECT function to create an array of R1C1-style addresses. The INDIRECT function takes a second optional argument, a logical value that specifies the type of reference (FALSE if an R1C1-style reference). The formula is

=INDIRECT("R"&UniqArrayRows&"C"&COLUMN(Range1),0)

This expression, when filled down in a range of cells, returns the unique items in the list, as shown in Figure 4-11.

The single worksheet formula (all in one cell, of course)

=INDIRECT("R"&LARGE(ROW(Range1)*(ROW(INDIRECT("1:"&COUNTA
(Range1)))=MATCH(Range1,Range1,0)),ROW(INDIRECT("1:"&SUM(1*
(ROW(INDIRECT("1:"&COUNTA(Range1)))=MATCH(Range1,Range1,0))))))
&"C"&COLUMN(Range1),0)

is equivalent to the preceding formula using names.

USING AN ARRAY FORMULA TO SORT A 1-D LIST

You can use the LARGE worksheet function in an array formula to sort a list of number values automatically by means of an array formula, rather than manually by using the **Sort...** command.

For example, let's imagine that you import a list of numerical data each day and paste them into Sheet1 of a workbook. The list must then be sorted in descending order. By means of the following array formula, the list can automatically appear on e.g., Sheet2 of the workbook, sorted in descending order:

=LARGE(List,ROW()-1)

In the preceding formula, the raw data in Sheet1 was named List; the formula was entered in row 2 of Sheet2 and filled down into sufficient rows to accommodate the sorted list.

A #NUM! error value is returned if the formula is filled into more rows than are required by the raw data; the formula

=IF(ISERROR(LARGE(List,ROW()-1)),"",LARGE(List,ROW()-1))

suppresses the display of this error.

USING AN ARRAY FORMULA TO SORT A 2-D LIST

Sorting a multi-column list using the LARGE function is only slightly more complicated, provided there are no duplicate entries in the column on which the sort is performed. You can use the MATCH and INDEX functions to return the values in the same row as the "sortkey" value. Figure 4-12 illustrates a portion of a list of polymer research samples and some of their physical properties. The columns in this raw data table were assigned the names SampleNumber, FormulationType, etc.

The column headings were copied and pasted into row 1 of another worksheet. In this second worksheet, the table of data was sorted by means of the following formulas, using the data in column J as the "sortkey".

The formula in cell J2

=LARGE(MeltingPoint,ROW()-1)

sorts the melting point values in ascending order; ROW()-1 must be used, since the table begins in row 2. The following formula in cell A2 (and similar formulas in cells B2 through I2)

=INDEX(SampleNumber,MATCH(J2,MeltingPoint,0))

returns the appropriate values in the same row as the sortkey value.

	A	B	C	D	E	F	G	H	I	J
1	Sample Number	Formulation Type	Molecular Weight (kD)	MW Dispersity (Mw/Mn)	Specific Gravity (ASTM D792)	Refractive Index	Dielectric Constant @ 100 Hz (ASTM D150)	Rockwell Hardness (ASTM D785)	Tensile Strength PSI/1000 (ASTM D638)	Melting Point (°C)
2	91976	T	14.6	2.4	1.135	1.497	2.88	110	15.4	232
3	91977	R	14.6	2.3	1.149	1.502	2.88	110	16.9	231
4	91978	R	11.5	2.6	1.105	1.500	2.88	107	16.9	209
5	91979	R	12.0	2.8	1.117	1.503	2.94	115	14.2	208
6	91980	R	12.8	2.9	1.181	1.496	2.89	108	13.8	207
7	91981	R	12.2	2.6	1.172	1.497	2.94	111	13.9	205
8	92007	A	13.6	2.2	1.174	1.501	2.88	114	14.4	215
9	92039	U	13.6	2.1	1.184	1.500	2.92	112	15.6	212
10	92071	R	14.5	3.4	1.135	1.496	2.91	113	16.3	185
11	92103	T	15.2	3.6	1.143	1.500	2.87	109	16.7	174
12	92135	T	11.5	3.7	1.148	1.504	2.89	110	17.4	218
13	92167	T	12.0	3.4	1.139	1.498	2.94	107	15.3	210
14	92199	R	12.8	3.0	1.178	1.495	2.89	106	16.7	215
15	92231	R	12.2	2.9	1.177	1.499	2.89	113	13.1	212

Figure 4-12. A multi-column list.

	A	B	C	D	E	F	G	H	I	J
1	Sample Number	Formulation Type	Molecular Weight (kD)	MW Dispersity (Mw/Mn)	Specific Gravity (ASTM D792)	Refractive Index	Dielectric Constant @ 100 Hz (ASTM D150)	Rockwell Hardness (ASTM D785)	Tensile Strength PSI/1000 (ASTM D638)	Melting Point (°C)
2	92351	U	11.0	2.3	1.165	1.498	2.90	110	16.7	255
3	91976	T	14.6	2.4	1.135	1.497	2.88	110	15.4	232
4	91977	R	14.6	2.3	1.149	1.502	2.88	110	16.9	231
5	92135	T	11.5	3.7	1.148	1.504	2.89	110	17.4	218
6	92135	T	11.5	3.7	1.148	1.504	2.89	110	17.4	218
7	92135	T	11.5	3.7	1.148	1.504	2.89	110	17.4	218
8	92007	A	13.6	2.2	1.174	1.501	2.88	114	14.4	215
9	92007	A	13.6	2.2	1.174	1.501	2.88	114	14.4	215

Figure 4-13. Using an array formula to sort a multi-column list.

This formula works fine as long as the table does not contain duplicate values of the sortkey. However, in this example there are many sets of duplicate melting point values, the first occurring in rows 5, 6 and 7. The MATCH function then returns the first occurrence of a match and as a result, all rows having duplicate sortkeys will display the same information in the associated cells, as illustrated in Figure 4-13.

Where the raw data table contains duplicate values of the sortkey, a more complicated formula is required. The following formula in cell A2 (and similar formulas in cells B2 through I2) returns the correct values in the same row as the sortkey value. Let's examine the formulas in row 5, the first row showing duplicate values of the sortkey in column J.

```
{=IF(COUNTIF($J$1:J4,J5),INDEX(SampleNumber,LARGE(ROW($B$2:$B$30)
*(MeltingPoint=J5),COUNTIF($J$1:$J$30,J5)-COUNTIF($J$1:J4,J5))1),
INDEX(SampleNumber,MATCH(J5,MeltingPoint,0)))}
```

Here's how this formula works. The expression

```
COUNTIF($J$1:J4,J5)
```

calculates the number of duplicate sortkey values (here MeltingPoint is the sortkey) that appear ahead of the current value; if zero, the final part of the formula is used; this is the *value_if_false* of the IF function. This expression

```
INDEX(SampleNumber,MATCH(J5,MeltingPoint,0))
```

is identical to the simple formula described earlier for the case of no duplicate sortkey values.

On the other hand, if there are duplicate sortkey values, we have to calculate which of the several rows of values to use. The expression

ROW($B$2:$B$30)*(MeltingPoint=J5)

returns the array of row numbers for which the melting point matches the sorted value in cell J5, i.e.,

{0;0;0;0;0;0;0;0;0;0;12;0;0;0;0;0;0;19;0;0;0;0;0;25;0;0;0;0;0}

The expression

COUNTIF($J$1:$J$30,J5)-COUNTIF($J$1:J4,J5)

calculates the total number of duplicates of the melting point value, minus the number of duplicates already encountered, and is thus the k value to use in the LARGE function (in this example, the expression returns 3 in row 5, 2 in row 6, 1 in row 7). Combining these two expressions returns the row number of the row that contains the appropriate information. We must subtract 1 to obtain the number to use in the INDEX function, since the array begins in row 2.

Finally, so that the formula can be filled down into many rows, where there may not be sufficient values to fill, an additional IF statement is used to suppress the #N/A error value.

{=IF(J5="","",IF(COUNTIF($J$1:J4,J5),INDEX(SampleNumber,LARGE(ROW ($B$2:$B$30)*(MeltingPoint=J5),COUNTIF($J$1:$J$30,J5)-COUNTIF ($J$1:J4,J5))-1),INDEX(SampleNumber,MATCH(J5,MeltingPoint,0))))}

5

ADVANCED CHARTING TECHNIQUES

This chapter shows several ways to plot multiple data series, and how to customize charts, create charts with multiple axes and link spreadsheet data to chart text.

GOOD CHARTS VS. BAD CHARTS

Most charts produced by chemists are XY charts. Unless you choose one of the "smoothed lines" options, Excel connects plotted points by straight-line segments; thus, a chart of data that should produce a smooth curve might look like the "bad chart" shown in Figure 5-1.

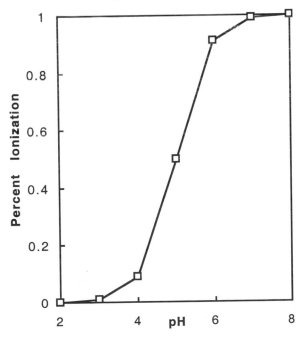

Figure 5-1. An example of a bad chart.

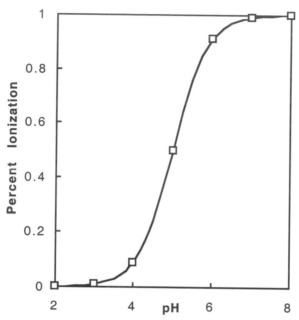

Figure 5-2. Smooth curves are produced by using small increments between data points.

To produce a smooth curve, you need to use small x increments between points. Figure 5-2 is an example of the use of small charting increments to produce smooth curves. If you want to minimize the number of calculations, you can limit your use of small plotting increments to regions where the curvature is changing rapidly.

CHARTS WITH MORE THAN ONE DATA SERIES

Excel can plot several data series in the same chart (the upper limit is 255 data series in a chart; see "Some Chart Specifications" at the end of this chapter). If the values in the data series are similar in magnitude, then plotting two or more sets of data is not any different from plotting one set. If the numbers are very different (for example, if one set of y values is in the range 1–10 and the other set is in the range 0.001–0.010), you must use a chart with a secondary axis (see "Charts with Secondary Axes" later in this chapter).

PLOTTING TWO DIFFERENT SETS OF Y VALUES IN THE SAME CHART

If more than two columns (or rows) of data are selected for plotting, Excel uses the leftmost column or uppermost row as the independent variable (plotted on the X Axis) and the remaining rows or columns as the dependent variables (plotted on the Y Axis). Figure 5-3 illustrates one column of x data and two columns of y data to be selected for a chart. If the data series are non-adjacent,

	A	B	C
3	X	Y1	Y2
4	0	0	0
5	1	3	5
6	3	9	15
7	5	15	25
8	7	21	35
9	9	27	45
10	11	33	55

Figure 5-3. Spreadsheet layout for two Y data series.

hold down the CONTROL key (Windows) or the COMMAND key (Macintosh) while you select the separated columns of data. (Excel will always use the leftmost column or uppermost row for the x values, though.)

Excel uses a different shape and color for the plotting symbols in each data series. You can change the plotting symbols or remove them if you customize the chart. Figure 5-4 illustrates the two data series from the spreadsheet of Figure 5-3.

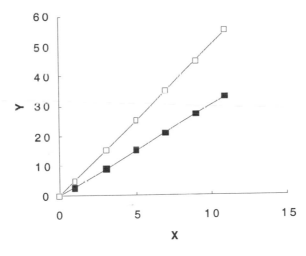

Figure 5-4. Chart with one X and two Y data series.

PLOTTING TWO DIFFERENT SETS OF X AND Y VALUES IN THE SAME CHART

To plot two sets of x, y data on a single chart, use the data layout shown in Figure 5-5.

A chart produced from the two data series in the spreadsheet of Figure 5-5 is shown in Figure 5-6.

	A	B	C
28	**X**	**Y1**	**Y2**
29	0	0	
30	1	3	
31	3	9	
32	5	15	
33	7	21	
34	9	27	
35	11	33	
36	0		0
37	2		10
38	4		20
39	6		30
40	8		40
41	10		50

Figure 5-5. Spreadsheet layout for two X and two Y data series.

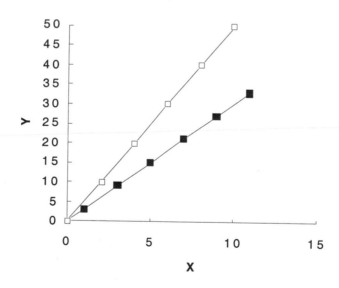

Figure 5-6. Chart with two X and two Y data series.

ANOTHER WAY TO PLOT TWO DIFFERENT SETS OF X AND Y VALUES IN THE SAME CHART

Another way to plot two sets of x, y data on a single chart is by means of the data layout shown in Figure 5-7. To use this approach, select the X and Y1 values (A2:B8) and use the Chart Wizard to create a chart. Next, select the X and Y2

	A	B
1	X	Y1
2	0	0
3	1	3
4	3	9
5	5	15
6	7	21
7	9	27
8	11	33
9		
10	X	Y2
11	0	0
12	2	10
13	4	20
14	6	30
15	8	40
16	10	50

Figure 5-7. Using **Paste Special**... for charts to add a data series.

values (A11:B16) and **Copy**, then click on the chart to activate it, then choose **Paste Special**... from the **Edit** menu to display the Paste Special dialog box for charts (Figure 5-8). Choose the Add Cells As New Series and X Values In First Column options, then press OK.

This approach has little to recommend it for simple charts, but it is useful to be able to **Copy** data ranges from widely separated areas of a worksheet and **Paste** them into a chart.

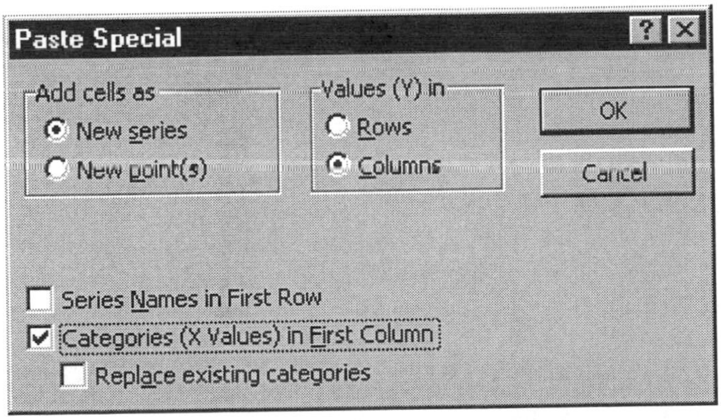

Figure 5-8. The Paste Special dialog box for charts.

EXTENDING A DATA SERIES OR ADDING A NEW SERIES

Occasionally you may want to add additional data points to a data series in a chart, or add a new series. Of course you can simply delete the chart and create a new one. But it's possible to add new data to an existing chart. You can use one of the following methods, depending on the type of chart and the layout of the data on the worksheet.

THE COPY AND PASTE METHOD
(FOR EMBEDDED CHARTS OR SEPARATE CHART SHEETS)

Highlight the range of data to be added to the chart and **Copy**. Switch to the chart sheet or activate an embedded chart by clicking on it. **Paste** the data. The new data will be added to the chart.

Excel usually does an excellent job of understanding how new data should be added to the chart, adding new data points to an existing series or adding a new series, as required. Excel will assume that the new data are additional data points in an existing series if you select additional cells in the same rows or columns as the original x values and y values; if you select new rows or columns that align with the original x values, the new data will be assumed to be new data series. Occasionally Excel will display the Paste Special dialog box, to ask you how the data should be added to the chart.

You can control how the data is added to the chart by choosing **Paste Special...** instead of **Paste** from the **Edit** menu. This allows you to add new data as new points or as a new series, independent of how the new selection aligns with the original data.

THE DRAG-AND-DROP METHOD
(FOR EMBEDDED CHARTS ONLY)

Highlight the range of data to be added to the chart. Using Drag-and-Drop, drag the new data to the chart. When the mouse pointer passes over the chart, the gray border indicating the selected cells will show that they will be added to the chart; a small plus sign will appear near the arrow pointer, indicating that you are adding a copy of the selected cells to the chart. Release the mouse button. The new data will be added to the chart.

THE COLOR-CODED RANGES METHOD
(FOR EMBEDDED CHARTS ONLY)

First, click on the outer border of the chart to select the Chart Area. All data series in the chart will be indicated on the worksheet by means of color-coded ranges: purple for x values, blue for y values, as shown in Figure 5-9. (If you selected cells containing labels for the columns, they will be color-coded green.)

To extend the range of all existing series, drag either of the handles down, as shown in Figure 5-10. To add a new Y data series, drag the *y* values handle to the right.

This method can be used only for embedded charts with data in adjacent ranges of cells. A chart with data in non-adjacent ranges does not display color-coded ranges when you click on the plot area.

To operate on a single data series, click on the *series*. The color-coded range of *x* values and a single color-coded range of *y* values will be displayed. To extend the data series, you have to drag the handle for the *x* values (the purple handle) down, then repeat for the *y* values. You can't add a new data series by dragging the y values handle (the blue handle) to the right.

	A	B	C	D
3	X	Y1	Y2	
4	0	0	0	
5	1	3	5	
6	3	9	15	
7	5	15	25	
8	7	21	35	
9	9	27	45	
10	11	33	55	
11	13	39	65	
12	15	45	75	
13	17	51	85	
14	19	57	95	
15				

Figure 5-9. Chart showing color-coded ranges.

	A	B	C	D
3	X	Y1	Y2	
4	0	0	0	
5	1	3	5	
6	3	9	15	
7	5	15	25	
8	7	21	35	
9	9	27	45	
10	11	33	55	
11	13	39	65	
12	15	45	75	
13	17	51	85	
14	19	57	95	
15				

Figure 5-10. Using the color-coded ranges to extend data series in a chart.

USING **SOURCE DATA...** IN THE **CHART** MENU
(FOR EMBEDDED CHARTS OR SEPARATE CHART SHEETS)

First, switch to the chart sheet or activate an embedded chart by clicking on it. Choose **Source Data...** from the **Chart** menu. Choose the Data Range tab if you want to change the range of several data series all at once; choose the Series tab if you want to operate on just one series, or add a new series. To change just one data series, select Series 1 (for example); cell references to the Name (the text that will be used as the legend), the x values and the y values are displayed in separate text input boxes. You can enter or edit these references by typing, or by selecting. For example, use the TAB key to select the X Values input box. The worksheet will be displayed with a marquee around the x values data series. Select the new range with the mouse. Tab to the Y Values input box, and repeat, then press OK.

EDITING THE SERIES FUNCTION IN THE FORMULA BAR
(FOR EMBEDDED CHARTS OR SEPARATE CHART SHEETS)

First, switch to the chart sheet or activate an embedded chart by clicking on it. Click on the desired data series in the chart. The definition of the data series, in the format =SERIES(*name*, *x-values_ref*, *y-values_ref*, *plot_order*, for example =SERIES("Series2", Sheet4!$A$1:$A$11, Sheet4!$C$1:$C$11, 2) will appear in the formula bar. Edit the references for both X and Y values to include the additional data points. Sometimes this is the fastest way.

CUSTOMIZING CHARTS

The following sections illustrate some ways to to improve the appearance or usefulness of a chart.

PLOTTING EXPERIMENTAL DATA POINTS
AND A CALCULATED CURVE

Plotting experimental data points and a smooth calculated curve is one of the most common applications of custom formatting. To do this you need to plot two y data series — the experimental data points and a series of points to describe the calculated curve. The y_{obsd} data should be formatted as a series of symbols with no connecting line, the y_{calc} data as a line with no symbols, as in Figure 5-11. To generate a smooth calculated curve, you'll need to have the y_{calc} points fairly close together. But since having too many points can slow recalculation of a worksheet, you should try to strike a balance between the two requirements.

Of course, to plot a calculated curve you need to have an equation that fits the data. It may be the least-squares straight line (obtained from LINEST) that best fits the data, or a curve produced by an equation appropriate for the data.

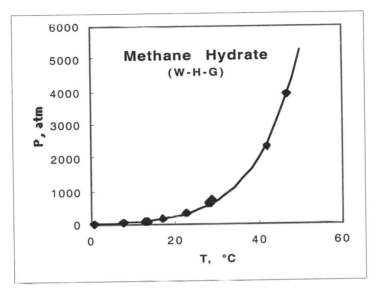

Figure 5-11. Chart with y_{obsd} and y_{calc}.

	B	C	D
8	T	P, atm	P(calc)
9	0.5	27	
10	7.7	58	
11	12.7	97	
12	13.3	105	
13	13.5	107	
14	17.0	157	
15	22.5	335	
16	27.8	640	
17	28.4	645	
18	28.8	765	
19	41.9	2344	
20	46.9	3918	
21			
22	0		22
23	2		28
24	4		36
25	6		46
26	8		59
27	10		76
28	12		96

Figure 5-12. Spreadsheet for plotting y_{obsd} and y_{calc}.

Figure 5-12 illustrates a portion of a data table showing experimental data points (temperature, pressure) for a phase diagram and part of the table of calculated pressure values, where the pressure was calculated using the theoretical relationship P = A * EXP(B/(T + 273)). Since the experimental temperature data points were in the approximate range 0–50 degrees, a series of temperature values (0–50 in increments of 2) for the calculated curve was created, beginning in row 22. The values of the parameters A and B were obtained from a least-squares fit of the experimental data to the straight-line relationship lnY = B/T + lnA.

To Plot y_{obsd} and y_{calc}

1. In the same column as the experimental x values, create a suitable range of x values (the best way is to use AutoFill) to calculate the theoretical curve. The x values can be appended either at the end or the beginning of the experimental x values.

2. Enter the expression for y_{calc} in a separate column and **Fill Down** to produce the calculated values.

3. Select the three data series (x, y_{obsd} and y_{calc}) and create an XY chart. If you created an embedded chart, you'll have to select it (click on it) before you can proceed to customize it.

4. The chart will have the experimental and calculated values more or less superimposed. Click on the experimental data series. It may take a bit of searching and clicking to find the experimental data series; if you can't find it, you can use the arrow keys to select the appropriate data series. You'll see the name of the data series (e.g., Series 1) displayed in the Name Box.

5. Choose **Selected Data Series...** from the **Format** menu and choose the Patterns tab. In the Patterns dialog box (Figure 2-5), choose Line = None and Marker = Automatic. If you want to change the style of data marker, select from the Style box. To change a solid symbol to an open one, choose Background = White. The symbol that you've selected is displayed in the Sample box in the lower right corner.

6. Now click on the data points for the theoretical curve. Again, you'll see the name of the appropriate data series displayed in the Name box.

7. Choose **Selected Data Series...** from the **Format** menu and choose the Patterns tab. Choose Marker = None and Line = Custom, then choose the style in the Style box.

ADDING ERROR BARS TO AN XY CHART

Error bars are often an important part of the graphical presentation of scientific data. You can add error bars to either the Y Axis, the X Axis or both axes of XY charts. The procedure for adding error bars in the Y Axis will be described, since this is what you'll almost always be doing.

	A	B	C	D
1	Kinetic data for an acid-catalyzed reaction			
2				
3	[H⁺]/10⁻³ M	k(obsd)/ sec⁻¹	std.dev	k(calc)
4	0.00			0.0006
5	0.06	0.00073	0.00018	0.00070
6	0.12	0.00085	0.00015	0.00076
7	0.30	0.00099	0.00019	0.00096
8	0.50	0.00119	0.0002	0.00118
9	0.75	0.00152	0.00018	0.00145
10	1.00	0.00144	0.00035	0.00173
11	1.38	0.00191	0.0005	0.00215
12	1.80	0.00291	0.00038	0.00261

Figure 5-13. Spreadsheet layout, with standard deviations, for a chart with error bars.

Figure 5-13 shows a spreadsheet containing kinetic data for an acid-catalyzed reaction, together with standard deviations of the k_{obsd} values.

Begin by creating and formatting an XY chart of the data (in this example, k_{obsd} and k_{calc} plotted vs. [H⁺]); then select the k_{obsd} data series, to which error bars are to be added, by clicking on the data series. Now choose **Selected Data Series...** from the **Format** menu and choose the Y Error Bars tab (Figure 5-14). The dialog box provides several ways to specify the magnitude of the error bars: a specified number of chart divisions, e.g., 5 (the same value for each data point), a specified percentage, e.g., 5%, of the value of each data point, a specified number

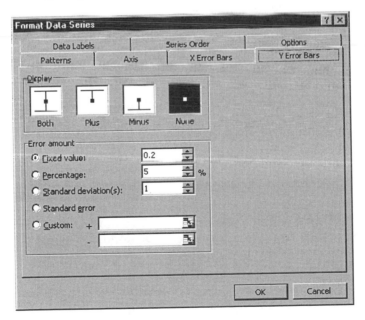

Figure 5-14. The Error Bars dialog box.

of standard deviations (but this standard deviation is the standard deviation calculated from the table of y values, not the standard deviations of individual data points), or Custom. Custom allows you to specify the magnitude of error bars according to values in a range of worksheet cells; this is the option to use if you've calculated standard deviations (or other measures of scatter) for each data point in your worksheet. Even though Both (above and below) is one of the four types of error bar options, if you choose Custom you'll have to specify the range of standard deviations twice, so don't even bother to click on the Both icon. Click in the Plus input box and then select the range of standard deviations in the worksheet. Repeat for the Minus input box, then click OK.

Figure 5-15 shows the formatted chart (k_{obsd} data points and k_{calc} line) with error bars corresponding to $\pm 1\,\sigma$.

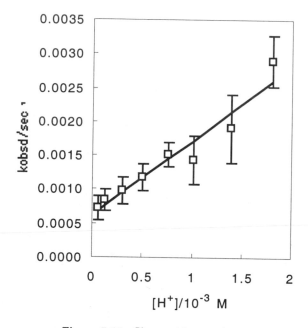

Figure 5-15. Chart with error bars.

ADDING DATA LABELS TO AN XY CHART

Data labels are text boxes associated with each data point of a chart series. You can choose to have either the label (the X value) or the value (the Y value) displayed as the data label. Neither of these two options is too useful, but you can manually edit the Data Label text to provide more useful information.

To add data labels to a particular series in a chart, activate the chart and click on the series. Choose **Selected Data Series...** from the **Format** menu and choose the Data Labels tab. Choose either the Value option or the Label option, then press OK.

To edit the text of a particular data label text, click once on the label to select all the data labels for a particular series, then a second time to select a single data label, then a third time to begin editing the text.

You can format data labels either singly or as a group. Click once on a label to select all the data labels for a particular series, then a second time to select a single data label. The Format Data Labels dialog box has tabs for Patterns, Font, Number and Alignment. The Alignment tab allows you to position the data label either Above, Below, Center, Left or Right.

CHARTS SUITABLE FOR PUBLICATION

To produce a suitable y_{calc} curve, you'll need to make sure that the x values are close enough together to produce a smooth curve. You can choose a sufficiently small increment for x to produce a smooth curve over the whole range of x, or you can manually insert smaller x increments only in regions whose curvature requires them.

Usually XY charts for publication in journals have a border around the chart area. To add a border, select the Plot area by clicking anywhere within it, then choose **Patterns** from the **Format** menu and choose Border = Automatic.

Publication-quality charts typically have data point symbols that are about 1/50th of the size of the chart. Since Excel's plotting symbols are 2 mm in size, irrespective of the size of the chart, it follows that the chart size should be about 100 mm, or 4 inches. The chart can be measured directly on the screen with a ruler and sized with the size box.

To copy a chart from Excel to Microsoft Word as a simple graphic object, first select the chart, then hold down the SHIFT key while clicking on the **Edit** menu. The **Copy** command will now be **Copy Picture...** When you choose the **Copy Picture...** command you can specify Size As Shown On Screen.

CHANGING THE DEFAULT CHART FORMAT

Excel 2000 uses the column chart type as the default chart format, but you can change the default to any of the other built-in chart formats (most likely you'd want the XY chart type to be the default). You can also set your own chart format as the default, and thus convert the Chart Wizard from a four-step wizard to a one-step wizard, so that by simply pressing the Chart Wizard's Finish button in the Chart Wizard Step 1 of 4 dialog box, you can go directly to the completed chart.

Use the procedure in the following box to define a custom chart format and make that the default format. The new chart format will remain in effect through all subsequent sessions of Excel.

You can add any number of different chart formats to the list of user-defined formats and convert a chart from one format to another.

To Make a Custom Chart Format the Default Chart Format

1. Create a chart in the desired format. Usually you'll want to make the format as general as possible (do not add title, format the axis scales, etc.), although in some instances you may want to make a chart with a specific format.

2. Activate the chart. Choose **Chart Type...** from the **Chart** menu and choose the Custom Types tab.

3. In the Select From box, press the User-defined button. A list of user-defined chart types will be displayed in the Chart Type box.

4. To add the selected chart's format to the list, press the Add button.

5. To add the selected chart's format to the list *and* make it the default chart format for all subsequent charts, press the Set As Default Chart button. In either case you'll be asked to give a name to the new custom chart type.

The procedure for deleting unwanted chart formats is shown in the box below.

To Delete a User-Defined Chart Format

1. Activate any chart.

2. Choose **Chart Type...** from the **Chart** menu and choose the Custom Types tab.

3. Press the User-Defined button in the Select From box.

4. Select the format to be deleted. The Delete button will become active.

5. Press the Delete button, then press Yes to delete the format and exit.

LOGARITHMIC CHARTS

In the Scale tab of the Format Axis dialog box, you will see an option to convert the axis (either the X Axis or the Y Axis) to a logarithmic scale. However, this option is limited; the Minimum and Maximum of the scale must be powers of 10, that is, values such as 0.01 or 1 or 10 or 1000. If your y values range from e.g., 90 to 115, and you choose the Logarithmic Scale option, the Y Axis will extend from 10 to 1000 and your data will be an almost horizontal line in the middle of the chart.

3-D CHARTS

When a dependent variable z depends on the values of two independent variables x and y, you can display the relationship graphically using a 3-D chart.

USING EXCEL'S BUILT-IN 3-D CHART FORMAT

The 3-D Surface chart type in Excel's gallery of charts is not a true 3-D chart, but rather a Line chart in two dimensions. The X and Y axes are Categories — only the Z axis is proportional to the data plotted. Thus if you attempt to produce a 3-D chart using Excel, you'll have to take this limitation into account and work within it. The values on the X and Y axes must be equally spaced. Since only one Z axis value can be charted for any pair of X and Y Axis values, you can't produce a plot of a closed surface, such as a sphere. But this type of chart can be useful, for example, to show the effect of changing two variables on the yield of a process.

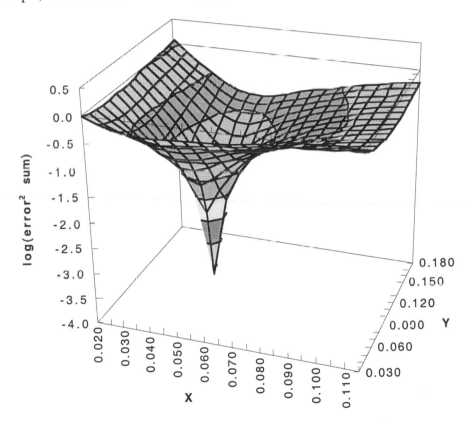

Figure 5-16. A 3-D chart.

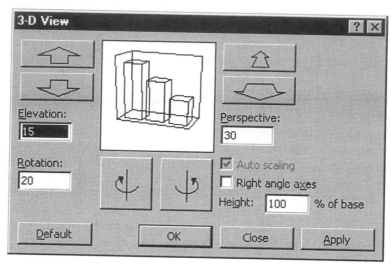

Figure 5-17. The 3-D View dialog box.

There are two useful format options within the subgallery of 3-D charts: the 3-D Surface chart and the 3-D Wireframe chart. The 3-D Surface chart uses colors to indicate areas on the surface with different ranges of Z axis values. The 3-D Wireframe chart is identical except that colors are not used. Figure 5-16 shows an example of a 3-D Surface chart.

You can change the view of a 3-D chart by choosing **3-D View...** from the **Chart** menu (Figure 5-17). Use the Rotation, Elevation and Perspective buttons to view the chart from different angles. Often a more pleasing or informative chart can be produced by changing the view. To return to the original viewing angle, press the Default button.

CHARTS WITH SECONDARY AXES

You've seen how Excel can plot more than one set of *y* data on a single chart. However, all the data series are plotted on a single Y Axis scale. Adding a secondary axis permits the graphing of sets of data with different X and/or Y Axis scales. For plotting scientific data, we'll be interested in producing charts with two different Y Axis scales plotted using the same X Axis, or with two different Y Axis scales and two different X Axis scales. The *secondary Y Axis* is plotted along the right side of the chart, the *secondary X Axis* along the top of the chart.

To produce an XY scatter chart with one X Axis scale and two Y Axis scales (a primary Y Axis along the left-hand side of the chart and a secondary Y Axis along the right-hand side, as illustrated in Figure 5-18), you first create the chart using the data for both *y* data series, then designate one of the *y* data series as the series to be plotted on the *secondary axis*. The procedure is described in the following box.

> ### To Create a Chart with a Secondary Y Axis
> **(two different Y Axis scales and the same X Axis)**
>
> 1. Select all data series to be plotted (the X Axis data series, two Y Axis data series).
> 2. Create an XY chart.
> 3. Click on the data series whose axis you want to change.
> 4. Choose **Selected Data Series...** from the **Format** menu and choose the Axis tab (see Figure 5-19).
> 5. Press the Secondary Axis button. A preview of the combination chart will be displayed. If the chart is suitable, press the OK button.

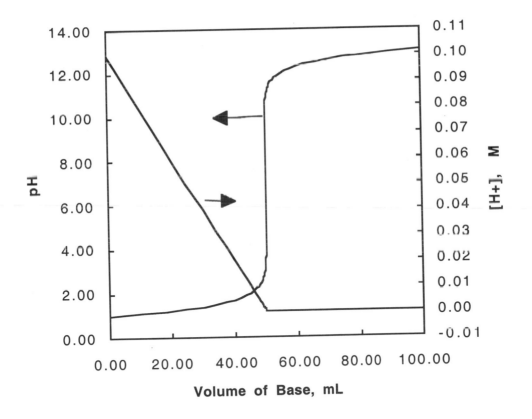

Figure 5-18. Chart with two Y Axis scales.

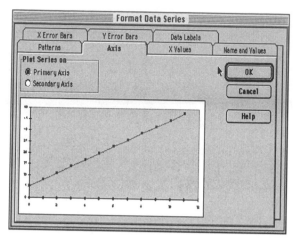

Figure 5-19. The Axis tab of the Format Data Series dialog box.

To produce an XY chart with two different X axes and two different Y axes, you must create a chart with one set of x and y data series, then paste the second set of data in the chart, then designate the series to be plotted on the secondary Y Axis and the secondary X Axis, as described in the following box.

To Create a Chart with a Secondary X Axis and a Secondary Y Axis
(two different X Axis scales and two different Y Axis scales)

1. Create the first chart in the usual way.

2. Select the second set of X and Y data series, and **Copy**.

3. Select the chart by clicking on it.

4. Choose **Paste Special...** from the **Edit** menu. In the Paste Special for charts dialog box (see Figure 5-8), press buttons for New Series and X-Values In First Column, then OK.

5. The new series will be added to the chart. Select the new series by clicking on it.

6. Choose **Selected Data Series...** from the **Format** menu. Choose the Axis tab and press the Plot Series on Secondary Axis button.

7. Choose **Axes...** from the **Insert** menu. Check the X Axis and Y Axis buttons in the Secondary Axis box, then OK. The chart will now have different axes on all four sides.

GETTING CREATIVE WITH CHARTS

Sometimes you may want to create a type of chart that is not "built in". With a little ingenuity you can create some useful charts. The example below illustrates how technical users can improvise to add more information to an XY plot in Excel.

A CHART WITH AN ADDITIONAL AXIS

Figure 5-20 shows an Arrhenius plot, in which kinetics data is plotted in the form $\ln k$ vs. $1/T$. It is common in such plots to provide, for clarity, an additional X-axis scale showing the actual temperatures used. Often this additional X-axis scale is displayed along the bottom of the chart, sometimes it is placed at the top, as shown here. It isn't possible to provide an additional scale like this using the tools provided by the Chart Wizard, but it can be done by "hand crafting". The upper scale is a "fake" scale — the Tick Marks are a data series, the Tick Mark Labels are data labels.

Figure 5-21 shows the area of a worksheet containing the chart data, the cells producing the calculated curve and the cells containing the data that create the upper X-axis tick marks (values in columns C and D in each case). The chart was created by selecting the experimental data points (C5:D10), then using the Chart Wizard to create the chart. This data series (Series 1) was formatted to display marker points and no line. Next the cells containing the values for the calculated curve were selected (C15:D16), copied, and pasted on the chart using **Paste Special**, specifying Add Cells As New Series and X Values In First Column. This data series (Series 2) was formatted to display a line with no marker points.

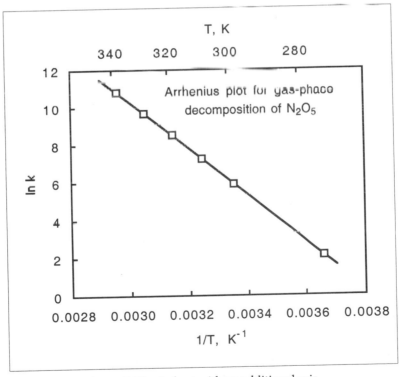

Figure 5-20. A chart with an additional axis.

The third data series in C20:D30 produces the tick marks. (To understand how these tick marks are created, remember that when Excel creates a line in an XY chart it simply connects numerical x,y coordinates with straight lines.) Since the maximum value in the Y-axis was 12, pairs of values were entered in column D (12 and 11.8, the latter having been found by trial and error to give a tick mark of suitable length). The empty rows between pairs are necessary, otherwise the tick marks would be connected to each other by a line, making a "sawtooth" pattern.

The tick marks were added to the chart by copying the table of tick mark values (C20:D30), then using **Paste Special** to add the new series (Series 3), then formatting the series to remove the marker points and change the color of the line to black.

	A	B	C	D
3			Data	
4	T, K	k	1/T	ln k
5	338	48700	0.002959	10.793
6	328	15000	0.003049	9.616
7	318	4980	0.003145	8.513
8	308	1350	0.003247	7.208
9	298	346	0.003356	5.846
10	273	7.87	0.003663	2.063
11				
12			Calculated line	
13	(slope m =	−12376	intercept b =	47.39)
14	T, K		1/T	mx + b
15	345		0.002899	11.519
16	270		0.003704	1.555
17				
18			Tick marks for upper scale	
19	T, K		1/T	
20	280		0.003571	12
21	280		0.003571	11.8
22				
23	300		0.003333	12
24	300		0.003333	11.8
25				
26	320		0.003125	12
27	320		0.003125	11.8
28				
29	340		0.002941	12
30	340		0.002941	11.8

Figure 5-21. Worksheet layout to produce a chart with an additional axis.

Finally, the tick mark labels were added by adding data labels, either values (y values) or labels (x values), to this series. (It doesn't matter whether you display values or labels, since you're going to change them anyway.) Now edit the data labels individually by clicking on them, then replacing the original text with the desired text. Change the Alignment of the data labels to Above.

When this approach is applied to Series 3 to provide Tick Mark Labels, you'll get two superimposed data labels at each x value (for example, 12 and 11.8 if you chose to display values.) This means that you'll have to select one of the two labels and delete it, then select the other label and edit it. A better way to add the data labels is to add a fourth data series with a single point for each tick mark; in this example, select C20:D20,C23:D23,C26:D26,C29:D29 and **Copy**, add this as a new series, Series 4, using **Paste Special**, then format the series to set both Line and Marker to None. Then add data labels, either values (Y values) or labels (X values), to this series. Change the Alignment of the data labels to Above. Edit the data labels individually, replacing them with the desired text.

A CHART WITH AN INSET

On the screen, you can easily create a chart with a smaller inset chart by superimposing the smaller chart on the main chart, but it's a little more difficult to transfer the result to paper. Once you've positioned the inset chart on the main chart, if you try to use Excel's **Copy** command, you can select and copy only one or the other of the two charts. There are two ways to produce a graphic containing both charts: you can either make a "screen shot" (see "Making a 'Screen Shot'" in Chapter 1) of the two superimposed charts, or you can copy the inset chart as a picture and paste it on the main chart.

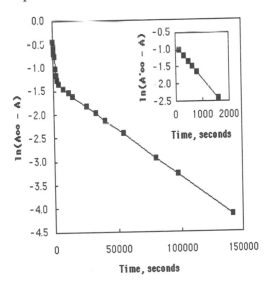

Figure 5-22 A chart with an inset.

Here's how to create a chart with an inset, similar to the example shown in Figure 5-22, which illustrates the separation of the two first-order rate constants of a consecutive process A → B → C.

To Create a Chart with an Inset

1. Create the main chart as an embedded chart or as a separate chart sheet. Customize it and size it as desired.

2. Create the inset chart as an embedded chart, and size it appropriately to fit in the main chart.

3. With the inset chart selected, hold down the SHIFT key while choosing the **Edit** menu, and choose **Copy Picture....** In the Copy Picture dialog box, choose As Shown On Screen, then press OK. Activate the main chart and **Paste** the picture. You can select the inset chart and move and size it as any graphic object. Position the inset chart on the main chart.

4. The black "handles" indicate the area of the main chart that will be obscured by the overlying inset chart. To remove the frame around the inset chart, select it, choose **Patterns** from the **Format** menu and set Border = None. If you set Background = None, you will be able to position the inset over the main chart without obscuring it.

LINKING CHART TEXT ELEMENTS TO A WORKSHEET

Any text element in a chart can be linked to a worksheet cell, causing the text in the worksheet to be displayed in the chart. (Some of this can be done automatically, as, for example, category labels.) But chart titles or unattached text boxes can also be linked to the worksheet. In this way titles can be generated automatically, or explanatory notes in text boxes can include information that will change as the data in the worksheet is modified.

Basically, to link a chart text element to a worksheet cell, you enter a formula as the chart text. The syntax of the formula is:

=worksheet_name!absolute_cell_reference.

Follow the procedure outlined in the box on the following page.

The formula entered as chart text must be only a cell reference. If you want the chart text to be text concatenated with a number, e.g. to produce a title such as Half-wave Potential = -0.74V, where the potential is a value in cell B2 of the worksheet, the complete formula ="Half-wave potential = "&B2&"V" must appear in the worksheet cell.

To Link Worksheet Text to a Chart

1. The worksheet must be open.

2. The chart must be activated. If the chart is an embedded chart, you must activate it by clicking on it.

3. With the chart as the active document, select the chart text element (Chart Title, X Axis Title, Y Axis Title or Unattached Text) by using the **Chart** menu or by clicking on the chart text element if it already exists in the chart.

4. In the formula bar, type =*worksheet_name!absolute_cell_reference*, e.g., ='Voltammetric curve'!$A$7. If the worksheet name contains spaces, it must be enclosed in single quotes.

5. Alternatively, you can have Excel supply the worksheet name and cell reference. In the formula bar, type "=". Activate the worksheet by clicking on it or by selecting it from the **Window** menu. The external reference to the worksheet will appear in the formula bar. Select the desired cell in the worksheet. The absolute reference to the cell will appear in the formula bar.

TO SWITCH PLOTTING ORDER IN AN XY CHART

The Chart Wizard always uses the leftmost column (of those that you selected) for the x values when it creates an XY chart. To change the plotting order, so that values from a column other than the leftmost one are used as the x values, you must first create a chart in the normal way, then either:

- Click on the desired data series in the chart. The series formula will appear in the formula bar, e.g.,

  =SERIES(,Demo1!$A$1:$A$11,Demo1!$B$1:$B$11,1).

  Manually edit the data series to reverse the X and Y values in the series, e.g.,

  =SERIES(,Demo1!$B$1:$B$11,Demo1!$A$1:$A$11,1)

- Choose **Source Data...** from the **Chart** menu and choose the Series tab. The Series dialog box displays the range of x values and y values for a selected series. Enter the new ranges for x and y and press OK.

SOME CHART SPECIFICATIONS (EXCEL 2000)

Maximum number of...

Charts linked to a worksheet	Limited by available memory
Worksheets referred to by a chart	255
Data series in one chart	255
Data points in a data series	32,000 for 2-D charts
	4,000 for 3-D charts
Data points for all data series in one chart	256,000

6

USING EXCEL'S DATABASE FEATURES

This chapter demonstrates how to create and use a database on an Excel worksheet. Although database software such as Filemaker or Access is superior to Excel, it is sometimes useful to be able to combine the features of a database with Excel's superior calculation and data analysis features.

THE STRUCTURE OF A LIST OR DATABASE

A rectangular range of data in an Excel worksheet can be used as a *list* or *database*. A database consists of a number of *records*, each of which can contain a number of *fields*. For example, a compilation of physical properties of organic compounds, such as the one in the *CRC Handbook of Chemistry and Physics*, is a database; the row of data for a particular compound is a record and the values for the melting point, boiling point, solubility, etc. are the data fields within the record. In Excel, a list or database must be arranged in tabular form, with row or column labels; that's the only requirement.

Most fields in a database will contain values that have been entered as text, numbers or dates. A database may also contain *calculated fields*, containing values that are calculated by using Excel formulas.

SORTING A LIST

One of the most common operations performed on a list is sorting it in increasing or decreasing order with respect to the value in one of its fields. Use the **Sort...** command in the **Data** menu to do this.

Figure 6-1 illustrates a portion of a list of the elements, together with their symbols, atomic weights and electronic configurations. The list (which was imported into Excel by using a scanner, as described in Chapter 7) is arranged in order of increasing atomic number. It may seem redundant to include a column of atomic numbers in the list, but this column is necessary if you want to return the sorted list to its original order.

To sort the list according to any of its fields, select the complete list (be sure to select all rows and columns, or the list may become irreversibly scrambled).

	A	B	C	D	E
1	Table of Atomic Weights and Electron Configurations				
2					
3	Name	Symbol	At. Num.	At.Wt.	Elec. Config.
4	Hydrogen	H	1	1.00797	1s1
5	Helium	He	2	4.0026	1s2
6	Lithium	Li	3	6.939	[He] 2s1
7	Beryllium	Be	4	9.0122	[He] 2s2
8	Boron	B	5	10.811	[He] 2s2 2p1
9	Carbon	C	6	12.01115	[He] 2s2 2p2

Figure 6-1. A portion of a list of atomic weights and symbols of the elements.

For a large list such as this one, it's convenient to use one of the methods for selecting a block of cells described in Chapter 1: select the first row of cells and press CONTROL+SHIFT+(down arrow) or place the mouse pointer on the bottom edge of the selected row, hold down the SHIFT key and double-click to select all cells to the bottom of the block. Then choose **Sort...** from the **Data** menu to display the Sort dialog box.

To choose the field to sort by (sometimes called a *sortkey*), choose the field name from the Sort By list box (Figure 6-2) or type a reference in the input box; for example, to sort by atomic symbol, you can either choose symbol (field names are not case-sensitive) or enter $B$3 in the Sort By input box. Then press OK. Figure 6-3 shows a portion of the list, sorted in ascending order according to column 1. If the list is not sorted as you want it, press the Undo button or choose **Undo Sort** from the **Edit** menu.

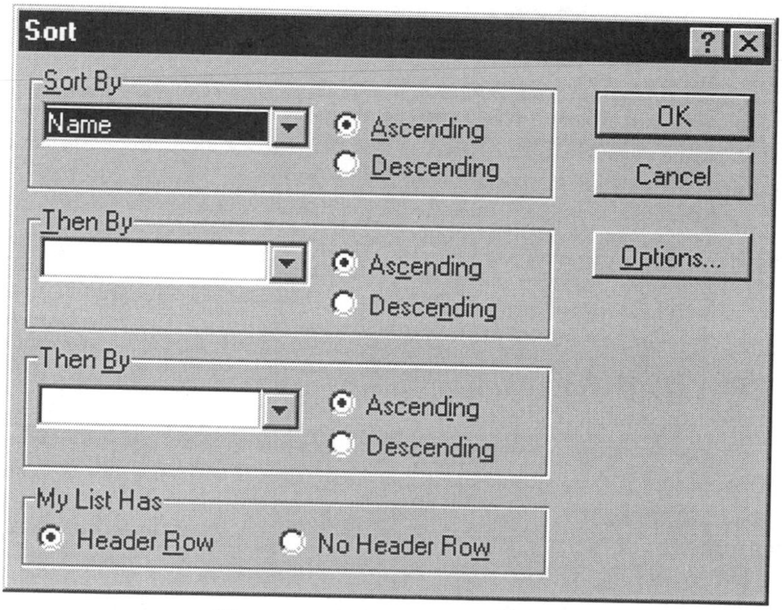

Figure 6-2. The Sort dialog box.

	A	B	C	D	E
3	**Name**	**Symbol**	**At. Num.**	**At.Wt.**	**Elec. Config.**
4	Actinium	Ac	89	[227]	[Rn] 7s2 6d1
5	Aluminum	Al	13	26.9815	[Ne] 3s2 3p1
6	Americium	Am	95	[243]	[Rn] 7s2 5f7
7	Antimony	Sb	51	121.75	[Kr] 5s2 4d10 5p3
8	Argon	Ar	18	39.948	[Ne] 3s2 3p6
9	Arsenic	As	33	74.9216	[Ar] 4s2 3d10 4p3

Figure 6-3. Portion of a list sorted according to element name.

SORTING ACCORDING TO MORE THAN ONE FIELD

As you can see from the Sort dialog box, it is possible to sort by up to three separate fields. For example, you can sort a list of chemistry students by descending order according to year of graduation (freshman, sophomore, junior, senior) and by ascending order alphabetically within each year.

SORT OPTIONS

Pressing the Options... button in the Sort dialog box displays the Sort Options dialog box and allows you to change the default sorting options.

Excel assumes that your data fields are in columns and sorts your list by rearranging rows. If you want to sort a list horizontally (i.e., to rearrange the columns of the list rather than the rows) press the Sort Left to Right button in the Orientation box (Figure 6-4).

You can choose case sensitive sorting, in which lowercase letters follow uppercase letters in an ascending sort (AaBbCc..., not ABC...abc...). If you choose this option, the sortkeys will also be case-sensitive.

The Sort Options dialog box contains four lists for custom sorts (other than ascending or descending). For example, you can sort in the order Jan, Feb, Mar, Apr, etc. You can create a custom sort order for your own specialized application (go to "Creating a custom sort order" in Excel's On-line Help for details).

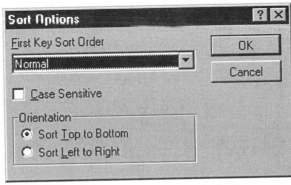

Figure 6-4. The Sort Options dialog box.

USING AUTOFILTER TO OBTAIN A SUBSET OF A LIST

Often you'll want to examine a subset of a list — only records for which one or more data fields match certain criteria. The process of abstracting a subset of a list is known as using a *data filter* or *querying* the database. The **AutoFilter** command makes it easy to obtain a subset of a list.

Figure 6-5 illustrates a portion of a list of polymer research samples and some of their physical properties. To use **AutoFilter**, first select the complete list or a partial area to be examined; if the list is separated from the rest of the worksheet data by blank cells, you need only select any cell within the list. Choose **Filter** from the **Data** menu and then choose **AutoFilter** from the submenu. Excel adds drop-down arrow buttons to row 1, which ideally should contain column labels.

To use a data filter on molecular weight (column B), click the arrow button in that column. Excel displays a drop-down list of all values in the column, plus "Custom...". To display all records that match one of the values in the selected field, you can select it from the list. To perform other logical comparisons, choose Custom... to display the Advanced Filter dialog box (Figure 6-6).

Click the comparison operator drop-down arrow and choose >. TAB over to the text box and enter 12, then press the OK button. The records in the list for samples with molecular weight greater than 12 kD are displayed (Figure 6-7).

	A	B	C	D	E	F	G	H	I
1	Sample Number	Molecular Weight (kD)	MW Dispersity (Mw/Mn)	Specific Gravity (ASTM D792)	Refractive Index	Dielectric Constant @ 100 Hz (ASTM D150)	Rockwell Hardness (ASTM D785)	Tensile Strength PSI/1000 (ASTM D638)	Melting Point (°C)
2	91976	14.6	2.4	1.135	1.497	2.88	110	15.4	232
3	91977	14.6	2.3	1.149	1.502	2.88	110	16.9	231
4	91978	11.5	2.6	1.105	1.500	2.88	107	16.9	209
5	91979	12.0	2.8	1.117	1.503	2.94	115	14.2	208
6	91980	12.8	2.9	1.181	1.496	2.89	108	13.8	207
7	91981	12.2	2.6	1.172	1.497	2.94	111	13.9	205
8	92007	13.6	2.2	1.174	1.501	2.88	114	14.4	215
9	92039	13.6	2.1	1.184	1.500	2.92	112	15.6	212
10	92071	14.5	3.4	1.135	1.496	2.91	113	16.3	185
11	92103	15.2	3.6	1.143	1.500	2.87	109	16.7	174
12	92135	11.5	3.7	1.148	1.504	2.89	110	17.4	218
13	92167	12.0	3.4	1.139	1.498	2.94	107	15.3	210
14	92199	12.8	3.0	1.178	1.495	2.89	106	16.7	215
15	92231	12.2	2.9	1.177	1.499	2.89	113	13.1	212

Figure 6-5. A portion of a list with AutoFilter buttons displayed.

Figure 6-6. The Advanced Filter dialog box.

To restore the complete list, simply choose **AutoFilter** from the **Data** menu again. Notice that the **AutoFilter** command is checked, indicating that the list has been filtered.

You can add a second data filter on data in the same field using the Advanced Filter dialog box. The second filter can be either an *And filter* (e.g., samples with molecular weight greater than 12 kD and less than 15 kD) or an *Or filter*.

	A	B	C	D	E	F	G	H	I
	Sample Number	Molecular Weight (kD)	MW Dispersity (Mw/Mn)	Specific Gravity (ASTM D792)	Refractive Index	Dielectric Constant @ 100 Hz (ASTM D150)	Rockwell Hardness (ASTM D785)	Tensile Strength PSI/1000 (ASTM D638)	Melting Point (°C)
1									
2	91976	14.6	2.4	1.135	1.497	2.88	110	15.4	232
3	91977	14.6	2.3	1.149	1.502	2.88	110	16.9	231
6	91980	12.8	2.9	1.181	1.496	2.89	108	13.8	207
7	91981	12.2	2.6	1.172	1.497	2.94	111	13.9	205
8	92007	13.6	2.2	1.174	1.501	2.88	114	14.4	215
9	92039	13.6	2.1	1.184	1.500	2.92	112	15.6	212
10	92071	14.5	3.4	1.135	1.496	2.91	113	16.3	185
11	92103	15.2	3.6	1.143	1.500	2.87	109	16.7	174
14	92199	12.8	3.0	1.178	1.495	2.89	106	16.7	215
15	92231	12.2	2.9	1.177	1.499	2.89	113	13.1	212
16	92263	13.6	3.9	1.190	1.497	2.92	109	13.1	211

Figure 6-7. A portion of a list that has been filtered.

Sample Number	Molecular Weight (kD)	MW Dispersity (Mw/Mn)	Specific Gravity (ASTM D792)	Refractive Index	Dielectric Constant @ 100 Hz (ASTM D150)	Rockwell Hardness (ASTM D785)	Tensile Strength PSI/1000 (ASTM D638)	Melting Point (°C)
91976	14.6	2.4	1.135	1.497	2.88	110	15.4	232
91977	14.6	2.3	1.149	1.502	2.88	110	16.9	231
92358	13.0	3.0	1.188	1.500	2.88	106	13.0	218

Figure 6-8. A list that has been filtered on two fields.

USING MULTIPLE DATA FILTERS

You can also add a second filter by querying data in a second field. For example, you can display all samples with melting point greater than 215°C. Figure 6-8 shows the subset of records having both molecular weight greater than 12 kD and melting point greater than 215°C.

To "undo" a particular filter, click the drop-down list button in that column and choose "(All)" from the list.

Although only three records are displayed, you can't calculate the average molecular weight of the three samples by entering the formula =AVERAGE() in a worksheet cell and selecting the displayed cells in column B that contain the molecular weight information. The range entered by selecting will be (in this example) B2:B25, and the average will not be the desired one. The most convenient way to apply SUM, AVERAGE, STDEV, etc., to a subset is to **Copy** the desired cells, **Paste** them in a convenient location and then perform the calculation.

DEFINING AND USING A DATABASE

Excel provides a number of menu commands and functions that permit you to use a list as a database. You can use menu commands to find records in the database that match criteria that you define, or you can use worksheet functions to extract numerical information from a database.

CREATING A DATABASE

To use Excel's database commands and functions, your list must meet the following requirements.

- The list must be in column format: each column in the list is a field in the database, each row in the list is a record in the database.

- The top row of the list must contain the field names, which you will use to identify the information stored in each field.

- If you leave at least one blank column and one blank row between the list and other data on the worksheet, Excel can detect and select the list automatically. If you want to use this automatic feature, however, you cannot have blank rows or columns within the list.

DEFINING A DATABASE

For Excel to recognize a list as a database, simply place the mouse pointer anywhere in the list. Alternatively, you can assign the range name **Database** to the list. This also permits Excel to detect the database automatically. Or you can use a reference to the desired range of cells, as illustrated in some of the examples that follow.

To assign the range name **Database** to the list, select the entire range of cells in the list, including the field names. Then choose **Set Database** from the **Data** menu (available in Excel 97 only). Excel assigns the name **Database** to the selected range. You can also define a database by assigning the name **Database** to the selected range of cells by using Define Name. You can define only one list in a worksheet as **Database**; if you have more than one list in a worksheet, you will have to redefine the database range in order to switch databases.

> *Excel Tip. The name* **Database** *is one of Excel's built-in names. Excel recognizes the name* **Database** *as the reference to use in database functions. Other built-in names include* **Criteria, Extract, Print_Area, Print_Titles** *Don't use any of these names as variable names except within the context of their normal use.*

ADDING OR DELETING RECORDS OR FIELDS

You can add new records to a database either by inserting new rows within the database or by entering new information below the last existing record. If you insert additional rows within the range, Excel updates the definition of the range name **Database** to conform to the new range. If you enter information below the last row of the defined range, you will have to redefine the range of the database by using the **Set Database** command (Excel 97) or by using Define Name.

If you prefer to keep the database records in the order of their entry, you can include a dummy row at the end of the database, and always insert new rows just above the dummy row. You can add new data fields (columns) in exactly the same way. Alternatively, you can use Excel's Data Form to add new records.

UPDATING A DATABASE USING DATA FORM

A convenient way to edit existing records or enter new data in a database is by using Excel's Data Form. The database range must already have been defined, either by using **Set Database** or by selecting a cell within the block of cells that comprise the database. Choose **Form...** from the **Data** menu. The Data Form dialog box will appear, with the name of the worksheet in the title bar, as in Figure 6-9.

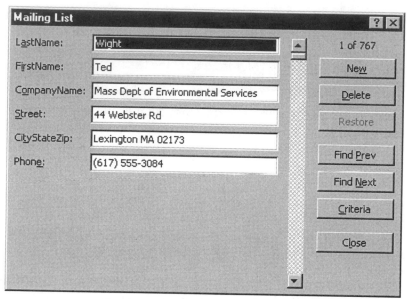

Figure 6-9. The Data Form dialog box.

Each field name in the database appears in the dialog box, along with the entries from the first record in the database. The current record and the number of records in the database are displayed in the upper right corner of the dialog box. Entries in *editable fields* display their values in a text entry box, while values that do not appear in a text box are either in calculated fields or in fields whose contents are protected.

To move from record to record, use either the Find Prev and Find Next buttons, the up and down arrow keys, or the scroll bar. The first field of each record is highlighted. Move forward from field to field within a record by using the TAB key. You can also select a field for editing by clicking the mouse pointer in it.

Edit characters in a text box in the usual way. Once a change has been made in a field, the Restore button is enabled; you can use it to restore the original data if you spot an error in the new text. To edit a single field in all records, tab forward to that field or click the mouse pointer in it; then use the up or down arrow keys to move to the same field in other records.

To add a new record, press the New button or drag the scroll button to the bottom. New Record will be displayed in the upper right corner of the dialog box. The new record will be added to the bottom of the database. Formulas of calculated fields are also included in the new record. If the new record will overwrite existing information outside the database, you'll get the "Cannot extend database" message.

The range name **Database** will be updated automatically when you add or remove records.

FINDING RECORDS THAT MEET CRITERIA

You can use **Advanced Filter...** from the **Filter...** submenu to display only records that match selection criteria, or to write them to another location in the worksheet.

To use Advanced Filter you must create a Criteria range somewhere on your worksheet, as shown in Figure 6-10, 6-11, 6-12 or 6-13. You can give it a range name if you wish, by using Define Name. Any name is acceptable, but Criteria is best because Excel will recognize it. You can also establish an Extract range (see "Extracting Records" later in this chapter) and give it the name Extract. You do not need to assign the name Database to the list to be filtered. Just select any cell within the list and Excel will recognize the range as a list.

Now choose **Advanced Filter...** from the **Filter...** submenu of the **Data** menu. Excel will display the Advanced Filter dialog box (Figure 6-10). If you selected a cell within the list, the reference to the list will appear in the List Range box. If you had assigned the name Criteria to the criteria table, the reference will appear in the Criteria Range box.

You can choose to Filter The List In-Place or Copy To Another Location. Filter The List In-Place hides the rows in the list that do not meet the criteria; Copy To Another Location copies the rows that meet the criteria to another location in the worksheet or to another worksheet (see "Extracting Records" later in this chapter).

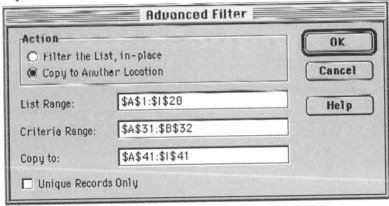

Figure 6-10. The Advanced Filter dialog box.

DEFINING AND USING SELECTION CRITERIA

You define selection criteria by setting up a table in the worksheet with one or more *field name*s in a row and the desired criteria below the field names. The Criteria field names must be the same as the database field names. For example, in Figure 6-11, the range K2:K3 is a criteria range to select all records having molecular weight greater than 12 kD. Now select the range of cells containing the field name and the criterion and use Define Name to assign the name

		K
2	**Molecular Weight (kD)**	
3	>12	

Figure 6-11. A simple criteria table.

Criteria to the selected range of cells. (Unfortunately, there is no **Set Criteria** command in the Excel 97 **Data** menu.) At any time, only one range in a worksheet can be defined as the Criteria range.

It's good practice to copy the complete row of database field names and use it to create a Criteria range, as in Figure 6-12. You can then enter single or multiple criteria in the appropriate cells.

USING MULTIPLE CRITERIA

You can select records on the basis of two or more criteria. Multiple criteria can be combined to produce a logical AND, a logical OR, or a combination of AND and OR.

If you enter two criteria in the same row, you have created an AND criterion (e.g., molecular weight >12 kD and melting point >215°C). If you enter two criteria in separate rows, you have created an OR criterion (e.g., molecular weight > 12 kD or melting point <200°C) as illustrated in Figure 6-13.

If the AND criteria apply to the same field (e.g., all samples with molecular weight >12 kD and <13 kD), you must duplicate the field name in the Criteria range (see Figure 6-14).

	A	B	C	D	E	F	G	H	I
30				*Criteria range*					
31	Sample Number	Molecular Weight (kD)	MW Dispersity(Mw/Mn)	Specific Gravity (ASTM D79	Refractive Index	Dielectric Constant @ 100 H:	Rockwell Hardness (ASTM	Tensile Strength PSI/1000	Melting Point (°C)
32		>12							

Figure 6-12. Criteria tables are best made by copying the row of database field names.

	A	B	C	D	E	F	G	H	I
34				*Criteria range*					
35	Sample Number	Molecular Weight	MW Dispersity(MW	Specific Gravity (	Refractive Index	Dielectric Constan	Rockwell Hardnes:	Tensile Strength	Melting Point (°C)
36		>12							
37									<200

Figure 6-13. A criteria table with OR logic.

	A	B
31	Molecular Weight (kD)	Molecular Weight (kD)
32	>12	<13

Figure 6-14. A criteria table with AND logic applied to the same field.

SPECIAL CRITERIA FOR TEXT ENTRIES

When you use a number value in a Criteria range, Excel returns only the records that match the field value exactly. But when you use a text value as a criterion, Excel returns all records that contain text values in the specified field that *begin* with the text value. For example, in a database of alumni information (LastName, FirstName, StreetAddress, City, State, ZIP, YearOfGraduation, etc.), using the letter N as criterion in the LastName criteria field will return all names that begin with N.

You can use wildcard characters in text criteria. The ? wildcard character represents any single character, the * character represents any number of characters. Thus, for example, the criterion *phen* will extract, from a database of chemicals, any entry that contains the character string "phen", such as phenol, 1,10-phenanthroline and benzophenone.

To extract only the records that have an exact match between a text criterion and a field value, enter the criterion in the form

="criterion_text"

where criterion_text is the text string you want to find.

EXTRACTING RECORDS

To extract a copy of all records that meet specified criteria, first define Database and Criteria ranges as described earlier. Then do the following to create a destination range for the extracted records: **Copy** the database field names and **Paste** them in a suitable area of the worksheet as labels for the extracted information. Select the row of field names and enough empty rows below the column labels for the extracted information; define this range as the Extract range by assigning the name Extract to the selected range of cells using Define Name.

Now choose **Advanced Filter...** from the **Filter...** submenu of the **Data** menu. Excel will display the Advanced Filter dialog box (Figure 6-10). Press the button for Copy To Another Location and, if you have assigned the name Extract, Excel will automatically enter the reference to the Extract range; otherwise, enter a reference in the input box. When you press OK, the records that meet the criteria will be copied to the Extract range (Figure 6-15). (Remember that all cells from the bottom of the Extract range to the bottom of the worksheet will be cleared.)

Excel Tip. *Be sure to select a range of empty cells as the Extract range. If you select only the field names as the Extract range, the* **Extract** *command will clear all cells below the Extract field names to the bottom of the worksheet, regardless of whether values are extracted into them, erasing any previous information. The* **Undo** *command cannot be used to reverse the action and restore the missing information.*

	A	B	C	D	E	F	G	H	I
40					*Extract range*				
41	Sample Number	Molecular Weight	MW Dispersity(M	Specific Gravity (	Refractive Index	Dielectric Constar	Rockwell Hardness	Tensile Strength	Melting Point (°C)
42	91976	14.6	2.4	1.135	1.497	2.88	110	15.4	232
43	91977	14.6	2.3	1.149	1.502	2.88	110	16.9	231
44	92358	13.0	3.0	1.188	1.500	2.88	106	13.0	218

Figure 6-15. An extract of a database.

	A	B	C	D	E
40			*Extract range*		
41	Sample Number	Molecular Weight	Specific Gravity	Refractive Index	Melting Point (°C)
42	91976	14.6	1.135	1.497	232
43	91977	14.6	1.149	1.502	231
44	92358	13.0	1.188	1.500	218

Figure 6-16. An extract containing only selected fields.

You can create an Extract range that contains only selected fields of each record. Only the fields included in the Extract range will be returned, as shown in Figure 6-16.

USING DATABASE FUNCTIONS

There are 12 *database functions* that return information about the records in a database: DAVERAGE, DCOUNT, DCOUNTA, DGET, DMAX, DMIN, DPRODUCT, DSTDEV, DSTDEVP, DSUM, DVAR and DVARP. Some of these are particularly useful and are described here. See *Microsoft Excel Worksheet Function Reference* or Excel's On-Line Help for further information.

DAVERAGE returns the average of the values in the specified field of all records that match the criteria.

DMAX or DMIN returns, respectively, the maximum or minimum value in the specified field of all records that match the criteria.

DSTDEV returns the standard deviation of the values in the specified field of all records that match the criteria.

All 12 database functions have the syntax (database, field, criteria). Database is either the reference to the database or the name assigned to it. Field is either the name of the field, as text, or a number indicating the position of the field within the database. Criteria is either the reference to the criteria range or the name assigned to it.

EXAMPLE. To obtain the average refractive index of all samples in the database that have molecular weight greater than 12 and melting point greater than 215, enter the database function DAVERAGE(Database, 5, Criteria). The formula returns the value 1.49948135, the average for the three samples that were extracted in the example shown in Figure 6-15.

You can use the formula =DAVERAGE(Database,"Refractive Index", Criteria) in place of the formula =DAVERAGE(Database, 5, Criteria). The field name must be identical to the field name in the database range. If the name used in the function is not identical to the name of one of the fields in the database, you'll get the #VALUE! error value.

7

IMPORTING DATA INTO EXCEL

Since you use Excel largely for the analysis of experimental data, the problem of how to get that data into Excel is crucial. Certainly you don't want to spend time transcribing data from a piece of paper into an Excel worksheet. There are several ways of transferring data to Excel. In order of decreasing preference, they are:

- directly from an instrument to an Excel worksheet
- from a data file on a diskette, or from a data file received electronically, to an Excel worksheet
- from data on paper, via a scanner, to an Excel worksheet

DIRECT INPUT OF INSTRUMENT DATA INTO EXCEL

Software programs are available to accept data from instruments and transfer it in real time to Excel. For example, SoftwareWedge (TAL Technologies, Inc., 2027 Wallace St., Philadelphia PA 19130) captures RS-232 serial I/O data, parses and filters it to user specifications, and transfers it to Excel for Windows via DDE (Dynamic Data Exchange). Data can be received from several serial ports simultaneously. The user can also define output strings, which can be sent to an instrument to control it directly from Excel. Discussion of this and similar software is beyond the scope of this book; specific information can be obtained from the manufacturer.

TRANSFERRING DATA FILES
FROM OTHER APPLICATIONS TO EXCEL

Most modern instruments are controlled by a computer and can write a data file to disk. Many are interfaced to external PCs. Here's how to import spectrophotometric data, saved to disk as a text file, into Excel.

USING THE TEXT IMPORT WIZARD

This example involves data from a Hewlett-Packard diode-array spectrophotometer interfaced to a PC; the values were collected between 350 and

	A
1	350,4.580689E-02
2	352,4.936218E-02
3	354,.0541687
4	356,6.047058E-02
5	358,6.842041E-02
6	360,7.719421E-02
7	362,8.996582E-02
8	364,.1021271

Figure 7-1. Portion of a comma-delimited text file imported into Excel without parsing.

820 nm at 2 nm intervals and saved to disk as a comma-delimited text file (Figure 7-1). Each data record consists of (wavelength, comma, absorbance). We'd like to read the data file into Excel and get the data fields into separate columns. (Data from a PC disk can be read by most Macintosh computers.)

If you use the **Open** command in Excel's **File** menu to open a text file, Excel will display the Text Import Wizard to allow you to parse the data. Either Delimited or Fixed Width text files can be parsed. The Fixed Width option is useful for columnar data separated by spaces, as illustrated in Figure 7-2.

	A
1	CH11101 1 L GENERAL CHEM LAB I M 2-5
2	CH11102 1 L GENERAL CHEM LAB I T 8 45-11 45
3	CH11103 1 L GENERAL CHEM LAB I T 1 30-4 30
4	CH11104 1 L GENERAL CHEM LAB I W 2-5
5	CH11105 1 L GENERAL CHEM LAB I TH 8 45-11 45

Figure 7-2. A fixed-width text file imported into Excel without parsing.

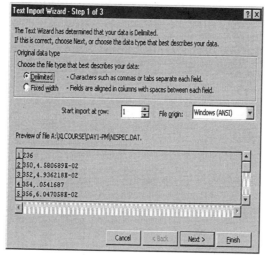

Figure 7-3. The Text Import Wizard Step 1 dialog box with preview of data to be parsed.

The Text Import Wizard usually recognizes whether the text file is Delimited or Fixed Width and displays the choice in the Step 1 dialog box, shown in Figure 7-3. You can override Excel's choice and manually select either the Delimited or the Fixed Width option.

If you chose Delimited in Step 1, the Step 2 dialog box allows you to select the type of delimiter and to see a preview of the parsed data (Figure 7-4).

If you chose Fixed Width in Step 1, the Step 2 dialog box (Figure 7-5) will display a ruler on which you can position vertical bars indicating the field widths. Click and drag on a delimiter bar to reposition it, double-click on a bar to remove it or click on the ruler to position a new delimiter bar.

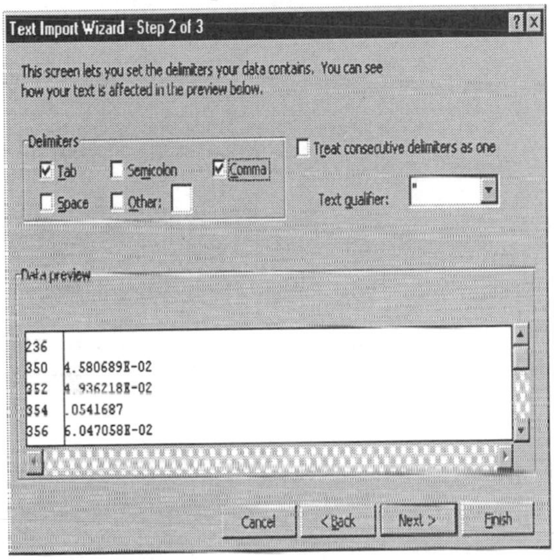

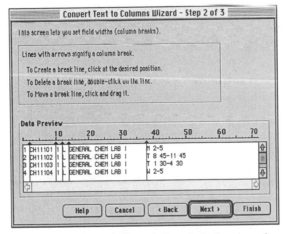

Figure 7-4. The Text Import Wizard Step 2 dialog box with preview of data fields

Figure 7-5. The Text Import Wizard Step 2 dialog box showing ruler and column break delimiters.

The Step 3 dialog box (Figure 7-6) offers several options in the Column Data Format category. If you press the Date option button, a column of text values representing dates will be converted into date serial numbers.

When you press Finish, a new Excel workbook will be created to contain the data. The parsed data is shown in Figure 7-7.

Excel Tip *The workbook is in text file format; be sure to* **Save** *it in Microsoft Excel Workbook format.*

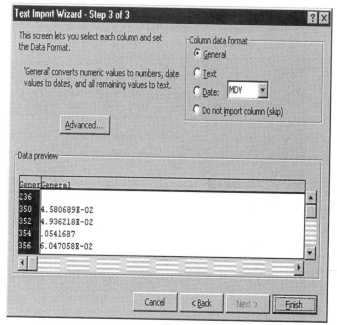

Figure 7-6. The **Text Import Wizard** Step 3 dialog box.

	A	B
1	350	4.58E-02
2	352	4.94E-02
3	354	0.0541687
4	356	6.05E-02
5	358	6.84E-02
6	360	7.72E-02
7	362	9.00E-02
8	364	0.1021271

Figure 7-7. The text file after parsing.

USING TEXT TO COLUMNS

The Convert Text to Columns Wizard allows you to parse text within Excel. The Wizard is actuated by means of the **Text to Columns...** command in the **Data** menu.

The Text to Columns dialog boxes are essentially identical to those of the Text Import Wizard (see Figure 7-8). The data to be parsed must be in a single column.

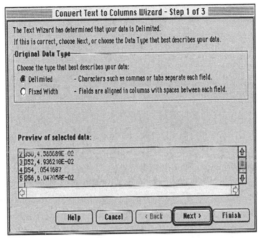

Figure 7-8 The Text to Columns Step 1 dialog box with preview of data to be parsed.

The Step 3 dialog box contains an input box (not present in the Text Import Wizard) that allows you to specify where you want the parsed data to be placed. The default option overwrites the original data column.

FROM HARD COPY (PAPER) TO EXCEL

Occasionally you'll have data in hard copy tabular form (from a book, report, correspondence, etc.). Unless the data set is very small, in which case it will be more efficient to enter the data manually, you'll want to use some semi-automatic way to import the data from paper copy to Excel.

USING A SCANNER TO TRANSFER NUMERIC DATA TO EXCEL

You can use a scanner and a software program that performs optical character recognition to create a data file from hard copy. Once impored into Excel, the data may have to be manipulated to get it into a useable form. The following example shows how to record a simple Excel macro for converting scanned data with blank lines and other undesired features into useful columnar data.

Figure 7-9 shows scanned spectral data (absorbance values at 10 nm intervals from 500 nm to 680 nm for a metal ion at varying concentrations of a ligand)

	A	B	C	D	E	F	G	H
1	2.51859E-4	0.0	1.21621E-4	1.42871E-2	0.0	10.0		
2								
3	0.465	0.489	0.506	0.517	0.522	0.519	0.502	0.473
4								
5	0.428	0.377	0.324	0.273	0.223	0.181	0.145	0.117
6								
7	0.093	0.074						
8								
9	2.51859E-4	0.0	1.21621E-4	2.85743E-2	0.0	10.0		
10								
11	0.569	0.581	0.576	0.555	0.529	0.496	0.459	0.416
12								

Figure 7-9. Spectral data imported from a monograph. (D. J. Leggett, *Computational Methods for the Determination of Formation Constants*, Plenum Press, New York, 1985)

copied from a monograph and imported into Excel. Because the data in the original copy was double-spaced, the scanner put the data in alternate rows of the spreadsheet. The data consists of (i) in Row 1, Row 9, etc., a row of data giving concentrations and path length, not necessary in our data table, (ii) in Row 3, Row 11, etc., eight values of absorbance measurements (for 500–570 nm) for one concentration of ligand, (iii) in Row 5, Row 13, etc., a second eight values of absorbance (580–660 nm), and in Row 7, Row 15, etc., the final two values of absorbance (670 and 680 nm).

Of the 120 lines in data in the sheet, more than half need to be deleted. With a little bit of experience in recording macros, you can manually fix up the first lines of the data set while recording a macro, then let the macro do the rest. Chapter 13 provides information on how to record a macro. Here's how to handle this example.

1. Before turning on the recorder, select the first row of data (Row 1).

2. Choose **Record New Macro...** from the **Macro** menu, and CONTROL+a (Windows) or OPTION+COMMAND+a (Macintosh) as the shortcut key.

3. Press the Relative Reference button.

4. Perform the following operations: (i) **Delete** the first two rows, (ii) **Cut** the second row of data and **Paste** it at the end of the first row of data, (iii) **Cut** the third row of data (two cells) and **Paste** at the end of the first row of data, (iv) select five rows and **Delete**, (v) select the next row of data, so that the previous operations will be repeated on the correct row when automated by using the macro.

5. Press the Stop Recording button or choose **Stop Recording** from the **Macro** submenu.

The recorded macro is shown in Figure 7-10.

```
Sub Macro1()
' Macro1 Macro
' Macro recorded 2/8/00 by Dr. Billo
' Keyboard Shortcut: Option+Cmd+a
    ActiveCell.Rows("1:2").EntireRow.Select
   Selection.Delete Shift:=xlUp
   ActiveCell.Offset(2, 0).Range("A1:H1").Select
   Selection.Cut
   ActiveCell.Offset(-2, 8).Range("A1").Select
   ActiveSheet.Paste
   ActiveCell.Offset(4, -8).Range("A1:B1").Select
   Selection.Cut
    ActiveCell.Offset(-4, 16).Range("A1").Select
   ActiveSheet.Paste
    ActiveCell.Offset(1, 0).Rows("1:5").EntireRow.Select
   Selection.Delete Shift:=xlUp
    ActiveCell.Rows("1:1").EntireRow.Select
End Sub
```

Figure 7-10. Recorded macro, with comments added.

	A	B	C	D	E	F	G	H	I	J
1	0.465	0.489	0.506	0.517	0.522	0.519	0.502	0.473	0.428	0.377
2	0.569	0.581	0.576	0.555	0.529	0.496	0.459	0.416	0.365	0.313
3	0.668	0.668	0.640	0.592	0.533	0.473	0.413	0.357	0.301	0.251

Figure 7-11. Reformatted data.

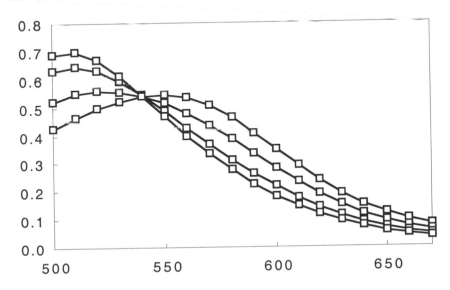

Figure 7-12. Chart created from imported data.

The macro can be actuated by using CONTROL+a. If the cursor is placed on the first line of the original data, and CONTROL+a pressed, the data is automatically formatted and the cursor moved down to the next row. Thus, repeatedly pressing CONTROL +a will successively format each line of data.

Alternatively, the macro can be modified to go through the data table until all rows have been processed. To achieve this result, simply add the line of code

```
If ActiveCell.Value <> "" Then GoTo Top
```

at the end and Top: at the beginning of the code. The complete data table is thus processed by a single CONTROL+a keystroke. A spreadsheet fragment showing some of the reformatted data is shown in Figure 7-11; Figure 7-12 is a chart of a portion of the imported data.

USING A SCANNER TO TRANSFER GRAPHICAL DATA TO EXCEL

You can use a scanner to convert graphs from strip chart recorders, published graphs or spectra, to digitized x, y data. Special software (e.g., Un-Scan-It from Silk Scientific, Inc., Orem UT 84059) is available for this purpose. First you scan the image using a scanner. Then you import the scanned image into the digitizing program. The program converts the scanned image into an x,y ASCII file.

SELECTING EVERY NTH DATA POINT

If you have imported a data file with a large number of data records (e.g., 8000 data points) you may wish to work with a reduced data set, say, every fiftieth point. The following sections describe three ways you can create such a list: by using AutoFill to **Fill Down** a pattern, by using Excel's Sampling tool, or by using a worksheet formula.

USING AUTOFILL

As you saw in Chapter 1, you can use AutoFill to **Fill Down** a pattern of selected cells. You can use this feature to select every Nth value in a range.

To create a list containing, for example, every tenth data point from the spectrophotometer data shown in Figure 7-7 (the complete data set covers the range 300–820 nm with an absorbance measurement every 2 nm), use the following procedure:

- In cell C1, enter the formula =A1, and in D1 enter the formula = B1. This will copy the first line of x,y data.

- Select (highlight) the range C1:D10 (1 row of values and 9 rows of empty cells).

- Use AutoFill to fill this pattern down to the end of the data, in row 236. This will return every 10th line of x,y values.

- While the range is still highlighted, **Copy** the values and **Paste Special** (Values). This is necessary because we are going to **Sort** the values in the next step.

- While the range is still highlighted, choose **Sort...** from the **Data** menu and choose Ascending. This will cause the values to rise to the top of the range. Note that this procedure succeeds only because the x values are increasing monotonically. If neither the x values nor the y values change monotonically, add an adjacent column of integers 1, 2, 3,... and use this as the sortkey.

USING THE SAMPLING TOOL

You can use Excel's Sampling tool, part of the Analysis ToolPak, to select every Nth value from a range. The Analysis ToolPak, which provides a range of statistical tools that you can use, is accessed by choosing **Data Analysis...** from the **Tools** menu. But since the Analysis ToolPak is an Add-In, the **Data Analysis...** command may not be present in the **Tools** menu (it's normally located at the bottom of the menu). If the command is not present, choose **Add-ins...** from the **Tools** menu to display the Add-ins dialog box, check the box for Analysis ToolPak and press OK (Figure 7-13). You should now see the **Data Analysis...** command in the **Tools** menu. Choosing the **Data Analysis...** command displays the Data Analysis dialog box (Figure 7-14). Choose Sampling from the list of statistical tools to display the Sampling dialog box (Figure 7-15).

In the Input Range box, enter the range of x values. A deficiency of the Sampling tool is that it can sample in only one range at a time; if you have a column of x values and a column of y values, you'll have to perform the sampling operation twice.

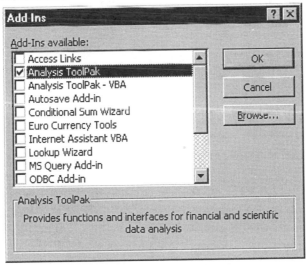

Figure 7-13. The Add-Ins dialog box.

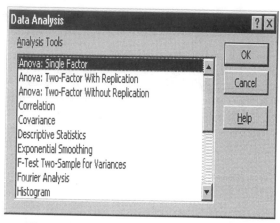

Figure 7-14. The Data Analysis dialog box.

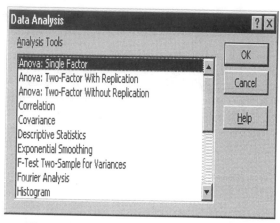

Figure 7-15. The Sampling dialog box.

The Sampling tool can perform either random or periodic sampling. Press the button for periodic sampling and enter 10 as the value for the period.

There are three output options: sending the results to a new worksheet ply, to a new workbook, or to a specified range of cells on the active sheet. For this example, we'll send the result to column C in the active sheet, so enter C1 as the start destination for the x values.

Finally, repeat the whole process to sample the range of y values.

USING A WORKSHEET FORMULA

Using the Sampling tool would be laborious if you needed to sample a data file with, for example, 17 columns: a column of x values and 16 columns of associated y values. You'd have to use the Sampling tool 17 times. In a case like this it would be preferable to have a formula, using relative references, that could be entered once, filled down to sample the column of x values, then filled right to create formulas that would sample each of the columns of y values. The following formulas illustrate how to do this.

Any of the four worksheet formulas described below can be used to select every Nth data point from a data table. All four utilize the INDEX(array, row_num,column_num) function to select an element from the array of data, but they use different methods to calculate the pointers row_num and column_num into the reference that contains the data.

- The first method uses names for the x and y data ranges; a separate worksheet formula must be entered for each column of data. The worksheet formula

  =INDEX(XData,Nth*(ROW()-ROW(XData))+1)

  when entered in cell T5 of Figure 7-16 and filled down yields the results shown in Figure 7-13. A similar formula is entered in cell U5 and filled down. XData and YData were defined as named ranges; the variable Nth contains the value 5.

- The second method uses a single worksheet formula that employs a mixed reference. The formula can then be copied into multiple columns. The formula

  =INDEX(R$5:R$18,Nth*(ROW()-ROW(R$5:R$18))+1)

  was entered in cell T5; then **Fill Right** was used to insert it into cell U5 to yield the formula

  =INDEX(S$5:S$18,Nth*(ROW()-ROW(S$5:S$18))+1)

  Upon application of **Fill Down**, the values shown in Figure 7-16 are produced.

- The third method uses an array formula that, because it involves both rows and columns, must be entered into the complete range of cells. The following array formula was entered in cells T5 :U7:

  {=INDEX(Data,Nth*(ROW()-MIN(ROW()))+1,COLUMN()-MIN(COLUMN())+1)}

• The fourth method uses an array formula that can be entered into a single row of cells (here T5:U5) and then copied into the complete range of cells using either **Fill Down** or AutoFill:

{=INDEX(Data,Nth*(ROW()-ROW(Data))+1,COLUMN()-MIN(COLUMN())+1)}

	R	S	T	U
3	<-- Data -->			
4	**XData**	**YData**		
5	350	0.0458	350	0.0458
6	351	0.0476	355	0.0574
7	352	0.0494	360	0.0772
8	353	0.0518	*etc.*	*etc.*
9	354	0.0542		
10	355	0.0574		
11	356	0.0605		
12	357	0.0645		
13	358	0.0684		
14	359	0.0728		
15	360	0.0772		
16	361	0.0836		
17	362	0.0900		
18	*etc.*	*etc.*		

Figure 7-16. Using a worksheet formula to select every Nth data point.

8

ADDING CONTROLS
TO A SPREADSHEET

You can create what Microsoft calls a "custom form". A custom form is a worksheet with one or more of the controls that are normally found in a dialog box — for example, option buttons to turn options on or off, or a list box to display a list from which the user can select an item. Thus you can create a user-friendly document that has many of the features usually found only in dialog boxes.

Most, but not all, of the controls that you have seen in Excel's dialog boxes can be installed on a worksheet. These controls return a value to a worksheet cell (the *cell link*). You can then use this value in worksheet formulas.

YOU CAN ADD OPTION BUTTONS, CHECK BOXES, LIST BOXES AND OTHER CONTROLS TO A WORKSHEET

The controls that can be installed on a worksheet to return a value to a worksheet cell are an option button, a check box, a list box, a drop-down list box (sometimes called a combo box), a scroll bar and a spinner. In addition, you can install a group box, which does not return a value but is used to group related controls, usually option buttons.

OPTION BUTTON

An option button is used to select one of a group of options. When related option buttons are placed within a group box, only one button of the group can be selected. Use option buttons when only one of several possibilities is allowed at any given time.

CHECK BOX

A check box is used to turn an option on or off. Use check boxes when several possibilities are allowed at any given time.

LIST BOX

A list box displays a list of items. Several of the items in the list are visible in the box; the user scrolls up or down to display the rest of the list, then chooses an item.

COMBO BOX

A drop-down list box (combo box) displays a list of items. Initially only one item is visible in the window; the user presses the drop-down button to display the rest of the list, then chooses an item.

SCROLL BAR

A scroll bar, with buttons and a slider, permits the user to increase or decrease a value in a worksheet cell. Maximum and minimum values can be set for the value returned by the scroll bar. A Scroll Bar can be installed either horizontally or vertically.

SPINNER

A spinner is a button with an up arrow and a down arrow that permits the user to increase or decrease a value in a worksheet cell. Maximum and minimum values can be set for the value returned by the spinner.

HOW TO ADD A CONTROL TO A WORKSHEET

Adding a control to a worksheet is a three-step process: first, you display the Forms toolbar (Figure 8-1); then you use the Forms toolbar to draw the desired control; finally, you set the control properties. For example, for a list box, the control properties are (i) the range of values to be displayed in the list and (ii) the cell link, the cell to which the list box returns a value. Follow the procedure in the following box.

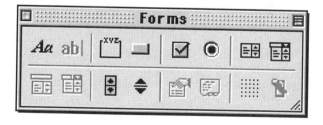

Figure 8-1. The Forms toolbar

To Add a Control to a Worksheet

1. Choose **Toolbars** from the **View** menu and choose **Forms** from the submenu.

2. Press the button for the control you want to add. Controls that can be used only in a Userform (a custom dialog box), not on a worksheet, are dimmed. On the worksheet, click and drag to draw the control. (You can move or re-size it later.)

3. While the control is still selected (has handles around it), choose **Control...** from the **Format** menu and enter the control properties, described below.

4. When the control properties have been entered, click anywhere on the worksheet to un-select and activate the control. Upon activation of the control, the mouse pointer becomes the "hand"' shape when the mouse pointer is positioned over the control.

5. To select the control for moving or re-sizing, hold down the CONTROL key while clicking on the control.

Each control has a title, e.g., Check Box 1, associated with it. The titles are numbered sequentially as you add them; for example, if you add a list box, then three option buttons, then a check box, they will have titles Option Button 2, Option Button 3, Option Button 4, Check Box 5 (the list box doesn't have a title that you can see). You can select the title text and change it.

Almost always you will place several related option buttons within a group box and link each option button to the same cell. (The best procedure is to add the group box first, then add the option buttons within the group box.)

Table 8-1 provides details on control properties.

CONTROL PROPERTIES

You set the properties of a control by choosing **Control...** from the **Format** menu or **Format Control...** from the shortcut menu to display the Format Control dialog box for the specific control, then choose the Control tab. Some control properties function only when the control is used with a macro. You can set these control properties but they will have no effect on a control installed on a worksheet.

Table 8-1. Control Properties

Check Box   Cell Link: a reference to a cell that returns the state of the check box: TRUE if the box is checked, FALSE if unchecked. This logical argument can be used in a formula.

Option Button   Cell Link: a reference to a cell that returns the number of the option button that is selected. Only one option button within a group box can be selected (pressed) at any time. The cell link returns the number of the option button within the group (e.g., 1, 2 or 3) not the number that appeared in the original title (e.g., Option Button 7, Option Button 8, Option Button 9).

List Box   Input Range: a reference to a range of cells whose values will be displayed in the list box.

Cell Link: a reference to a cell that returns a value that is the relative position of the selected item in the list. This value can then be used in a formula to return a result based on the selected item in the list.

Selection Type: you must use Single when a list box is installed on a worksheet.

Combo Box   Input Range: a reference to a range of cells whose values will be displayed in the drop-down list box.

Cell Link: a reference to a cell that returns the relative position of the selected item in the list.

Scroll Bar   Minimum Value: the minimum value that the scroll bar can return (corresponds to the top of a vertical scroll bar or the left end of a horizontal scroll bar).

Maximum Value: the maximum value that the scroll bar can return (corresponds to the bottom of a vertical scroll bar or the right end of a horizontal scroll bar).

Incremental Change: the amount the scroll slider button moves when an arrow at either end of the scroll bar is clicked. The default value is 1.

Page Change: the amount the scroll slider button moves when you click between the scroll button and one of the scroll arrows. The default value is 10.

Cell Link: a reference to a cell that returns the current value of the scroll bar. This number can be used in a formula to return a result based on the position of the scroll slider button in the scroll bar.

Spinner   The control properties for a spinner are the same as those for a scroll bar, but spinners do not have a Page Change property.

A LIST BOX ON A WORKSHEET

As a simple first exercise, we'll create a list box that displays a list of items. The list that will be displayed consists of the names of the months of the year. When you select a particular month from the list, the list box returns a number from 1 to 12, the relative position of the selected item in the list.

First, create a range of cells containing the list to be displayed. Fill the range A1:A12 (use a different worksheet than the one where you're going to install the list box) with the text January, February, etc. You can use **AutoFill** to do this.

To install the list box on a worksheet, follow the procedure outlined earlier in this chapter. First, display the Forms toolbar. Click on the List Box button (second from the right in the top row in Figure 8-1) and use the mousepointer to draw the outline of a list box. While the list box is still selected (has handles around it), choose **Control...** from the **Format** menu. In the Input Range box, enter the external reference to the range of cells containing the list to be displayed. In the Cell Link box, enter a cell reference, e.g., C5, then press OK.

To make the list box active, click on any cell in the worksheet. The handles around the list box will disappear.

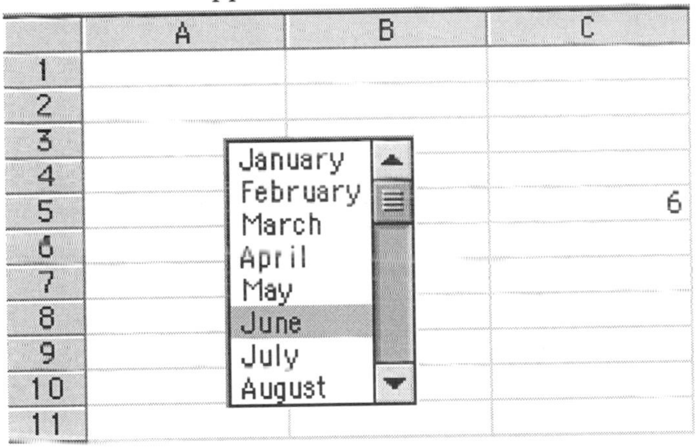

Figure 8-2. A list box installed on a worksheet. The cell link is cell C5.

The list box should now display the months of the year, as shown in Figure 8-2. When you select an item from the list, the Cell Link will return the number of the selected item. You can use the cell link value in a formula.

To select the list box for moving or re-sizing, click on it once while holding down the CONTROL key.

A DROP-DOWN LIST BOX ON A WORKSHEET

Figure 8-3 shows a drop-down list box that is used to display a subset of a list. This drop-down list box displays, from a list of names and telephone numbers, the subset of names that begin with a particular letter or string. The user can display all the names that begin with the letter "D", or "McD", for example. When the user chooses a name from the subset displayed in the drop-down list box, that person's telephone number is displayed in a cell.

The advantage of using the drop-down list box rather than a simple list box is that the box automatically sizes to fit the number of values displayed.

In this example names, addresses and telephone numbers are listed in columns A, B and C of the worksheet. The range A2:A139 was assigned the name Names, C2:C139 the name Phones. The list of names was sorted in alphabetical order.

Figure 8-3 shows the list box displaying the sublist of names beginning with the letter D. To select the subset to be displayed, the user enters a letter or string in cell F2.

The reference SubListStart, in cell G2, contains the formula
        =MATCH(Letter&"*",Names,0)
This formula returns the relative position of the first item in the list that matches the input string. In our example, the first name that begins with D is the 35th item in the list in column A. The reference SubListLength, in cell H2, contains the array formula
        {=SUM(1*(LEFT(Names,LEN(Letter))=Letter))}
This formula returns the number of items in the list that match the input string. These two formulas are combined in the formula
        =OFFSET($A$1,SubListStart,0,SubListLength,1)
to return a reference to the cells containing the sublist. This formula was entered as a named formula (using Define Name) and given the name DropList. In fact, all the formulas in this example can be entered as named formulas.

	F	G	H	I
1	Letter	SubListStart	SubListLength	Cell Link (relative)
2	D	35	10	2
3		Daly, Josephine		
4		Deckster, Meghan K		
5		Dell, Daniel M		
6		Dickson, Martin J		
7		Dimatti, Peter A		
8		Ditolla, Maria L		
9		Dominguez, Christina D		
10		Dopfer, David E		
11		Drabanti, Brad		
12		Drozdowski, Christina		

Figure 8-3. A drop-down list box displaying a subset of a list. The cell link is cell I2.

Finally, using **Control...** from the **Format** menu, DropList was entered as the reference for the Input Range of the list box and cell I2 as the Cell Link.

The formula to return the telephone number corresponding to the name selected from the list box

=INDEX(Telephones,SubListStart+CellLink-1)

was entered in cell H6. Figure 8-4 illustrates the final result.

	F	G	H	I
1	Letter	SubListStart	SubListLength	Cell Link (relative)
2	D	35	10	2
3				
4		Deckster, Meghan K	⬍	
5				
6		Telephone #:	555-5618	
7				

Figure 8-4. A drop-down list box used to return a telephone number.

OPTION BUTTONS AND A DROP-DOWN LIST BOX

This example illustrates the use of option buttons in combination with the drop-down list box of the preceding example. The option buttons are used to specify whether to display the address or telephone number corresponding to the selected name.

Two option buttons were positioned inside a group box (see Figure 8-5) and the titles "Option Button 2" and "Option Button 3" were changed to "Address" and "Phone", respectively. The Cell Link for the two buttons was cell J2, which was given the name ColumnOffset.

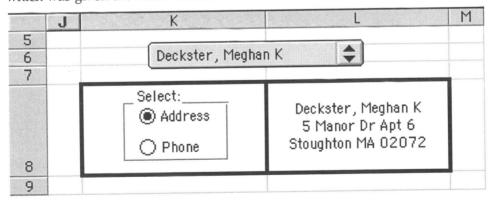

	J	K	L	M
5				
6		Deckster, Meghan K ⬍		
7				
8		Select: ⦿ Address ◯ Phone	Deckster, Meghan K 5 Manor Dr Apt 6 Stoughton MA 02072	
9				

Figure 8-5. A drop-down list box used to return address or telephone number.

ColumnOffset is used in the following formula, in cell L8, to specify whether to return either the address (in column B of the range DirectoryTable, or the telephone number in column C of DirectoryTable:

=INDEX(DirectoryTable,SubListStart+CellLink-1,ColumnOffset+1)

PART III

SPREADSHEET MATHEMATICS

9

SOME MATHEMATICAL TOOLS FOR SPREADSHEET CALCULATIONS

This chapter describes some mathematical methods that are useful for spreadsheet calculations. For the most part, they are methods that are applicable to tables or arrays of data.

LOOKING UP VALUES IN TABLES

The VLOOKUP(*lookup_value, array, column_index_num, match_type_logical*) worksheet function is useful for obtaining values from tables. This section shows how to use VLOOKUP to obtain a value both from a one-way table and a two-way table.

GETTING VALUES FROM A ONE-WAY TABLE

The spreadsheet shown in Figure 9-1, containing a list of element symbols and atomic weights, is part of a one-way table (the table entries extend down to row 110). In a one-way table, the lookup values occupy a single row or column of the table (here, the lookup values are found in column B).

We want to obtain the atomic weight of a particular element from the table, in order to use it in a formula. The formula =VLOOKUP(symbol,$B$2:$D$110,2,0) returns the atomic weight corresponding to symbol. The formula =VLOOKUP(symbol,$B$2:$D$110,3,0) returns the electron configuration.

	B	C	D
1	**Symbol**	**At. Wt.**	**Elec. Config.**
2	H	1.00797	1s1
3	He	4.0026	1s2
4	Li	6.939	[He] 2s1
5	Be	9.0122	[He] 2s2
6	B	10.811	[He] 2s2 2p1
7	C	12.01115	[He] 2s2 2p2

Figure 9-1. Portion of a data table of the elements.

	A	B	C	D
1	Element	Symbol	At. Wt.	Elec. Config.
2	Hydrogen	H	1.00797	1s1
3	Helium	He	4.0026	1s2
4	Lithium	Li	6.939	[He] 2s1
5	Beryllium	Be	9.0122	[He] 2s2
6	Boron	B	10.811	[He] 2s2 2p1
7	Carbon	C	12.01115	[He] 2s2 2p2

Figure 9-2. Portion of a data table of the elements.

When you use VLOOKUP, you must always "look up" in the first column of the defined database and retrieve associated information from a column to the right in the same row; you cannot VLOOKUP to the left, for example. In the table shown in Figure 9-2, you cannot use VLOOKUP to return the element name corresponding to symbol. If you want to perform a lookup to the left of, or above, the lookup value, you can either construct your own lookup-type function using the MATCH function (see later in this chapter) or use the LOOKUP function.

GETTING VALUES FROM A TWO-WAY TABLE

A two-way table has two lookup values, usually in the leftmost column and in the top row. The value to be returned from the table is the value located at the intersection of the row and column containing the two lookup values.

Once again we can use VLOOKUP to obtain the value from the table. In the preceding example, the value used for the *column_index_num* argument of VLOOKUP was the fixed value 2, because we are always going to be returning a value from column 2 of array. In the two-way table example shown in Figure 9-3, *column_index_num* is a variable; we must determine by means of an expression the column from which the desired value is to be returned. The MATCH expression can be used to obtain the relative column number in the table corresponding to the lookup value.

To find the value in the table for $T = 950°F$ and $P = 5000$ psia, the expression

=VLOOKUP(B44,$A$4:$P$34,MATCH(B45,$B$3:$P$3,1)+1,1)

where the T and P lookup values are entered in cells B44 and B45 respectively, returns the value 0.035 for the viscosity. The same formula, using names instead of references,

=VLOOKUP(T,T_Table,MATCH(P,P_Row,1)+1,1)

returns the value for the viscosity.

	A	B	C	D	▨	M	N	O	P
1		Viscosity of Steam and Water (cP)							
2		psia							
3	T, °F	1	2	5		5000	7500	10000	12000
17	650	0.022	0.022	0.022		0.082	0.088	0.092	0.096
18	700	0.023	0.023	0.023		0.071	0.079	0.085	0.086
19	750	0.024	0.024	0.024		0.057	0.071	0.078	0.081
20	800	0.025	0.025	0.025		0.040	0.062	0.071	0.075
21	850	0.026	0.026	0.026		0.035	0.052	0.064	0.070
22	900	0.028	0.028	0.028		0.035	0.045	0.057	0.064
23	950	0.029	0.029	0.029		0.035	0.042	0.052	0.059
24	1000	0.030	0.030	0.030		0.035	0.041	0.049	0.055
25	1050	0.031	0.031	0.031		0.036	0.040	0.047	0.052
26	1100	0.032	0.032	0.032		0.037	0.040	0.045	0.050
27	1150	0.034	0.034	0.034		0.037	0.041	0.045	0.049
28	1200	0.034	0.034	0.034		0.038	0.041	0.045	0.048
29	1250	0.035	0.035	0.035		0.039	0.042	0.045	0.048
30	1300	0.037	0.037	0.037		0.040	0.043	0.045	0.048
31	1350	0.038	0.038	0.038		0.041	0.044	0.046	0.049
32	1400	0.039	0.039	0.039		0.042	0.044	0.047	0.049
33	1450	0.040	0.040	0.040		0.043	0.045	0.047	0.049
34	1500	0.041	0.041	0.041		0.044	0.046	0.048	0.050

Figure 9-3. Portion of a two-way data table.

INTERPOLATION METHODS: LINEAR

Often it's necessary to estimate the value of y for some value of x intermediate between two adjacent data points — the process of *interpolation*. If the separation between x values is small, and the y values do not change rapidly with x, then linear interpolation may be adequate. To calculate the value of y at a value x_i that is intermediate between x_0 and x_1, use the linear interpolation formula (equation 9-1):

$$y_i = y_0 + \frac{x_i - x_0}{x_1 - x_0}(y_1 - y_0) \tag{9-1}$$

TABLE LOOKUP WITH LINEAR INTERPOLATION

You can use the MATCH and INDEX worksheet functions in a formula to perform linear interpolation in a table. The syntax of the MATCH function is MATCH(*lookup_value,lookup_array,match_type*). If *match_type* = 1, MATCH returns the position of the largest array value that is less than or equal to *lookup_value*. The x values in the array must be in ascending order. Thus, for any lookup value, you can use MATCH to find the pointer to the position of x_0, the value of x in the table that is less than or equal to the lookup value, and then use this pointer in the INDEX function to return values for x_0, x_1, y_0 and y_1, as in the following formulas:

position	=MATCH(LookupValue,XValues,1)
x_0	=INDEX(XValues,position)
x_1	=INDEX(XValues,position+1)
y_0	=INDEX(YValues,position)
y_1	=INDEX(YValues,position+1)

The preceding formulas were applied in the example shown in Figure 9-4, to perform table lookup with linear interpolation in the simple data table in A5:B8. The intermediate values shown in B13:B17 in Figure 9-5 were used to develop, in a stepwise manner, the following interpolation formula used in B18.

=y_0+(LookupValue-x_0)*(y_1-y_0)/(x_1-x_0)

A chart of the data with some interpolated values is shown in Figure 9-6.

	A	B
3	Table data	
4	XValues	YValues
5	1.1	100
6	2.1	200
7	3.1	400
8	4.1	750
9		

Figure 9-4. Data table for interpolation.

	A	B	C
12	LookupValue	Value	
13	3.5	3	(position of lookup value)
14		3.1	(x_0)
15		4.1	(x_1)
16		400	(y_0)
17		750	(y_1)
18		540	(interpolated value)

Figure 9-5. Intermediate values for stepwise development of interpolation formula.

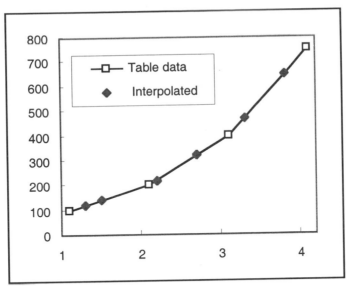

Figure 9-6. Chart showing some interpolated values

The formulas for the intermediate values in B13:B17 can be copied and pasted into the interpolation formula in B18 to create a single "megaformula" (all in one cell, of course):

=INDEX(YValues,MATCH(LookupValue,XValues,1))+(LookupValue-INDEX(
XValues,MATCH(LookupValue,XValues,1)))'(INDEX(YValues,MATCH(
LookupValue,XValues,1)+1) INDEX(YValues,MATCH(LookupValue,
XValues,1)))/(INDEX(XValues,MATCH(LookupValue,XValues,1)+1)-
INDEX(XValues,MATCH(LookupValue,XValues,1)))

If the *x* values in the data table are in descending order, the *match_type* argument in the MATCH function must be changed to –1.

INTERPOLATION METHODS: CUBIC

Simple linear interpolation is not always adequate for data tables with extensive curvature. *Cubic interpolation* uses the values of four adjacent data points to evaluate the coefficients of the cubic equation $y = a + bx + cx^2 + dx^3$.

A compact and elegant implementation of cubic interpolation in the form of an Excel 4.0 Macro Language custom function was provided by Orvis[*]. A slightly modified version, in VBA (Visual Basic for Applications) language, is provided here (Figure 9-7). The syntax of the custom function is CubicInterp(*known_xs, known_ys, x_value*).

---

[*]   William J. Orvis, *Excel 4 for Scientists and Engineers*, Sybex Inc., Alameda, CA, 1993.

```
Function CubicInterp(XValues, YValues, X)
'  Performs cubic interpolation, using an array of XValues, YValues.
'  The XValues must be in ascending order.
'  Based on XLM code from "Excel for Chemists", page 239,
'  which was based on W. J. Orvis' code.
Row = Application.Match(X, XValues, 1)
   If Row < 2 Then Row = 2
   If Row > XValues.Count - 2 Then Row = XValues.Count - 2
For I = Row - 1 To Row + 2
   Q = 1
For J = Row - 1 To Row + 2
   If I <> J Then Q = Q * (X - XValues(J)) / (XValues(I) - XValues(J))
Next J
   Y = Y + Q * YValues(I)
Next I
CubicInterp = Y
End Function
```

Figure 9-7. Cubic interpolation function macro.

The cubic interpolation function can be used to produce a smooth curve through data points. Figure 9-8 illustrates a portion of a spreadsheet with experimental spectrophotometric data taken at 5 nm intervals in columns A and B (the data are in rows 6–86), and a portion of the interpolated values, at 1 nm intervals, in columns C and D. The formula in cell D24 is

= CubicInterp ($A$6:$A$86,$B$6:$B$86,C24).

The smoothed curve through the data points is illustrated in Figure 9-9.

	A	B	C	D
4	Original Data		Interpolated Data	
5	x	y	x	y
24	390	0.552	390	0.552
25	395	0.582	391	0.559
26	400	0.598	392	0.566
27	405	0.600	393	0.572
28	410	0.586	394	0.577
29	415	0.559	395	0.582
30	420	0.521	396	0.586
31	425	0.473	397	0.590
32	430	0.419	398	0.593
33	435	0.362	399	0.596
34	440	0.305	400	0.598

Figure 9-8. Interpolation by using a cubic interpolation function macro.

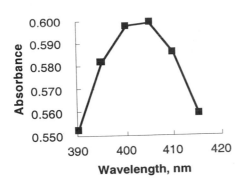 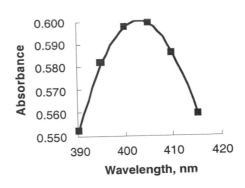

Figure 9-9. (Left) Chart created using data points only. (Right) Chart with smooth curve interpolated using a cubic interpolation function.

The cubic interpolation function forces the curve to pass through all the known data points. If there is any experimental scatter in the data, the result will not be too pleasing. A better approach for data with scatter is to find the coefficients of a least-squares line through the data points, as described in Chapter 11 or 12.

NUMERICAL DIFFERENTIATION

The process of finding the derivative or slope of a function is the basis of *differential calculus*. Since you will be dealing with spreadsheet data, you will be concerned not with the algebraic differentiation of a function, but with obtaining the derivative of a data set or the derivative of a worksheet formula by numeric methods.

Often a function depends on more than one variable. The *partial derivative* of the function $F(x,y,z)$, e.g., $\delta F/\delta x$, is the slope of the function with respect to x, while y and z are held constant.

FIRST AND SECOND DERIVATIVES OF A DATA SET

The simplest method to obtain the first derivative of a function represented by a table of x, y data points is to calculate $\Delta y/\Delta x$. The first derivative or slope of the curve at a given data point x_n, y_n can be calculated using either of the following formulas:

$$\text{slope} = \frac{\Delta y}{\Delta x} = \frac{y_{n+1} - y_n}{x_{n+1} - x_n} \tag{9-2}$$

$$\text{slope} = \frac{y_n - y_{n-1}}{x_n - x_{n-1}} \tag{9-3}$$

2	A V/mL	B pH	C ΔV	D ΔpH	E V(avge)	F ΔpH/ΔV
22	1.90	4.981	0.100	0.229	1.850	2.29
23	1.95	5.157	0.050	0.176	1.925	3.52
24	2.00	5.389	0.050	0.232	1.975	4.64
25	2.05	5.928	0.050	0.539	2.025	10.78
26	2.08	7.900	0.030	1.972	2.065	65.73
27	2.10	9.115	0.020	1.215	2.090	60.75
28	2.15	9.604	0.050	0.489	2.125	9.78
29	2.20	9.856	0.050	0.252	2.175	5.04
30	2.30	10.125	0.100	0.269	2.250	2.69

Figure 9-10. First derivative of titration data, near the end-point.

The second derivative, d^2y/dx^2, of a data set is calculated in a similar manner, namely by calculating $\Delta(\Delta y/\Delta x)/\Delta x$.

Calculation of the first or second derivative of a data set tends to emphasize the "noise" in the data set; that is, small errors in the measurements become relatively much more important.

Points on a curve of x, y values for which the first derivative is either a maximum, a minimum or zero are often of particular importance and are termed *critical points*.

The spreadsheet shown in Figure 9-10 uses pH titration data to illustrate the calculation of the first derivative of a data set .

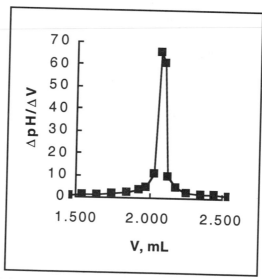

Figure 9-11. First derivative of titration data, near the end-point.

	E	F	G	H	I	J
2	V(avge)	ΔpH/ΔV	ΔV	Δ(ΔpH)	V(avge)	Δ(ΔpH)/ΔV
22	1.850	2.29	0.100	0.57	1.800	5.7
23	1.925	3.52	0.075	1.23	1.888	16.4
24	1.975	4.64	0.050	1.12	1.950	22.4
25	2.025	10.78	0.050	6.14	2.000	122.8
26	2.065	65.73	0.040	54.95	2.045	1373.8
27	2.090	60.75	0.025	-4.98	2.078	-199.3
28	2.125	9.78	0.035	-50.97	2.108	-1456.3
29	2.175	5.04	0.050	-4.74	2.150	-94.8
30	2.250	2.69	0.075	-2.35	2.213	-31.3

Figure 9-12. Second derivative of titration data, near the end-point.

Since the derivative has been calculated over the finite volume $\Delta V = V_{n+1} - V_n$, the most suitable volume to use when plotting the $\Delta pH/\Delta V$ values, as shown in Figure 9-10, is:

$$V_{average} = \frac{V_{n+1} + V_n}{2} \tag{9-4}$$

The maximum in $\Delta pH/\Delta V$ indicates the location of the inflection point of the titration (Figure 9-11). The second derivative, $\Delta(\Delta pH/\Delta V)/\Delta V$, which is calculated by means of the spreadsheet shown in Figure 9-12, can be used to locate the inflection point more precisely. The second derivative passes through zero at the inflection point. Linear interpolation can be used to calculate the point at which the second derivative is zero (Figure 9-13).

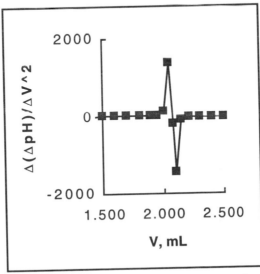

Figure 9-13. Second derivative of titration data, near the end-point.

There are more sophisticated equations for numerical differentiation. These equations use three, four or five points instead of two points to calculate the derivative. Since they usually require equal intervals between points, they are of less generality. Their main advantage is that they minimize the effect of "noise".

DERIVATIVES OF A FUNCTION

The first derivative of a formula in a worksheet cell can be obtained with a high degree of accuracy by evaluating the formula at x and at $x + \Delta x$. Since Excel carries 15 significant figures, Δx can be made very small. Under these conditions $\Delta F/\Delta x$ approximates dF/dx very well.

The spreadsheet fragment shown in Figure 9-14 illustrates the calculation of the first derivative of a function ($F = x^3 - 3x^2 - 130x + 150$) by evaluating the function at x and at $x + \Delta x$. Here a value of Δx of 1×10^{-9} was used; alternatively Δx could be obtained by using a worksheet formula such as =1E−9*x. For comparison, the first derivative was calculated from the expression from differential calculus: $F' = 3x^2 - 6x - 130$.

The Excel formulas in cells B12, C12, D12 and E12 are

= t*x^3+u*x^2+v*x + w

= t*(x+delta)^3 +u*(x+delta)^2 +v*(x+delta) + w

	A	B	C	D	E
1		Numerical Differentiation			
2		F(x) = tx^3 + ux^2 + vx + w			
3		First derivative F'(x) = 3tx^2 + 2ux + v			
5		t	1		
6		u	-3		
7		v	-130		
8		w	150		
9		Δ (delta)	1.00E-09		
10				Using differences	Using calculus
11	x	F(x)	F(x+Δ)	F'(x)	F'(x)
12	-10	150	150	230.0	230.0
13	-9	348	348	167.0	167.0
14	-8	486	486	110.0	110.0
15	-7	570	570	59.0	59.0
16	-6	606	606	14.0	14.0
17	-5	600	600	-25.0	-25.0
18	-4	558	558	-58.0	-58.0
19	-3	486	486	-85.0	-85.0

Figure 9-14. Calculating the first derivative of a function.

=(C12-B12)/delta

=3*t*x^2+2*u*x+v

Figure 9-15 shows a chart of the function and its first derivative.

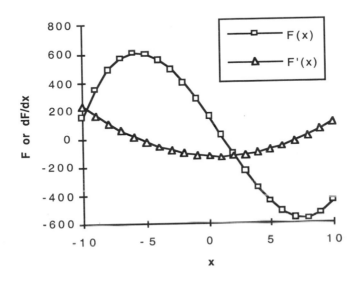

Figure 9-15. The function $F = x^3 - 3x^2 - 130x + 150$ and its first derivative.

NUMERICAL INTEGRATION

A common use of numerical integration is to determine the area under a curve. We will describe three methods for determining the area under a curve: the rectangle method, the trapezoid method and Simpson's method. Each involves approximating the area of each portion of the curve delineated by adjacent data points; the area under the curve is the sum of these individual segments.

The simplest approach is to approximate the area by the rectangle whose height is equal to the value of one of the two data points, illustrated in Figure 9-16.

As the x increment (the interval between the data points) decreases, this rather crude approach becomes a better approximation to the area. The area under the curve bounded by the limits $x_{initial}$ and x_{final} is the sum of the individual rectangles, as given by equation 9-5.

$$\text{area} = \Sigma \, y_i(x_{i+1} - x_i) \tag{9-5}$$

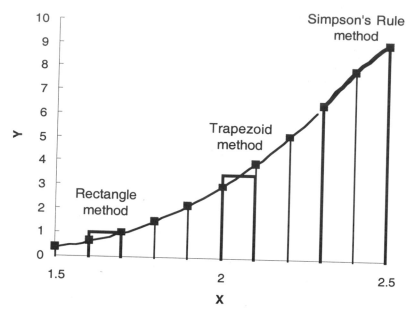

Figure 9-16. Graphical illustration of methods of calculating the area under a curve.

For a better approximation you can use the average of the two y values as the height of the rectangle. This is equivalent to approximating the area by a trapezoid rather than a rectangle. The area under the curve is given by equation 9-6.

$$\text{area} = \sum \frac{y_i + y_{i+1}}{2}(x_{i+1} - x_i) \qquad (9\text{-}6)$$

Simpson's rule approximates the curvature of the function by means of a quadratic equation. To evaluate the coefficients of the quadratic requires the use of the y values for three adjacent data points. The x values must be equally spaced.

$$\text{area} = \sum \frac{y_i + 4y_{i+1} + y_{i+2}}{6}(x_{i+1} - x_i) \qquad (9\text{-}7)$$

AN EXAMPLE: FINDING THE AREA UNDER A CURVE

The curve shown in Figure 9-17 is the sum of two Gaussian curves, with position and standard deviation $\mu = 90$, $\sigma = 10$ and $\mu = 130$, $\sigma = 20$, respectively. The equation used to calculate each Gaussian curve is

$$y = \frac{\exp[-(x - \mu)^2/2\sigma^2]}{\sigma\sqrt{2\pi}} \qquad (9\text{-}8)$$

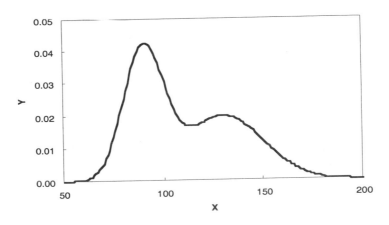

Figure 9-17. A curve that is the sum of two Gaussian curves.

The area of each Gaussian curve is equal to 1.000; thus the total area under the curve shown in Figure 9-17 is 2.000. The Excel formula used to calculate y is as follows (m is the position μ and s is the standard deviation σ):

=EXP(-0.5*((x-curv1 m)/curv1 s)^2)/(SQRT(2*PI())*curv1 s)+EXP(-0.5* ((x-curv2 m)/curv2 s)^2)/(SQRT(2*PI())*curv2 s)

The area under the curve, between the limits $x = 50$ and $x = 200$, was calculated by using each of the preceding equations (9-5, 9-6 and 9-7): the rectangular approximation, the trapezoidal approximation and Simpson's rule. In each case a constant x increment of 10 was used. A portion of the spreadsheet is shown in Figure 9-18.

The formulas in row 9, used to calculate the area increment, are as follows:

=10*F9   (rectangular approximation)

=10*(F8+F9)/2   (trapezoidal approximation)

=10*(F8+4*F9+F10)/6   (Simpson's rule)

	E	F	G	H	I
6			Rectangular	Trapezoidal	Simpson's
7	x	Y	Approximation	Approx.	Rule
8					
9	50	0.00002	0.00020	0.00010	0.00095
10	60	0.00049	0.00487	0.00253	0.01265
11	70	0.00562	0.05621	0.03054	0.08007
12	80	0.02507	0.25073	0.15347	0.24751
13	90	0.04259	0.42594	0.33834	0.37687
14	100	0.03067	0.30673	0.36633	0.30464

Figure 9-18. Portion of a spreadsheet for calculating the area under a curve.

The area increments were summed and the area under the curve, calculated by the three methods, is shown in Figure 9-19. All three methods of calculation appear to give acceptable results in this case.

	G	H	I
6	Rectangular	Trapezoidal	Simpson's
7	Approximation	Approx.	Rule
25			
26	1.9999	1.9997	1.9998

Figure 9-19. Area under a curve, calculated by three different methods.

DIFFERENTIAL EQUATIONS

Certain chemical problems, such as those involving chemical kinetics, can be expressed by means of differential equations. For example, the coupled reaction scheme

$$A \underset{k_2}{\overset{k_1}{\rightleftharpoons}} B \underset{k_4}{\overset{k_3}{\rightleftharpoons}} C$$

results in the simultaneous equations

$$\frac{d[A]}{dt} = -k_1[A] + k_2[B]$$

$$\frac{d[B]}{dt} = k_1[A] - k_2[B] - k_3[B] + k_4[C]$$

$$\frac{d[C]}{dt} = k_3[B] - k_4[C]$$

To "solve" this system of simultaneous equations, we want to be able to calculate the value of [A], [B] and [C] for any value of t. For all but the simplest of these systems of equations, obtaining an exact or analytical expression is difficult or sometimes impossible. Such problems can always be solved by numerical methods, however. Numerical methods are completely general. They can be applied to systems of differential equations of any complexity, and they can be applied to any set of initial conditions. Numerical methods require extensive calculations but this is easily accomplished by spreadsheet methods.

In this chapter we will consider only ordinary differential equations, that is, equations involving only derivatives of a single independent variable. As well, we will discuss only initial-value problems — differential equations in which information about the system is known at $t = 0$. Two approaches are common: Euler's method and the Runge–Kutta (RK) methods.

EULER'S METHOD

Let us use as an example the simulation of the first-order kinetic process $A \rightarrow B$ with initial concentration $C_0 = 0.2000$ mol/L and rate constant $k = 5 \times 10^{-3}$ s^{-1}. We'll simulate the change in concentration vs. time over the interval from $t = 0$ to $t = 600$ seconds, in increments of 20 seconds.

The differential equation for the disappearance of A is $d[A]/dt = -k[A]$. Expressing this in terms of finite differences, the change in concentration $\Delta[A]$ that occurs during the time interval from $t = 0$ to $t = \Delta t$ is $\Delta[A] = -k[A]\Delta t$. Thus, if the concentration of A at $t = 0$ is 0.2000 M, then the concentration at $t = 0 + \Delta t$ is $[A] = 0.2000 - (5 \times 10^{-3})(0.2000)(20) = 0.1800$ M. The formula in cell B7 is

=B6-k*B6*DX.

The concentration at subsequent time intervals is calculated in the same way.

The advantage of this method, known as *Euler's method*, is that it can be easily expanded to handle systems of any complexity. Euler's method is not particularly useful, however, since the error introduced by the approximation $d[A]/dt = \Delta[A]/\Delta t$ is compounded with each additional calculation. Compare the Euler's method result in column B of Figure 9-20 with the analytical expression for the concentration, $[A]_t = [A]_0 e^{-kt}$, in column C. At the end of approximately one half-life (seven cycles of calculation in this example), the error has already increased to 3.6%. Accuracy can be increased by decreasing the size of Δt, but only at the expense of increased computation. A much more efficient way of increasing the accuracy is by means of a series expansion. The Runge–Kutta methods, which we describe next, comprise the most commonly used approach.

	A	B	C	D
1	Simulation of first order kinetics by Euler's Method			
2		rate constant =	5.0E-03	(k)
3		time increment =	20	(DX)
4				
5	t	A(t)	e^(-kt)	
6	0	0.2000	0.2000	
7	20	0.1800	0.1810	
8	40	0.1620	0.1637	
9	60	0.1458	0.1482	
10	80	0.1312	0.1341	
11	100	0.1181	0.1213	
12	120	0.1063	0.1098	
13	140	0.0957	0.0993	

Figure 9-20. Euler's method.

THE RUNGE–KUTTA METHODS

The Runge–Kutta methods for numerical solution of the differential equation $dy/dx = F(x, y)$ involve, in effect, the evaluation of the differential function at intermediate points between x_i and x_{i+1}. The value of y_{i+1} is obtained by appropriate summation of the intermediate terms in a single equation. The most widely used Runge–Kutta formula involves terms evaluated at x_i, $x_i + \Delta x/2$ and $x_i + \Delta x$. The *fourth-order Runge–Kutta* equations for $dy/dx = F(x, y)$ are

$$y_{i+1} = y_i + \frac{T_1 + 2T_2 + 2T_3 + T_4}{6} \qquad (9\text{-}9)$$

where

$$T_1 = F(x_i, y_i)\,\Delta x \qquad (9\text{-}10)$$

$$T_2 = F\left(x_i + \frac{\Delta x}{2}, y_i + \frac{T_1}{2}\right)\Delta x \qquad (9\text{-}11)$$

$$T_3 = F\left(x_i + \frac{\Delta x}{2}, y_i + \frac{T_2}{2}\right)\Delta x \qquad (9\text{-}12)$$

$$T_4 = F(x_i + \Delta x, y_i + T_3)\,\Delta x \qquad (9\text{-}13)$$

If more than one variable appears in the expression, then each is corrected by using its own set of T_1 to T_4 terms.

The spreadsheet in Figure 9-21 illustrates the use of the RK method to simulate the first-order kinetic process A → B with initial concentration $[A]_0 = 0.2000$ and rate constant $k = 5 \times 10^{-3}$. The differential equation is $d[A]_t/dt = -k[A]_t$. This equation is one of the simple form $dy/dx = F(y)$, and thus only the y_i terms of T_1 to T_4 need to be evaluated. The RK terms (note that T_1 is the Euler method term) are shown on the following page.

	A	B	C	D	E	F	G
1	Runge-Kutta simulation of first order kinetics						
2			rate constant =		5.0E-03	(k)	
3			time increment =		20	(DX)	
4							
5	t	TA1	TA2	TA3	TA4	A(t)	e^(-kt)
6	0					0.2000	0.2000
7	20	-0.0200	-0.0190	-0.0191	-0.0181	0.1810	0.1810
8	40	-0.0181	-0.0172	-0.0172	-0.0164	0.1637	0.1637
9	60	-0.0164	-0.0156	-0.0156	-0.0148	0.1482	0.1482
10	80	-0.0148	-0.0141	-0.0141	-0.0134	0.1341	0.1341
11	100	-0.0134	-0.0127	-0.0128	-0.0121	0.1213	0.1213
12	120	-0.0121	-0.0115	-0.0116	-0.0110	0.1098	0.1098
13	140	-0.0110	-0.0104	-0.0105	-0.0099	0.0993	0.0993

Figure 9-21. The Runge–Kutta method.

$$T_1 = -k[A]_t\, \Delta x \tag{9-14}$$
$$T_2 = -k([A]_t + T_1/2)\, \Delta x \tag{9-15}$$
$$T_3 = -k([A]_t + T_2/2)\, \Delta x \tag{9-16}$$
$$T_4 = -k([A]_t + T_3)\, \Delta x \tag{9-17}$$

The RK equations in cells B7, C7, D7, E7 and F7, respectively, are:

=-k*F6*DX

=-k*(F6+TA1/2)*DX

=-k*(F6+TA2/2)*DX

=-k*(F6+TA3)*DX

=F6+(TA1+2*TA2+2*TA3+TA4)/6.

If you use the names TA1,..., TA4, TB1,..., TB4, etc., you'll find that (i) the nomenclature is expandable to systems requiring more than one set of Runge–Kutta terms, (ii) the names are accepted by Excel, whereas T1 is not a valid name, and (iii) you can use **AutoFill** to generate the column labels TA1,..., TA4.

Compare the RK result in column F of Figure 9-21 with the analytical expression for the concentration, $[A]_t = [A]_0 e^{-kt}$, in column G. After one half-life (row 13) the RK calculation differs from the analytical expression by only 0.00006%. (Compare this with the 3.6% error in the Euler method calculation at the same point.) Even after 10 half-lives (not shown), the RK error is only 0.0006%.

If the spreadsheet is constructed as shown in Figure 9-21, you can't use a formula in which a name is assigned to the concentration array in column F. This is becasue the formula in B7, for example, will use the concentration in F7 instead of the required F6. An alternative arrangement that permits using a name for the concentration $[A]_t$ is shown in Figure 9-22. Each row contains the concentration at the beginning and at the end of the time interval.   The name A_t

	A	B	C	D	E	F	G	H
1		Runge–Kutta simulation of first order kinetics						
2			rate constant =		5.0E-03	(k)		
3			time increment =		20	(DX)		
4								
5	t	A(t)	TA1	TA2	TA3	TA4	A(t+DX)	e^(-kt)
6	0	0.2000	-0.0200	-0.0190	-0.0191	-0.0181	0.1810	0.1810
7	20	0.1810	-0.0181	-0.0172	-0.0172	-0.0164	0.1637	0.1637
8	40	0.1637	-0.0164	-0.0156	-0.0156	-0.0148	0.1482	0.1482
9	60	0.1482	-0.0148	-0.0141	-0.0141	-0.0134	0.1341	0.1341
10	80	0.1341	-0.0134	-0.0127	-0.0128	-0.0121	0.1213	0.1213

Figure 9-22. Alternative spreadsheet layout for the Runge–Kutta method.

can now be assigned to the array of values in column B; the former formulas (now in cells C7–G7) contain A_t in place of F6, and cell B7 contains the formula =G6.

In essence, the fourth order Runge–Kutta method performs four calculation steps for every time interval. In the solution by Euler's method, decreasing the time increment to 5 seconds, to perform four times as many calculation steps, still only reduces the error to 0.9% after 1 half-life.

Summary of Steps to Implement the RK Method

1. Write down the differential equations that describe the system.

2. From the set of differential equations, enter formulas in spreadsheet cells using Euler's method, **Fill Down** and make sure that the results make sense.

3. Insert five columns to the right of each variable. Label them TA1,..., TA4 and A(t), TB1,...,TB4 and B(t), etc.

4. **Copy** the Euler's method formula from the formula bar and **Paste** it into the four cells for TA1,...,TA4. The Euler's method formula should be of the form $= y_0 + F^*y_0^*\Delta x$. Edit it to leave only the $= F^*y_0^*\Delta x$ part. Modify the formulas to produce the formulas for TA1,···, TA4. TA1 remains $F^*y_0^*\Delta x$, TA2 becomes $F^*(y_0 + TA1/2)^*\Delta x$, TA3 becomes $F^*(y_0 + TA2/2)^*\Delta x$, TA4 becomes $F^*(y_0 + TA3)^*\Delta x$.

5. Enter the formula for $y(m)$: $ym + (TA1 + 2^*TA2 + 2^*TA3 + TA4)/6$

6. **Fill Down** the TA1,...,TA4 and $y(m)$ cells. $y(m)$ should agree approximately with the Euler's method column.

7. **Delete** the columns containing the Euler's method values.

ARRAYS, MATRICES AND DETERMINANTS

Spreadsheet calculations lend themselves almost automatically to the use of arrays of values. As you've seen, arrays in Excel can be either one- or two-dimensional. For the solution of many types of problem, it is convenient to manipulate an entire rectangular array of values as a unit. Such an array is termed a *matrix*. (In Excel, the terms "range", " array " and "matrix" are virtually interchangeable.) An $m \times n$ matrix (m rows and n columns) of values is illustrated on the following page.

$$\begin{vmatrix} a_{11} & a_{12} & \cdots & a_{1n} \\ a_{21} & a_{22} & \cdots & a_{2n} \\ \vdots & \vdots & \vdots & \vdots \\ a_{m1} & a_{m2} & \cdots & a_{mn} \end{vmatrix}$$

The values comprising the array are called *matrix elements*. Mathematical operations on matrices have their own special rules.

A *square matrix* has the same number of rows and columns. If all the elements of a square matrix are zero except those on the main diagonal (a_{11}, a_{22},..., a_{nn}), the matrix is termed a *diagonal matrix*. A diagonal matrix whose diagonal elements are all 1 is a *unit matrix*.

A matrix which contains a single column of m rows or a single row of n columns is called a *vector*.

A *determinant* is simply a square matrix. There is a procedure for the numerical evaluation of a determinant, so that an $N \times N$ matrix can be reduced to a single numerical value. The value of the determinant has properties that make it useful in certain tests and equations. (See, for example, "Solving Sets of Simultaneous Linear Equations" in Chapter 10.)

AN INTRODUCTION TO MATRIX ALGEBRA

Matrix algebra provides a powerful method for the manipulation of sets of numbers. Many mathematical operations — addition, subtraction, multiplication, division, etc. — have their counterparts in matrix algebra. Our discussion will be limited to the manipulations of square matrices. For purposes of illustration, two 3 × 3 matrices will be defined, namely

$$\mathbf{A} = \begin{vmatrix} a & b & c \\ d & e & f \\ g & h & i \end{vmatrix} = \begin{vmatrix} 2 & 3 & 4 \\ 3 & 2 & -1 \\ 4 & 3 & 7 \end{vmatrix}$$

and

$$\mathbf{B} = \begin{vmatrix} r & s & t \\ u & v & w \\ x & y & z \end{vmatrix} = \begin{vmatrix} 2 & 0 & 2 \\ 0 & 3 & -3 \\ -3 & -2 & 1 \end{vmatrix}$$

The following examples illustrate addition, subtraction, multiplication and division using a constant.

Addition or subtraction of a constant: $\mathbf{A} + q = \begin{vmatrix} a+q & b+q & c+q \\ d+q & e+q & f+q \\ g+q & h+q & i+q \end{vmatrix}$

Multiplication or division by a constant: $q\mathbf{A} = \begin{vmatrix} qa & qb & qc \\ qd & qe & qf \\ qg & qh & qi \end{vmatrix}$

Addition or subtraction of two matrices (both must contain the same number of rows and columns):

$$\mathbf{A} + \mathbf{B} = \begin{vmatrix} a & b & c \\ d & e & f \\ g & h & i \end{vmatrix} + \begin{vmatrix} r & s & t \\ u & v & w \\ x & y & z \end{vmatrix} = \begin{vmatrix} a+r & b+s & c+t \\ d+u & e+v & f+w \\ g+x & h+y & i+z \end{vmatrix}$$

Performing matrix algebra with Excel is very simple. Let's begin by assuming that the matrices **A** and **B** have been defined by selecting the 3R × 3C arrays of cells containing the values and naming them by using **Define Name**. To add a constant (e.g., 3) to matrix **A**, simply select a range of cells the same size as the matrix, enter the formula =A+3, then press COMMAND+RETURN or CONTROL+SHIFT+RETURN (Macintosh) or CONTROL+SHIFT+ENTER (Windows). Subtraction of a constant, multiplication or division by a constant, or addition of two matrices also is performed by using standard Excel algebraic operators.

Multiplication of two matrices can be either *scalar* multiplication or *vector* multiplication. Scalar multiplication of two matrices consists of multiplying corresponding elements, i.e.,

$$\mathbf{A} + \mathbf{B} = \begin{vmatrix} a & b & c \\ d & e & f \\ g & h & i \end{vmatrix} + \begin{vmatrix} r & s & t \\ u & v & w \\ x & y & z \end{vmatrix} = \begin{vmatrix} a+r & b+s & c+t \\ d+u & e+v & f+w \\ g+x & h+y & i+z \end{vmatrix}$$

Thus it's clear that both matrices must have the same dimensions $m \times n$. Scalar multiplication is commutative, that is, $\mathbf{A}*\mathbf{B} = \mathbf{B}*\mathbf{A}$

The vector multiplication of two matrices is somewhat more complicated:

$$\mathbf{A}\,\mathbf{B} = \begin{vmatrix} a & b & c \\ d & e & f \\ g & h & i \end{vmatrix} \begin{vmatrix} r & s & t \\ u & v & w \\ x & y & z \end{vmatrix} = \begin{vmatrix} ar+bu+cx & as+bv+cy & at+bw+cz \\ dr+eu+fx & ds+ev+fy & dt+ew+fz \\ gr+hu+ix & gs+hv+iy & gt+hw+iz \end{vmatrix}$$

Vector multiplication is not commutative, that is $\mathbf{A}\,\mathbf{B} \neq \mathbf{B}\,\mathbf{A}$.

Vector multiplication can be accomplished easily by the use of one of Excel's worksheet functions for matrix algebra, MMULT(**matrix1, matrix2**). For the matrices **A** and **B** defined above,

$$\mathbf{A}\,\mathbf{B} = \begin{vmatrix} -8 & 1 & -1 \\ 9 & 8 & -1 \\ -13 & -5 & 6 \end{vmatrix} \qquad \mathbf{B}\,\mathbf{A} = \begin{vmatrix} 12 & 12 & 22 \\ -3 & -3 & -24 \\ -8 & -10 & -3 \end{vmatrix}$$

Two matrices can be vector-multiplied when **A** has the same number of rows as **B** has columns.

Vector multiplication of two matrices is possible only if the matrices are *conformable*, that is, if the number of columns of **A** is equal to the number of rows of **B**. The opposite condition, if the number of *rows* of **A** is equal to the number of *columns* of **B**, is not equivalent. The following examples, involving multiplication of a matrix and a vector, illustrate the possibilities:

MMULT (4 × 3 matrix, 3 × 1 vector) = 3 × 1 result vector

MMULT (4 × 3 matrix, 1 × 4 vector) = **#VALUE!**

MMULT (1 × 4 vector, 4 × 3 matrix) = 1 × 4 result vector

The *transpose* of a matrix, indicated by a prime ('), is produced when rows and columns of a matrix are interchanged, i.e.,

$$\mathbf{A}' = \begin{vmatrix} a & d & g \\ b & e & h \\ c & f & i \end{vmatrix}$$

The transpose is obtained by using the worksheet function TRANSPOSE(*array*) or the Transpose option in the **Paste Special...** menu command (see "Using Paste Special to Transpose Rows and Columns" in Chapter 1).

The process of *matrix inversion* is analogous to obtaining the reciprocal of a number a. The matrix relationship that corresponds to the algebraic relationship $a \times (1/a) = 1$ is

$$\mathbf{A}\,\mathbf{A}^{-1} = \mathbf{I}$$

where $\mathbf{A}^{-1}$ is the inverse matrix and **I** is the unit matrix. The process for inverting a matrix "manually" (i.e., using pencil, paper and calculator) is complicated, but the operation can be carried out readily by using Excel's worksheet function MINVERSE(*array*). The inverse of the matrix **A** above is:

$$\mathbf{A}^{-1} = \begin{vmatrix} -0.25 & -0.3333 & -0.5 \\ 0.75 & 0.6667 & 0.5 \\ 0.75 & 0.3333 & 0.5 \end{vmatrix}$$

The "pencil-and-paper" evaluation of a determinant of N rows × N columns is also complicated, but it can be done simply by using the worksheet function MDETERM(*array*). The function returns a single numerical value, not an array, and thus you do not have to use CONTROL+SHIFT+ENTER. The value of the determinant of **A**, represented by $|\mathbf{A}|$, is 12.

POLAR TO CARTESIAN COORDINATES

You may occasionally need to chart a function that involves angles. Instead of using the familiar Cartesian coordinate system (x, y and z coordinates), such functions often use the polar coordinate system, in which the coordinates are two angles, θ and ϕ, and a distance r. The two coordinate systems are related by the equations $x = r \sin\theta \cos\phi$, $y = r \sin\theta \sin\phi$, $z = r \cos\theta$. Angle θ is the angle between the vector r and the Cartesian z axis; ϕ is the angle between the projection of r on the x, y plane and the x axis. Since Excel's trigonometric functions only consider x- and y axes, the simplified relationships are, for angles in the x, y plane: $x = r \cos\phi$, $y = r \sin\phi$.

As an example of transformation of polar to Cartesian coordinates, we'll graph the wave function for the d_{x2-y2} orbital in the x, y plane. The angular component of the wave function in the x, y plane is

$$\Phi = \sqrt{\frac{15}{16p}} \cos 2\phi \tag{9-18}$$

and Φ can be equated to the radial vector r for the conversion of polar to Cartesian coordinates.

In the spreadsheet fragment shown in Figure 9-23, column A contains angles from 0 to 360 in 2-degree increments. Column B converts the angles to radians (required by the COS worksheet function) using the relationship =A4*PI()/180 in row 4. The formulas in cells C4, D4 and E4 are:

=SQRT(15/(16*PI()))*COS(2*B4)

=C4*COS(B4)

=C4*SIN(B4).

The chart of the x and y values is shown in Figure 9-24.

	A	B	C	D	E
1	Representation of Angular Wave Function of dx2–y2 Orbital				
2					
3	Angle, deg	Angle, rad	d	x coord	y coord
4	0	0.0000	0.5463	0.5463	0.0000
5	2	0.0349	0.5449	0.5446	0.0190
6	4	0.0698	0.5410	0.5396	0.0377
7	6	0.1047	0.5343	0.5314	0.0559
8	8	0.1396	0.5251	0.5200	0.0731
9	10	0.1745	0.5133	0.5055	0.0891
10	12	0.2094	0.4990	0.4881	0.1038

Figure 9-23. Converting from polar to Cartesian coordinates

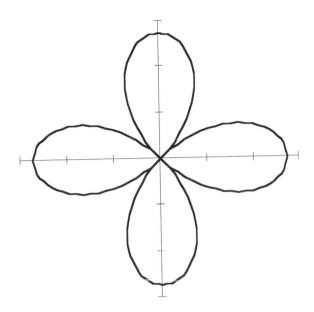

Figure 9-24. Angular wave function for $d_{x^2 - y^2}$ orbital.

USEFUL REFERENCE

K. Jeffrey Johnson, *Numerical Methods in Chemistry*, Marcel Dekker, New York, 1980.

GRAPHICAL AND NUMERICAL METHODS OF ANALYSIS

In this chapter you'll learn how to use graphical and numerical methods to solve chemical problems. The methods described range from the simple (finding the roots of a polynomial from a graph of the function) to the complex (using matrix methods to solve sets of simultaneous equations). Excel does most of the work, though, by virtue of its graphing capabilities and built-in numerical analysis capabilities.

FINDING ROOTS OF EQUATIONS

Sometimes a chemical problem can be reduced algebraically, by pencil and paper, to a polynomial expression for which the solution to the problem is one of the roots of the polynomial. Almost everyone remembers the quadratic formula for the roots of a quadratic equation, but finding the roots of a more complicated polynomial is more difficult. We begin by describing three methods for finding the real roots of a polynomial

THE GRAPHICAL METHOD

The roots of a polynomial $y = F(x)$ are the values of x that make $F(x) = 0$. One simple way to find those values is to create a spreadsheet table of x and corresponding y values, then create a chart from the values. The x *values* where the curve crosses the x axis (where $y = 0$) are the roots of the equation. Figure 10-1 is a graph of the function $y = x^3 - 3x^2 - 13x + 15$, for which the roots are 5, 1 and –3. The roots can't be read from this chart with any degree of accuracy, but it's a simple matter to create a chart of the region immediately around an intersection point and get a much more precise value. In general, though, the main use of this method is to gain an idea of the approximate value of the roots.

How many real roots does a given polynomial have? Generally, in chemical problems, x represents a quantity that can only be positive, such as a concentration or an equilibrium constant. *Descartes' Rule of Signs* states that for any polynomial $F(x)$ written in decreasing powers of x, the number of positive

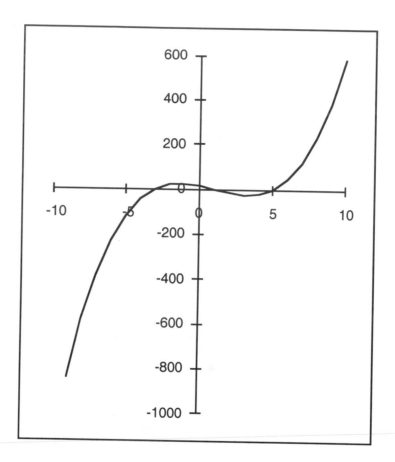

Figure 10-1. Graph of the equation $y = x^3 - 3x^2 - 13x + 15$. The roots are 5, 1 and −3.

roots of the equation $F(x) = 0$ is equal to the number of changes in sign between the coefficients of adjacent terms, or is less than this number by some positive even integer. For example, for the cubic equation $x^3 + (2.29 \times 10^2)x^2 - (2.29 \times 10^1)x - (9.14 \times 10^{-1}) = 0$, there is one change in sign, between the x^2 term and the x term. Thus there is only one positive root of the equation. Almost always, in chemical problems, such a polynomial will have only one positive real root, and this one root will be the solution to the problem.

THE METHOD OF SUCCESSIVE APPROXIMATIONS

The graphical method can give us only an approximate idea of the roots of a polynomial. To obtain a more accurate numerical result, the roots of $y = F(x)$ can easily be obtained by trial and error, finding the values of x that make the function y equal to zero.

As an example, let's calculate the solubility of barium carbonate in pure water. Barium carbonate, $BaCO_3$, is a sparingly soluble salt with $K_{sp} = 5.1 \times 10^{-9}$:

$$BaCO_3(s) = Ba^{2+} + CO_3^{2-} \qquad\qquad K_{sp} = [Ba^{2+}][CO_3^{2-}] \qquad (10\text{-}1)$$

If the hydrolysis of the carbonate ion to form bicarbonate is taken into account (equation 10-2, $K_b = 2.1 \times 10^{-4}$), solving the system of mass- and charge-balance

$$CO_3^{2-} + H_2O = OH^- + HCO_3^- \qquad\qquad K_b = \frac{[OH^-][HCO_3^-]}{[CO_3^{2-}]} \qquad (10\text{-}2)$$

equations leads to the following pseudo-quadratic expression, in which x is the solubility of barium carbonate.

$$x^2 - \sqrt{K_b\,K_{sp}}\ x - K_{sp} = 0 \qquad (10\text{-}3)$$

Inserting the values of the constants, you can write

$$x^2 - 1.0 \times 10^{-6}\,x^{1/2} - 5.1 \times 10^{-9} = 0 \qquad (10\text{-}4)$$

You can locate the positive root in a systematic fashion by changing the value of x in increments and observing the value of the function y. When y exhibits a sign change, you'll know that it has passed through zero. Then decrease the size of the increment and repeat the process. In this way you can determine the value of x to any desired degree of accuracy.

You can start with any value of x. Eventually you'll get the right answer. But if you use an educated guess, you'll get there sooner. Since the K_{sp} of $BaCO_3$ is

	A	B
	X	Y
1		
2	7.00E-05	-8.57F-09
3	8.00E-05	-7.64E-09
4	9.00E-05	-6.49E-09
5	1.00E-04	-5.10E-09
6	1.10E-04	-3.49E-09
7	1.20E-04	-1.65F-09
8	1.30E-04	3.98E-10
9	1.40E-04	2.67E-09
10	1.50E-04	5.15E-09
11	1.60E-04	7.85E-09
12	1.70E-04	1.08E-08
13	1.80E-04	1.39E-08

Figure 10-2. Stage 1 in solution by successive approximations

	C	D
7	1.20E-04	-1.65E-09
8	1.21E-04	-1.46E-09
9	1.22E-04	-1.26E-09
10	1.23E-04	-1.06E-09
11	1.24E-04	-8.60E-10
12	1.25E-04	-6.55E-10
13	1.26E-04	-4.49E-10
14	1.27E-04	-2.40E-10
15	1.28E-04	-2.97E-11
16	1.29E-04	1.83E-10
17	1.30E-04	3.98E-10

Figure 10-3. Stage 2 in solution by successive approximations

	E	F
15	1.280E-04	-2.97E-11
16	1.281E-04	-8.52E-12
17	1.282E-04	1.27E-11

Figure 10-4. Stage 3 in solution by successive approximations

5.1×10^{-9}, and the solubility can be approximated by $\sqrt{K_{sp}}$, use 7×10^{-5} as the initial guess. The solubility, as carbonate hydrolyzes to bicarbonate, will be higher than this, so increase it in increments of 1×10^{-5}.

Open a new worksheet and label cell A1 "X". Label cell B1 "Y". Enter 7E-5 in cell A2, 8E-5 in cell A3, then use Autofill to **Fill Down** about 10 cells. In cell B2 enter the formula =A2^2-1.0E-6*SQRT(A2)-5.1E-9, and use AutoFill to **Fill Down**. The result is shown in Figure 10-2. The function y changes sign between rows 7 and 8 (between $x = 1.2E-04$ and $1.3E-04$). Now repeat the process beginning with $x = 1.2E-04$ and using a smaller increment. To do this, **Copy** cells A7:B7 and **Paste** into cell C7:D7. In cell C8 enter 1.21E-04. Select C7:C8 and use AutoFill to **Fill Down** about 10 cells, then use AutoFill to **Fill Down** column D. The result is shown in Figure 10-3. The sign change occurs between rows 15 and 16.

THE NEWTON–RAPHSON METHOD

Instead of simply changing the value of x in regular increments until the function y approaches zero, the *Newton–Raphson method* uses the slope of the function at an initial estimate of the root, x_1, to obtain an improved estimate of the root, x_2. Figure 10-5 illustrates a chart of the function $y = 3x^3 + 2.5x^2 - 5x - 11$, between $x = 0$ and $x = 6$. The value $x = 5$ has been chosen as an initial estimate for finding a root of the function. Here's how an improved estimate for the root is calculated. The slope of the curve at x_1 is the first derivative of the function, dy/dx. The improved estimate of the root is given by $x_2 = x_1 - (y_1/\text{slope})$. The process is repeated until no appreciable change in x occurs.

In pencil-and-paper calculations, the slope of the function would be obtained by calculating its first derivative, i.e., $dy/dx = 9x^2 + 5x - 5$ in this particular case. In spreadsheet calculations, however, the slope is conveniently obtained by numerical differentiation, i.e., slope $= \Delta y/\Delta x$. The procedure is as follows: (i) for an initial guess value of x, the increment Δx is obtained by multiplying x by a suitable small number, e.g., 0.0000001; (ii) the function y increases by the corresponding amount Δy; (iii) the slope of the function at a particular x value is thus $\Delta y/\Delta x$; (iv) this slope m is used with the coordinates of the initial guess value to calculate the x value where the slope straight line intersects the x axis at $y=0$. Steps (ii)–(iv) are repeated, using the new value of x, until convergence is reached.

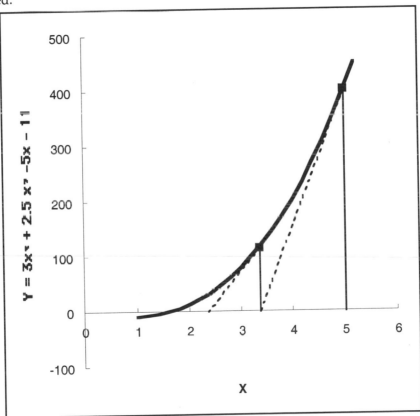

Figure 10-5. The Newton–Raphson method used to obtain a root of the equation $y = 3x^3 + 2.5x^2 - 5x - 11$.

	A	B	C	D	E	F	G	H
27								
28		X	Y	X+ΔX	Y+ΔY	slope	new X	
29		5	401.5	5.000001	401.50012	245.00002	3.36122	
30								
31		3.36122	114.3621	3.36122	114.36214	113.48661	2.35351	
32		2.35351	30.18833	2.35351	30.18834	56.61865	1.82032	
33		1.82032	6.27767	1.82032	6.27767	33.92381	1.63527	
34		1.63527	0.62762	1.63527	0.62762	27.24337	1.61223	
35		1.61223	0.0091	1.61223	0.00911	26.45485	1.61189	
36		1.61189	0.	1.61189	0.00001	26.44315	1.61189	
37		1.61189	0.	1.61189	0.	26.44314	1.61189	
38								

Figure 10-6. Calculation of a root of a function by the Newton–Raphson method.

The method is illustrated in Figure 10-5 and in the spreadsheet fragment shown in Figure 10-6. The formulas in cells C29, D29, E29, F29 and G29 are, respectively,

C29: =3*B29^3+2.5*B29^2-5*B29-11   (the function y)

D29: =B29+0.0000001*B29   (increment x by a small amount Δx)

E29: =3*D29^3+2.5*D29^2-5*D29-11   (this is $y + \Delta y$)

F29: =(E29-C29)/(D29-B29)   ($m = \Delta x / \Delta y$)

G29: =(F29*B29-C29)/F29   ($x_{new} = (m^* x_{old} - y_{old}) / m$)

Since Δx can be made very small, the slope calculated this way will be an excellent approximation to the true slope. The Newton-Raphson method usually converges rapidly.

If a function has more than one real root, the particular root to which the Newton-Raphson method converges will depend on the initial estimate chosen. Thus, to obtain a particular root, some guidance must be provided by the user.

SOLVING A PROBLEM USING GOAL SEEK...

Excel provides a built-in way to perform Newton–Raphson approximations: the **Goal Seek** command in the **Tools** menu. **Goal Seek** varies the value of a selected cell (the *changing cell*) to make the value of another cell (the *target cell*) reach a desired value.

To illustrate, let's repeat the calculation of the solubility of barium carbonate. **Goal Seek** allows you to obtain the same result much more easily. Open a new worksheet and in cell A1 enter the value 1. In cell B1, enter the formula =A1^2 - 1.0E-6*SQRT(A1)-5.1E-9. The task now will be to use **Goal Seek...** to find the value in cell A1 that makes the function (in B1) equal to zero.

The accuracy of the result will depend on the magnitude of the Maximum Change parameter, which you can adjust by choosing **Options...**(Windows) or **Preferences...** (Macintosh) from the **Tools** menu and then choosing the Calculation tab (Figure 10-7). The default value is 0.001 You'll see in a moment that adjusting the Maximum Change parameter is critical when you are using **Goal Seek.** But for now, set the value of Maximum Change to 1E-12.

Choose **Goal Seek** from the **Tools** menu. As shown in Figure 10-8, enter $B$1 in the Set Cell box (the cell reference will appear there if you selected that cell before choosing **Goal Seek...**). Put the cursor in the To Value box and enter the desired value, zero. Put the cursor in the By Changing Cell box and enter $A$1 by selecting the cell or by typing. Then click on OK.

After a few iteration cycles the Goal Seek Status dialog box (Figure 10-9) is displayed.

Adjusting the Maximum Change parameter is critical when using **Goal Seek.** That's because Excel stops iterating when the change in the result is less than the Maximum Change parameter. Therefore the Maximum Change parameter needs to be adjusted to be much less than the value of the function. Table 10-1 illustrates the importance of adjusting the Maximum Change parameter. For most chemical calculations it's a good idea to set Maximum Change to 1E-12 or 1E-15 as a matter of course.

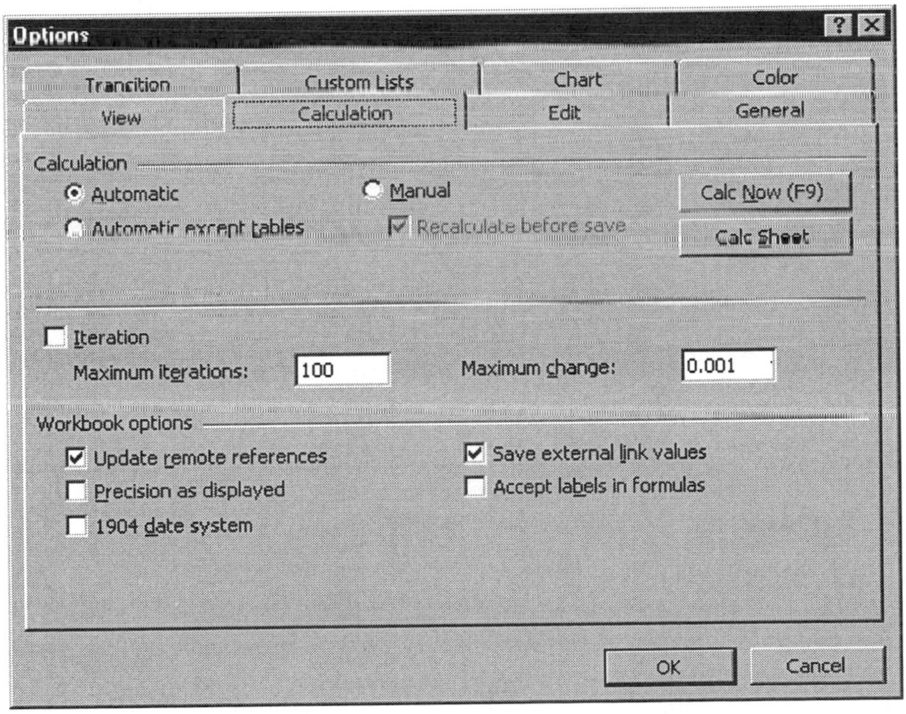

Figure 10-7. The Calculation Options dialog box.

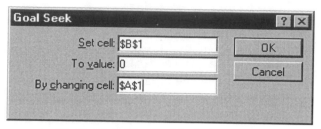

Figure 10-8. The Goal Seek dialog box.

Goal Seek Status ? X

Goal Seeking with Cell B1
found a solution.

Target Value: 0

Current Value: 1.18346E-13

OK

Cancel

Step

Pause

Figure 10-9. The Goal Seek Status dialog box.

Table 10-1. The Effect of Setting the Maximum Change Parameter

Trial	Maximum change	Final value of $F(x)$*	Value of x
1	1.00E-03	5E-04	2.22E-02
2	1.00E-06	6E-07	7.82E-04
3	1.00E-09	3E-10	1.30E-04
4	1.00E-12	2E-13	1.281E-04
5	1.00E-15	4E-17	1.281E-04

* The final value of the function depends to some extent on the starting value of x.

For problems requiring the variation of two or more parameters, that is, varying the values of several cells to make the value of another cell reach a desired value, you must use the Solver, which is described in detail in Chapter 12.

SOLVING A PROBLEM BY INTENTIONAL CIRCULAR REFERENCE

When a formula refers to itself, either directly or indirectly, it creates a *circular reference*. If a circular reference occurs, Excel issues a "Cannot resolve circular references" message and displays a zero value in the cell.

Usually circular references occur unintentionally, because the user incorrectly entered a cell reference in an equation. But occasionally a problem can be solved by intentionally creating a circular reference.

To illustrate the use of an intentional circular reference, let's return to an earlier example in this chapter: finding one of the roots of a cubic equation by using the Newton–Raphson method. In Figure 10-6 the successive approximations were generated in successive rows in the worksheet by filling down the formulas, and the calculations were deemed to have reached convergence when the cell values no longer changed.

The whole process can be set up to happen automatically by creating an intentional circular reference. The calculation is illustrated in Figure 10-11. The cells in rows 28 and 29 of Figure 10-6 were copied and pasted into rows 39–40 of Figure 10-11. Then the initial value 5 in cell B40 was replaced by the formula =G40. In this way the improved estimate of x was entered as the start value of the process.

	A	B	C	D	E	F	G	H
38								
39		x	Y	X+ΔX	Y+ΔY	slope	new X	
40		5.	401.5	5.000001	401.50012	245.00002	3.36122	
41								
42								

Figure 10-11.   Calculation of a root of a function by the Newton–Raphson method (before creating intentional circular reference).

When you press ENTER, the "Cannot resolve circular references" message is displayed. To force Excel to evaluate the circular reference, using the results of the previous calculation cycle as start values for the next cycle, choose **Options...**from the **Tools** menu and choose the Calculation tab. Check the Iteration box and enter 0.000001 in the Maximum Change box. (The default settings are Maximum Iterations = 100 and Maximum Change = .001.) When you press the OK button the circular reference will be evaluated. The results of the calculations are shown in Figure 10-12.

	A	B	C	D	E	F	G	H
38								
39		X	Y	X+ΔX	Y+ΔY	slope	new X	
40		1.61189	0.	1.61189	0.	26.44314	1.61189	
41								
42								

Figure 10-12.   Finding a root by the Newton-Raphson method and circular reference.

Another example of a problem that can be solved by intentional circular reference and iteration is the calculation of the pH of a solution of a weak acid. Combining the equations for mass balance and K_a, we obtain the following equation for the $[H^+]$ of a weak acid of concentration C mol/L.

$$[H^+] = \sqrt{K_a (C - [H^+])} \qquad (10\text{-}5)$$

Provided the extent of dissociation is small (i.e., $[H^+] \ll C$), the $[H^+]$ term can be neglected and the equation reduces to:

$$[H^+] = \sqrt{K_a C} \qquad (10\text{-}6)$$

The majority of weak acid problems can be solved using the approximate equation. However, if the extent of dissociation is relatively large (typically, greater than 10%), the $[H^+]$ term cannot be neglected and equation 10-5, when multiplied and rearranged, yields a quadratic equation. These problems are usually solved by using the quadratic formula, but an alternative approach is to use the method of successive approximations. An initial estimate of $[H^+]$, obtained from equation 10-6, is subtracted from C, as in equation 10-5, to obtain an improved value of $[H^+]$. The process is repeated until there is no further significant change in $[H^+]$. Each cycle of calculation is referred to as an iteration.

As an example of a weak acid problem that can readily be solved by circular reference and iteration, consider the calculation of the pH of a 0.1000 M solution of sodium bisulfate, $NaHSO_4$. The bisulfate ion is a relatively strong weak acid with $K_a = 1.2 \times 10^{-2}$. (Roughly speaking, weak acids with K_a values greater than 10^{-4} will require the quadratic or successive approximations approach.) The spreadsheet fragment in Figure 10-13 illustrates a case of a cell that contains a circular reference. Cell C5 contains the formula =SQRT($C$3*(A5-C5)).

An equivalent circular reference, in which a cell refers to itself indirectly, is shown in Figure 10-14. Cell C5 contains the formula =SQRT($C$3*B5) and cell B5 contains the formula =A5-C5.

	A	B	C
3	Bisulfate ion	Ka =	1.20E-02
4	[HA](initial)		[H⁺]
5	0.1		0.02916

Figure 10-13. Circular reference. Cell C5 refers to itself.

	A	B	C
3	Bisulfate ion	Ka =	1.20E-02
4	[HA](initial)	[HA](final)	[H⁺]
5	0.1	0.07084	0.02916

Figure 10-14. Circular reference. Cell C5 refers to a cell that refers to C5.

If the expression =SQRT($C$3*(A5-C5)) is typed into cell C5 in Figure 10-14, the "Cannot resolve circular references" message is displayed; when you click on OK in the dialog box, a zero value is displayed in the cell.

Choose **Options...** from the **Tools** menu and choose the Calculation tab. Check the Iteration box and enter 0.000001 in the Maximum Change box.

When you press the OK button, the worksheet will be recalculated until (in this case) the change in the cells is less than 0.00001.

For many chemistry calculations, this limit will not be sufficiently small. You will often need to reduce the Maximum Change parameter; you will rarely have to increase the Maximum Iterations parameter.

The current iteration number is displayed in the formula bar. To terminate calculations, press COMMAND+(period) or ESC (Macintosh) or ESC (Windows).

SOLVING SETS OF SIMULTANEOUS LINEAR EQUATIONS

Sometimes a chemical system can be represented by a set of n linear equations in n unknowns, i.e.,

$$a_{11}x_1 + a_{12}x_2 + a_{13}x_3 + \cdots = c_1$$
$$a_{21}x_1 + a_{22}x_2 + a_{23}x_3 + \cdots = c_2$$
$$\cdot$$
$$a_{n1}x_1 + a_{n2}x_2 + a_{n3}x_3 + \cdots = c_n \qquad (10\text{-}7)$$

where x_1, x_2, x_3, ... are the experimental unknowns, c is the experimentally measured quantity, and the a_{ij} are coefficients. The equations must be linearly independent; in other words, no equation is simply a multiple of another. These equations can be represented in matrix notation by

$$\mathbf{A}\,\mathbf{X} = \mathbf{C} \qquad (10\text{-}8)$$

A familiar example is the spectrophotometric determination of the concentrations of a mixture of n components by absorbance measurements at n different wavelengths. The coefficients a_{ij} are the ε, the molar absorptivities of the components at different wavelengths (for simplicity, the cell path length, usually 1.00 cm, has been omitted from these equations). For example, for a mixture of three species P, Q and R, where absorbance measurements are made at λ_1, λ_2 and λ_3, the equations are:

$$\varepsilon_{\lambda_1}^P[P] + \varepsilon_{\lambda_1}^Q[Q] + \varepsilon_{\lambda_1}^R[R] = A\lambda_1$$

$$\varepsilon_{\lambda_2}^P[P] + \varepsilon_{\lambda_2}^Q[Q] + \varepsilon_{\lambda_2}^R[R] = A\lambda_2$$

$$\varepsilon_{\lambda_3}^P[P] + \varepsilon_{\lambda_3}^Q[Q] + \varepsilon_{\lambda_3}^R[R] = A\lambda_3$$

Thus nine coefficients are required for the determination of three unknown concentrations.

CRAMER'S RULE

According to Cramer's rule, a system of simultaneous linear equations has a unique solution if the determinant D of the coefficients is non-zero.

$$D = \begin{vmatrix} a_{11} & a_{12} & a_{13} \\ a_{21} & a_{22} & a_{23} \\ a_{31} & a_{32} & a_{33} \end{vmatrix}$$

Thus, for example, for the set of equations

$$2x + y - z = 0$$
$$x - y + z = 6$$
$$x + 2y + z = 3$$

the determinant is

$$\begin{vmatrix} 2 & 1 & -1 \\ 1 & -1 & 1 \\ 1 & 2 & 1 \end{vmatrix}$$

The coefficients and constants lend themselves readily to spreadsheet solution, as illustrated in Figure 10-15.

Using the formula =MDETERM(A2:C4), the value of the determinant is found to be –9, indicating that the system is soluble.

	A	B	C	D
1	Coefficients			Constants
2	2	1	- 1	0
3	1	- 1	1	6
4	1	2	1	3

Figure 10-15. Spreadsheet data for three equations in three unknowns.

	A	B	C
8	0	1	- 1
9	6	- 1	1
10	3	2	1

Figure 10-16. The determinant for obtaining x.

The x values that comprise the solution of the set of equations can be calculated in the following manner: x_k is given by a quotient in which the denominator is D and the numerator is obtained from D by replacing the kth column of coefficients by the constants c_1, c_2, ... The unknowns are obtained readily by copying the coefficients and constants to appropriate columns in another location in the sheet. For example, to obtain x, the determinant is shown in Figure 10-16 and $x = 2$ is obtained from the formula

=MDETERM(A8:C10)/MDETERM(A2:C4)

$y = -1$ and $z = 3$ are obtained from appropriate forms of the same formula.

SOLUTION USING MATRIX INVERSION

If equation 10-8 is multiplied by the inverse of A, we obtain the relationship

$$X = A^{-1} C \tag{10-9}$$

In other words, the solution matrix is obtained by multiplying the matrix of constants by the inverse matrix of the coefficients. To return the solution values shown in Figure 10-17, the array formula

{=MMULT(MINVERSE(A2:C4),D2:D4)}

was entered in cells E2:E4.

	A	B	C	D	E
1	Coefficients			Constants	Solution
2	2	1	-1	0	2
3	1	-1	1	6	-1
4	1	2	1	3	3

Figure 10-17. Solving a set of simultaneous equations by means of matrix methods.

11

LINEAR REGRESSION

Excel provides several ways to find the coefficients that provide the best fit of a function to a set of data points — a process sometimes referred to as *curve fitting*. The "best fit" of the curve is considered to be found when the sum of the squares of the deviations of the data points from the calculated curve is a minimum. In the field of statistics, finding the least-squares best-fit parameters that describe a data set is known as *regression analysis*. In this chapter you'll learn how to perform simple and multiple linear regression

LEAST-SQUARES CURVE FITTING

Regression analysis is a statistical technique used to determine whether experimental variables are interdependent and to express quantitatively the relationship between them, and the degree of correlation. For most chemical systems, the mathematical form of the equation relating the dependent and independent variables is known; the goal is to obtain the values of the coefficients in the equation — the *regression coefficients*. In other cases, the data may be fitted by an empirical fitting function such as a power series, simply for purposes of graphing or interpolation. In any event, you must provide the form of the equation; regression analysis merely provides the coefficients.

A secondary but no less important goal is to obtain the standard deviations[*] of the regression parameters, and an estimate of the goodness of fit of the data to the model equation.

The *method of least squares* yields the parameters that minimize the sum of squares of the residuals (the deviation of each measurement of the dependent variable from its calculated value).

$$SS_{resid} = \sum_{n=1}^{N}(y_{obsd} - y_{calc})^2 \qquad (11\text{-}1)$$

---

[*] In Chapters 11 and 12, the symbol σ is used for the population standard deviation (i.e., when the sample size is large) and the symbol s for the sample standard deviation (when the sample size is small).

Linear regression is not restricted merely to straight-line relationships, but refers to any relationship that is *linear in the coefficients*, that is, any relationship of the form $y = a_0 + a_1x_1 + a_2x_2 + a_3x_3 + \dots$. The x_i can be different independent variables (e.g., pressure, temperature, time) or functions of a single independent variable (e.g., $[H^+]$, $[H^+]^2$, $[H^+]^3$).

LEAST-SQUARES FIT TO A STRAIGHT LINE

One of the most common methods of treating a set of x, y data points is to draw a straight line through them. Although it is relatively easy to draw a straight line through a series of points if they all fall on or near the line, it becomes a matter of judgment if the data are scattered. The method of least-squares provides the best method for objectively determining the best straight line through a series of points. In its simplest form, the least-squares approach assumes that all deviations from the line are the result of error in the measurement of the dependent variable y. Excel provides several worksheet functions that return regression parameters of the least-squares best fit of the straight line $y = mx + b$ to a data set.

THE SLOPE, INTERCEPT AND RSQ FUNCTIONS

The worksheet functions SLOPE(***known_ys, known_xs***) and INTERCEPT(***known_ys, known_xs***) return the slope, m, and intercept, b, respectively of the least-squares straight line through a set of data points. For example, Figure 11-1 illustrates some spectrophotometric calibration curve data (concentration of potassium permanganate standards in column B, absorbance of the standards in column C). The formula =SLOPE(C4:C8,B4:B8) in cell C10 was used to obtain the slope of the straight-line calibration curve. The SLOPE and INTERCEPT functions should be used with some caution, since they do not provide a measure of how well the data conforms to a straight line relationship.

	A	B	C	D	E
1	Calibration curve of potassium permanganate solutions				
2					
3		C, M	Abs	A(calc)	
4		0.000E+00	0.000	0.002	
5		1.029E-04	0.257	0.258	
6		2.058E-04	0.518	0.513	
7		3.087E-04	0.771	0.769	
8		4.116E-04	1.021	1.025	
9					
10		slope =	2.484E+03		
11		intercept =	0.0022		
12					

Figure 11-1. Using SLOPE and INTERCEPT functions.

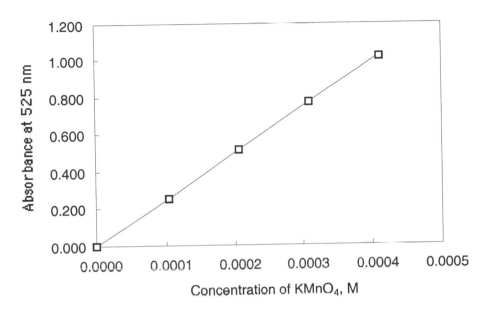

Figure 11-2. Least-squares best fit line through data points of a calibration curve.

At the least, a chart of the data should be produced for visual inspection of the fit, as illustrated in Figure 11-2. You can also use the RSQ(**known_ys, known_xs**) worksheet function to return R^2, the square of the correlation coefficient, to provide information about the goodness of fit to the straight line.

Column D in Figure 11-1 contains the absorbance values calculated from the slope and intercept; for example, cell D4 contains =$C$10*B4+$C$11.

LINEAR REGRESSION USING LINEST

The worksheet function LINEST performs linear regression analysis on a set of x,y data points. (LINEST stands for **LIN**ear **EST**imation, not **LINE ST**raight.) The general form of the linear equation that can be handled by LINEST is

$$y = m_1x_1 + m_2x_2 + m_3x_3 + \dots + b \qquad (11\text{-}2)$$

LINEST returns the array of regression parameters $m_n, \dots, m_2, m_1, b$. The syntax is LINEST(**known_ys**, known_xs, const_logical, stats_logical). If const_logical is TRUE, 1 or omitted, the regression parameters include an intercept b; if const_logical is FALSE or 0, the fit does not include the intercept b. If stats_logical is TRUE or 1, LINEST returns an array of regression statistics in addition to the regression coefficients $m_n, \dots, m_1$ and b. The layout of the array of returned values is shown in Figure 11-3. A one-, two-, three-, four-, or five-row array may be selected.

$m(n)$	$m(n-1)$	...	$m(2)$	$m(1)$	b
std.dev(n)	std.dev($n-1$)	...	std.dev(2)	std.dev(1)	std.dev(b)
R^2	SE(y)				
F	df				
SS(reg)	SS(resid)				

Figure 11-3. Layout of regression results and statistics returned by **LINEST**.

Mathematical relationships between the regression parameters are given below (N = number of data points, k = number of regression coefficients to be determined):

$$df \text{ (degrees of freedom)} = N - k \tag{11-3}$$

$$SS_{regression} = \Sigma(y_{mean} - y_{calc})^2 \tag{11-4}$$

$$SS_{residuals} = \Sigma(y_{obsd} - y_{calc})^2 \tag{11-5}$$

$$R^2 = 1 - \frac{SS_{resid}}{SS_{regression}} \tag{11-6}$$

$$F = \frac{SS_{regression}}{SS_{resid}/df} \tag{11-7}$$

$$SE(y) = \sqrt{\frac{SS_{resid}}{N - k}} \tag{11-8}$$

The coefficient of determination, R^2 (or the correlation coefficient, R), is a measure of the goodness of fit of the data to (in this case) a straight line. If x and y are perfectly correlated (i.e., the difference between y_{obsd} and y_{calc} is zero) then $R^2 = 1$. In contrast, an R^2 value of zero means that there is no correlation between x and y. A value of R^2 of less than 0.9 corresponds to a rather poor fit of data to a straight line.

The SE(y) parameter, the standard error of the y estimate, is sometimes referred to as the RMSD (root-mean-square deviation).

The F-statistic is used to determine whether the proposed relationship is significant (that is, whether y does in fact vary with respect to x). For most relationships observed in chemistry, a relationship will unquestionably exist. If it is necessary to determine whether the variation of y with x is statistically significant, or merely occurs by chance, you can consult a book on statistics.

The regression coefficients, the standard deviations of the coefficients and R^2, the coefficient of determination, are the statistical parameters of most interest to chemists.

LEAST-SQUARES FIT OF $y = mx + b$

A common application of LINEST is to find the best straight line through a set of data points, i.e., to find the regression parameters m and b of the best straight line $y = mx + b$ through the data points.

To use the LINEST function for this example, select an array two columns wide and one to five rows deep (depending on the regression parameters that you want returned). The array of regression parameters and statistics returned by LINEST when a 3R × 2C array is selected is shown in Figure 11-4. Then enter the LINEST function, followed by the appropriate values for known_ys and known_xs. Enter TRUE or 1 for constant_logical and for stats_logical. Since LINEST is an array function, after typing the closing parenthesis of the function, press COMMAND+RETURN (Macintosh) or CONTROL+SHIFT+ENTER (Windows). The regression parameters and statistics are returned, as shown in Figure 11-5. The returned values were formatted to display an appropriate number of significant figures.

m	b
std.dev of m	std.dev of b
R^2	SE(y)

Figure 11-4. Table of regression statistics returned by LINEST for slope and intercept of a straight line.

	B	C	D
13		*From LINEST:*	
14		m	b
15	parameters	2.484E+03	0.0022
16	std. devs	1.17E+01	0.0030
17	R², SE(y)	0.99993	0.00381

Figure 11-5. Slope and intercept of a straight line, with regression statistics.

REGRESSION LINE WITHOUT AN INTERCEPT

If the LINEST argument const_logical is set to FALSE, the m coefficient of the line of the form $y = mx$ that best fits the data is returned. Applying the LINEST function to the permanganate data yields the parameters shown in Figure 11-6.

	B	C	D
13		*From LINEST:*	
14		m	b
15	parameters	2.491E+03	0
16	std. devs	6.38E+00	#N/A
17	R², SE(y)	0.99992	0.00359

Figure 11-6. Slope of a straight line through the origin, with regression statistics.

WEIGHTED LEAST-SQUARES

If the y values to be fitted range over several orders of magnitude, it is sometimes advisable to use a *weighting factor* in the regression. Otherwise, if the error in the measurement is proportional to the magnitude of the measurement, the residuals of the largest measurements will have an overwhelming effect on the sum of squares. The weighting factor W is applied to each data point, so that the sum of squares is calculated according to equation 11-9.

$$\text{weighted } SS_{\text{resid}} = \sum_{n=1}^{N} W_n (y_n - y_{\text{calc}})^2 \qquad (11\text{-}9)$$

Some weighting functions that are often used include $W_n = 1/y_n$ and $W_n = 1/y_n^2$.

Weighting of the regression data can also be used if each y_n is an average of J_n observations. In this case the weighting factor is $W_n = J_n$.

MULTIPLE LINEAR REGRESSION

As is clear from equation 11-2, LINEST can be used to find the least-squares coefficients for equations involving more than one x variable. In the example that follows, LINEST was used to find the correlation between the wavelength of the visible absorption band of square-planar copper(II) complexes of peptide, amino acid and polyamine ligands and the identity of the four ligand donor groups coordinated to the copper(II) ion. According to the "rule of average environment", ν_{max}, the frequency or energy of the absorption band (as wavenumber), can be expressed as the sum of the ligand-field contributions of the four individual donor atoms in the complex. The wavenumber, in reciprocal centimeters, is used instead of wavelength since ν_{max} is proportional to the ligand-field splitting caused by the donor atoms; $\nu_{\text{max}} = 10^8/\lambda_{\text{max}}$. The donor atoms considered in the study consisted of N(amino), N(peptide), O(carboxylate), O(peptide), H_2O and OH^-. Preliminary analysis indicated that the ligand-field strengths of O(peptide), H_2O and OH^- are identical, and these were combined in the treatment that follows. Thus

$$\nu_{\text{max}} = \sum_{i=1}^{4} n_i \nu_i \qquad (11\text{-}10)$$

and the regression equation is

$$\nu_{\text{calc}} = \nu_1 n_1 + \nu_2 n_2 + \nu_3 n_3 + \nu_4 n_4 \qquad (11\text{-}11)$$

	A	B	C	D	E	F	G	H
1		*Analysis of Copper(II) Chromophores*						
2		#N	#N	#O	#O			
3		(amino)	(peptide)	(carboxy)	(H_2O)	λ_{max},nm	v_{max},cm^{-1}	λ_{max},nm
4	**Complex**	n_1	n_2	n_3	n_4	*Observed*	*Observed*	*Predicted*
5	CuGG+	1	0	0	3	735	13.61	738
6	CuGGG+	1	0	0	3	730	13.70	738
8	CuGa^{2+}	1	0	0	3	725	13.79	738
9	CuGGa^{2+}	1	0	0	3	725	13.79	738
11	Cu(Ga)$_2$	2	0	0	2	665	15.04	662
12	Cu gly+	1	0	1	2	715	13.99	713
13	Cu(gly)$_2$	2	0	2	0	617	16.21	623
14	CuH$_{-1}$GG	1	1	1	1	635	15.75	630
15	CuH$_{-1}$GGG	1	1	0	2	660	15.15	650
17	CuH$_{-1}$GGa+	1	1	0	2	660	15.15	650
19	Cu(Ga)(H$_{-1}$Ga)	2	1	0	1	605	16.53	590
20	Cu en^{2+}	2	0	0	2	660	15.15	662
22	CuH$_{-2}$GGG$^-$	1	2	1	0	555	18.02	564
24	CuH$_{-2}$GGa	1	2	0	1	580	17.24	580
26	CuH$_{-2}$GGGOH$^-$	1	2	0	1	578	17.30	580
27	CuH$_{-2}$GGaOH	1	2	0	1	575	17.39	580
28	CuH$_{-1}$GGOH	1	1	1	1	650	15.38	630
29	Cu dien^{2+}	3	0	0	1	611	16.37	600
30	Cu(H$_{-1}$Ga)$_2$	2	2	0	0	540	18.52	532
31	CuH$_{-3}$GGGG	1	3	0	0	520	19.23	524
34	Cu(en)$_2$$^{2+}$	4	0	0	0	549	18.21	549
35	Cu(1,2-DAP)$_2$$^{2+}$	4	0	0	0	545	18.35	549
37	Cu(tetmeen)$_2$$^{2+}$	4	0	0	0	547	18.28	549

Figure 11-7. Data table for multiple linear regression analysis of spectra of copper(II) complexes by LINEST. Ligand abbreviations: GG, glycylglycinate; Ga, glycinamide; gly, glycinate; en, ethylenediamine; dien, diethylenetriamine; 1,2 DAP, 1,2-diaminopropane; tetmeen, tetramethylethylenediamine. Data from E. J. Billo, *Inorg. Nucl. Chem. Lett.* **1974**, 10, 613.

where the n_i are the numbers of each type of donor atom in the complex (n_4 = total number of O(peptide), H_2O, OH$^-$ donor atoms). The n_i are the independent variables and the v_i are the regression coefficients to be determined.

The data table (Figure 11-7) has been compressed by hiding rows containing some of the less interesting data. To obtain the regression parameters, a 3R × 4C array was selected on the spreadsheet and the formula =LINEST(G5:G37,B5:E37,0,1) was entered. Since an array is to be returned, the array formula was entered by using COMMAND+ENTER (Macintosh) or CONTROL+SHIFT+ENTER (Windows). The array of returned values is shown in Figure 11-8.

The regression parameters can be used to predict the absorption maxima of other complexes, as illustrated in Figure 11-9.

		Results from LINEST		
		v_4	v_3	v_2	v_1
		$O(H_2O)$	$O(carboxy)$	$N(peptide)$	$N(amino)$
coeff.s		3.00	3.47	4.85	4.55
std.dev.s		0.03	0.09	0.04	0.03
R^2, SE(y.)		0.985	0.24	#N/A	#N/A

Figure 11-8. Regression results and statistics returned by LINEST.

Predicted Absorption Maxima of Some Copper(II) Complexes (nm)						
	n_1	n_2	n_3	n_4	calc	obsd
$Cu(H_2O)_4{}^{2+}$	0	0	0	4	834	808
$Cu(ala)_2$	2	0	2	0	623	620
$Cu(en)(Ga)^{2+}$	3	0	0	1	600	600
$Cu(en)(H_{-1}Ga)^+$	3	1	0	0	540	550
$Cu(en)(ala)^+$	3	0	1	0	584	582
$Cu(H_{-2}biuret)_2$	0	4	0	0	516	505

Figure 11-9. Using LINEST results to predict other values.

MULTIPLE LINEAR REGRESSION USING A POWER SERIES

LINEST can also be used to find the regression coefficients for equations of higher order. It is sometimes convenient, in the absence of a suitable equation, to fit data to a power series. The equation can then be used for data interpolation. Often a power series $y = a + bx + cx^2 + dx^3$ is sufficient to fit data of moderate curvature. The lowest order polynomial that produces a satisfactory fit should be used; if there are N data points, the highest order polynomial that can be used is of order $(N - 1)$.

The example presented in Figure 11-10 fits the solubility of oxygen in water as a function of temperature over the range 0–100°C. Columns for the T^2 and T^3 independent variables were inserted and the solubility data in column D were fitted to a cubic equation using LINEST.

The regression parameters returned by LINEST are shown in Figure 11-11. All four parameters seem to be significant, since the standard deviations are, at the most, less than 5% of the parameter value.

Figure 11-12 shows the fit of the polynomial to the data points. Except for the data points for 80°C and above, the deviation of the calculated line from the data points is less than 2% .

MULTIPLE LINEAR REGRESSION USING TRENDLINE

You can also fit a least-squares line to data points such as those shown in Figure 11-12 by adding a *trendline* to a chart. You can choose from a menu of

	A	B	C	D
1	Solubility of Oxygen in Water			
2	(g O$_2$/100 g H$_2$O)			
3				
4	T, deg C	T^2	T^3	S, g
5	0	0	0	0.006945
6	5	25	125	0.006072
7	10	100	1000	0.005368
8	15	225	3375	0.004802
9	20	400	8000	0.004339
10	25	625	15625	0.003931
11	30	900	27000	0.003588
12	35	1225	42875	0.003315
13	40	1600	64000	0.003082
14	45	2025	91125	0.002858
15	50	2500	125000	0.002657
16	60	3600	216000	0.002274
17	70	4900	343000	0.001856
18	80	6400	512000	0.001381
19	90	8100	729000	0.00079
20	100	10000	1000000	0

Figure 11-10. Fitting O$_2$ solubility in water by a power series. (Data reprinted with permission from *CRC Handbook of Chemistry and Physics*. Copyright CRC Press, Boca Raton FL.)

22	Results from LINEST			
23	-1.3E-08	2.21E-06	-0.00016	0.00685682
24	5.65E-10	8.39E-08	3.4E-06	3.6700E-05
25	0.999456	4.93E-05	#N/A	#N/A

Figure 11-11. Least-squares parameters of a power series for O$_2$ solubility in water.

mathematical functions — linear, logarithmic, polynomial, power, exponential — as curve-fitting functions.

To add a trendline, select the chart by clicking on it, then choose **Add Trendline...** from the **Chart** menu. If the chart has several data series, either select the desired data series before choosing **Add Trendline...** or choose the desired data series from the Based On Series box.

Choose the Type tab and then choose the appropriate fitting function from the gallery of functional forms. (Depending on the data in the series, the exponential, power or logarithmic choices may not be available.) If you choose the polynomial form you can select the order of the polynomial by using the spinner. If you choose 3, for example, Excel will fit a polynomial of order three

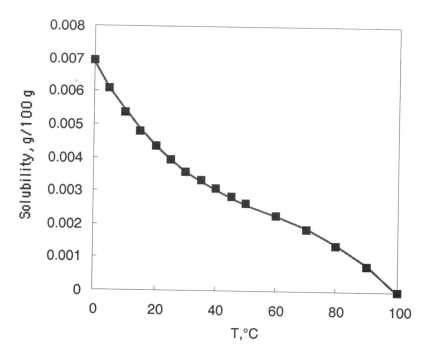

Figure 11-12. Fitting O_2 solubility in water by a power series.

(i.e., a cubic equation) to the data points. The maximum order is a polynomial of order six.

Now choose the Options tab and check the boxes for Display Equation On Chart and Display R-squared Value On Chart; then press OK. Excel displays the trendline on the chart as a heavy solid line and the equation (with the least squares coefficients) and R^2 value as a title on the chart.

If you want to use the coefficients for calculations, you'll have to copy them from the chart and paste them into worksheet cells. Usually the coefficients as displayed in the chart are not precise enough for calculations, but you can number-format the text to display more figures before copying the coefficients.

MULTIPLE LINEAR REGRESSION USING THE ANALYSIS TOOLPAK

Linear regression can also be performed using the Add-In package called the Analysis ToolPak. Choose **Data Analysis...** from the **Tools** menu; if the **Data Analysis...** command is not present in the **Tools** menu, you must use the **Add-Ins...** command in the **Tools** menu to install it.

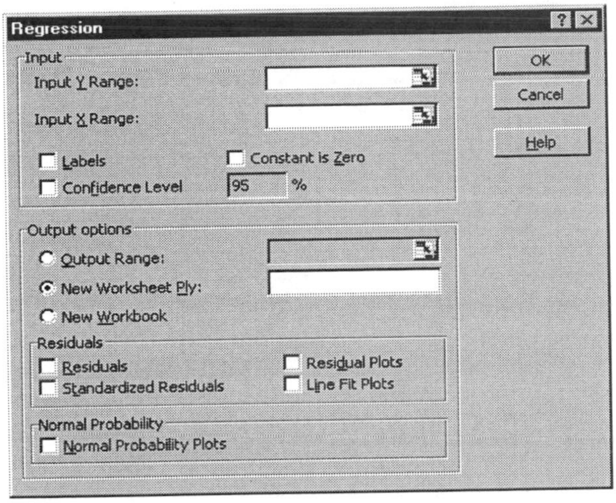

Figure 11-13. The Regression dialog box.

After you choose **Data Analysis...** from the Tools menu, choose Regression from the Analysis Tools list box. The Regression dialog box (Figure 11-13) will prompt you to enter the range of dependent variable (y) values and the range of independent variable (x) values, as well as whether the constant is zero, whether the first cell in each range is a label, and the confidence level desired in the output summary. Then select a range for the summary table. You need select only a single cell for this range; it will be the upper left corner of the range. You can also request a table of residuals and a normal probability plot. If you select a cell or range such that the summary table would over-write cells containing values, you will get a warning message.

In contrast to the results returned by LINEST, the output is clearly labeled, and additional statistical data is provided. Regression data for the example shown in Figure 11-1 is shown in the three tables of Figure 11-14. Three tables are produced: regression statistics, analysis of variance, and regression coefficients. (The coefficients table has been broken into two parts to fit the page.)

The major limitation of the regression tool is that, unlike LINEST, it is not a function. With LINEST, the returned values are dynamically linked to the original data and are updated if the raw data is changed. With the Regression tool, the values are calculated from the raw data and entered into worksheet cells; they do not change if you change the input data.

Regression Statistics	
Multiple R	0.999966633
R Square	0.999933268
Adjusted R Square	0.999911024
Standard Error	0.003812261
Observations	5

Analysis of Variance	df	Sum of Squares	Mean Square	F	Significance F
Regression	1	0.6533136	0.6533136	44952.7706	2.3137E-07
Residual	3	4.36E-05	1.4533E-05		
Total	4	0.6533572			

	Coefficients	Standard Error	t Statistic
Intercept	0.0022	0.00295296	0.74501401
x1	2483.96501	11.715673	212.020684

P-value	Lower 95%	Upper 95%
0.49765621	-0.0071977	0.01159766
2.9688E-09	2446.68048	2521.24955

Figure 11-14. Data obtained by using Regression from the Analysis ToolPak: (from top) Regression Statistics, Analysis of Variance, Regression Coefficients and Statistics.

USING THE REGRESSION STATISTICS

Regression statistics can be used to make decisions concerning the value or significance of regression coefficients — for example, whether the least-squares slopes of two lines are identical. This statistical procedure is known as *hypothesis testing*. The most common procedure is to set up a null hypothesis: for example, that the difference between the two slopes is zero. The answer — yes or no — must be stated within the context of a stated level of uncertainty. A confidence level of 95% is most commonly used; if the confidence level chosen (e.g., 99%) is too stringent, then a significant difference may be missed, while if the confidence level is too low (e.g., 50%) an insignificant difference may be accepted as real.

The *t*-test is used to test a null hypothesis. The computed *t*-statistic (calculated using one of the relationships given below) is compared with the value found in a table of *t* values corresponding to the appropriate number of degrees of freedom and at the desired confidence level. If the computed *t*-statistic exceeds the value from the table, the null hypothesis is rejected.

TESTING WHETHER AN INTERCEPT
IS SIGNIFICANTLY DIFFERENT FROM ZERO

For the permanganate calibration curve example examined earlier in this chapter, the intercept was found to be 0.0022 absorbance units with a standard

t-Table for Various Levels of Probability			
df	90%	95%	99%
1	6.31	12.70	63.70
2	2.92	4.30	9.92
3	2.35	3.18	5.84
4	2.13	2.78	4.60
5	2.02	2.57	4.03
10	1.81	2.23	3.17
15	1.75	2.13	2.95
20	1.72	2.09	2.84
30	1.70	2.04	2.75
60	1.67	2.00	2.66
∞	1.64	1.96	2.58

Figure 11-15. An abbreviated t-table.

deviation of 0.0030. Here it is obvious that the postulated zero intercept lies within the confidence interval for the intercept. But consider another case, where LINEST returns $b = 0.0011$ and $\sigma_b = 0.0005$; this calibration curve also contained $N = 5$ data points. Is the intercept statistically different from zero?

To test whether the intercept b is equal to a given value b_0, compute the t-statistic

$$t = \frac{b - b_0}{\sigma_b} \qquad (11\text{-}12)$$

and compare it to the t distribution with df degrees of freedom (for a straight line, $df = N - 2$ degrees of freedom).

For the second calibration curve

$$t = \frac{0.0011 - 0}{0.0005} = 2.2 \qquad (11\text{-}13)$$

If you consult a table of t values in a statistics text (a portion of such a table is shown in Figure 11-15), you will find, for 95% probability and 3 degrees of freedom, that $t = 3.18$. Since the calculated t-statistic is less than t-critical from the table, the intercept value of 0.0011 does not differ significantly from zero.

TESTING WHETHER TWO SLOPES ARE SIGNIFICANTLY DIFFERENT

To test whether the slopes of two different straight lines are equal, use the following equation:

$$t = \frac{m_2 - m_1}{\sigma_m(\text{pooled})} \qquad (11\text{-}14)$$

The pooled estimate of the standard deviation of the slope is given as follows, provided the number of data points is large:

$$\sigma_m(\text{pooled}) = \sqrt{\sigma^2_{m1} + \sigma^2_{m2}} \tag{11-15}$$

If the numbers of data points for each line, n_1 and n_2, are not large, then you must use the following equation to calculate the pooled estimate of the standard deviation:

$s_m(\text{pooled}) =$

$$\sqrt{\frac{(n_1 - 2)\, SE^2_{y1} + (n_2 - 2)\, SE^2_{y2}}{(n_1 - 2) + (n_2 - 2)}} \ \sqrt{\frac{1}{(n_1 - 1)S^2_{x1}} + \frac{1}{(n_2 - 1)S^2_{x2}}} \tag{11-16}$$

where S^2_{x1}, the variance of the x-values in data set 1, is given by:

$$S^2_{x1} = \frac{\Sigma x_1^2 - ((\Sigma x_1)^2 / n_1)}{(n_1 - 1)} \tag{11-17}$$

The variance can easily be obtained using the Excel function **VAR**. Equation 11-16 for the pooled estimate of the standard deviation is implemented, for example, by using the Excel formula

```
=((N.1-2)*SEY1^2+(N.2-2)*SEY2^2)/(N.1+N.2-4)*(1/((N.1-
1)*var1)+1/((N.2-1)*var2))
```

where N.1, SEY1 and var1 are the number of data points, the standard deviation of y and the variance of the x values of data set number 1, respectively.

TESTING WHETHER A REGRESSION COEFFICIENT IS SIGNIFICANT

In a multiple linear regression model using an empirical equation, you may want to know whether individual coefficients are significant, that is, whether they are useful in predicting the dependent variable. As in the preceding examples, if the standard deviation of the coefficient is small relative to the coefficient, the coefficient is clearly significant, while if the standard deviation of the coefficient is larger than the coefficient itself, it is clear that the coefficient is much less significant.

If you divide the coefficient by its standard deviation, the result is the t-statistic for that coefficient. The t-statistic is used to test hypotheses about the value of the coefficient. In a multiple regression model, the value of a_n/σ_n shows the relative importance of each term in the model.

TESTING WHETHER REGRESSION COEFFICIENTS ARE CORRELATED

Occasionally, in a multiple linear regression model using an empirical equation, two or more independent variables are correlated among themselves, that is, one independent variable is a linear function of another independent variable. Intercorrelated variables are said to exhibit *multicollinearity*. A non-chemical example of multicollinearity might exist in the relationship between sales of a product and the demographics of the customer base. Perhaps it is found that both educational level and family income are good predictors of sales of the product, yet it is clear that these two independent variables are strongly correlated.

If two independent variables are perfectly correlated (i.e., one is an exact linear function of the other) then LINEST will return #NUM! in all cells. If they are only approximately correlated, then LINEST will return regression parameters that can be very misleading: R^2 can be very close to 1, but the coefficients may have no meaning. The standard errors of one or more of the coefficients will probably be fairly large.

The Analysis ToolPak provides a convenient method for determining the correlation between independent variables. Choose Correlation from the Analysis Tools list box. The dialog box will prompt you to enter the range of independent variables; the dependent variables are not included. The output from Correlation, applied to the independent variable data from Figure 11-9, is shown in Figure 11-16. The output is a symmetrical matrix of correlation coefficients. The diagonal elements (correlation of a variable with itself) are all 1. The off-diagonal elements indicate the degree of correlation of two different independent variables. Unless a value is very close to 1.0, there is no significant correlation between independent variables.

	A	B	C	D	E
1		(amino)	(peptide)	(carboxy)	(H2O)
2	(amino)	1			
3	(peptide)	-0.4132	1		
4	(carboxy)	-0.1193	-0.0863	1	
5	(H2O)	-0.5072	-0.4790	-0.2327	1

Figure 11-16. Degree of correlation of independent variables from the Analysis ToolPak.

CONFIDENCE INTERVALS FOR SLOPE AND INTERCEPT

The confidence limits for the slope of a straight line are given by $m \pm t s_m$, where t is obtained from the t-table for the desired confidence level and $(n - 2)$ degrees of freedom. Similarly the confidence limits for the intercept are $b \pm t s_b$. Using the linest results for the permanganate calibration curve, shown in figure 11-5, and using $t = 3.18$ for 95% confidence level, the 95% confidence limits for the slope are $2.484 \times 10^3 \pm (3.18)(1.17 \times 10^1)$ or $(2.484 \pm 0.037) \times 10^3$.

CONFIDENCE LIMITS AND PREDICTION LIMITS
FOR A STRAIGHT LINE

The upper and lower *confidence limits* for y_{calc} at a particular value of x are given by equation 11-18. The confidence limits for y_{calc} are calculated by using a value of the t-statistic for a particular number of degrees of freedom and a particular probability level. Often the 90% probability level is used.

$$y_{calc} \pm t_{N-2} \sqrt{\frac{SS_{resid}}{N-2}} \sqrt{\frac{1}{N} + \frac{(x-\bar{x})^2}{(N-1)S^2_x}} \qquad (11\text{-}18)$$

The ± limits can be calculated using the Excel formula

=SQRT(SUM(d.sq)/(N-2))*SQRT(1/N+((X-X_bar)^2)/((N-1)*VAR(X)))

where d.sq is the residual squared, N is the number of observations, X_bar is the mean of the x-values, and VAR(X) returns the variance of the x-values.

The upper and lower *prediction limits* are given by the equation

$$y_{calc} \pm t_{N-2} \sqrt{\frac{SS_{resid}}{N-2}} \sqrt{1 + \frac{1}{N} + \frac{(x-\bar{x})^2}{(N-1)S^2_x}} \qquad (11\text{-}19)$$

which differs from equation 11-18 only in the extra 1 within the square root term. The prediction limits describe the confidence limits for predicting a single y-value for a particular x-value. The prediction limits contain two sources of error: the error in estimating y_{calc} and the error associated with a single measurement.

Very commonly, as in the calibration curve example, we use the measured y-value to estimate an x-value. Once the slope and intercept of a straight-line calibration curve have been established, it is easy to calculate an x-value (e.g., a concentration) from a measured y-value. The *estimation limits* of the estimated x-value are given by the equation

$$x_{meas} \pm t_{N-2} \frac{1}{b} \sqrt{\frac{SS_{resid}}{N-2}} \sqrt{1 + \frac{1}{N} + \frac{(y_{meas} - \bar{y})^2}{b^2 \Sigma(x_i - \bar{x})^2}} \qquad (11\text{-}20)$$

If y_{meas} is an average of M readings, then the equation becomes

$$x_{meas} \pm t_{N-2} \frac{1}{b} \sqrt{\frac{SS_{resid}}{N-2}} \sqrt{\frac{1}{M} + \frac{1}{N} + \frac{(y_{meas} - \bar{y})^2}{b^2 \Sigma(x_i - \bar{x})^2}} \qquad (11\text{-}21)$$

USEFUL REFERENCES

David G. Kleinbaum and Lawrence L. Kupper, *Applied Regression Analysis and Other Multivariable Methods*, Duxbury Press, Belmont, CA, 1978.

J. C. Miller and J. N. Miller, *Statistics for Analytical Chemistry*, 3rd ed., Ellis Horwood, New York , 1993.

12

NON-LINEAR REGRESSION
USING THE SOLVER

In this chapter you'll learn how to use the Solver, Excel's powerful optimization package, to perform non-linear least-squares curve fitting.

NON-LINEAR FUNCTIONS

The function

$$y = F(x_1, x_2, ..., a_0, a_1, a_2, ...)$$ (12-1)

where y and $x_1, x_2, ...$ are the dependent and independent variables, respectively, and $a_0, a_1, a_2, ...$ are coefficients, can be either *linear* or *non-linear* in the coefficients. A linear function is one of the type

$$y = a_0 + a_1 Z_1 + a_2 Z_2 + a_3 Z_3 + \cdots$$ (12-2)

where the Z's are functions of the independent variables x_i. The word "linear" in linear regression does not mean that the function is a straight line, but that the partial derivatives with respect to each coefficient are not functions of other coefficients.

An example of a function that is linear in the coefficients is

$$y = a_0 + a_1 x + a_2 x^2 + a_3 x^3$$ (12-3)

and it should be clear from reading Chapter 11 that LINEST can be used to obtain the regression coefficients for a data set that can be described by such an equation. However, if the function is one such as

$$y = \exp(a_0 + a_1 x)$$ (12-4)

or

$$y = \exp(-a_1 x) - \exp(-a_2 x)$$ (12-5)

then it is not linear in the coefficients, since the equation cannot be rearranged to obtain an expression containing separate $a_i Z_i$ terms. Some non-linear equations can be transformed into a linear form. Equation 12-4, for example, can be transformed by taking the logarithm to the base e of each side, to yield the equation

$$\ln y = a_0 + a_1 x \qquad (12\text{-}6)$$

which is linear. Others, such as equation12-5, cannot be converted into a linear form and are said to be *intrinsically non-linear*.

USING THE SOLVER TO PERFORM NON-LINEAR LEAST-SQUARES CURVE FITTING

There are many published computer programs and commercial software packages that perform non-linear regression analysis, but you can obtain the same results very easily by using the Solver. When applied to the same data set, the Solver gives the same results as commercial software packages.

USING THE SOLVER FOR OPTIMIZATION

The Solver is an optimization package that finds a maximum, minimum or specified value of a *target cell* by varying the values in one or several *changing cells*. It accomplishes this by means of an iterative process, beginning with trial values of the coefficients. The value of each coefficient is changed by a suitable increment, the new value of the function is calculated and the change in the value of the function is used to calculate improved values for each of the coefficients. The process is repeated until the desired result is obtained. The Solver uses gradient methods or the simplex method to find the optimum set of coefficients.

With the Solver you can apply constraints to the solution. For example, you can specify that a coefficient must be greater than or equal to zero, or that a coefficient must be an integer. Solutions to chemical problems will rarely use the integer option, and although the ability to apply constraints to a solution may be tempting, it can sometimes lead to an incorrect solution.

The Solver is an Add-In, a separate software package. To save memory, it may not automatically be opened whenever you start Excel. If the Solver Add-in has already been opened, you will see the **Solver...** command in the **Tools** menu. If not, open the Solver Add-In by choosing **Add-ins...** from the **Tools** menu, check the box for the Solver, then press OK.

The Solver code is not written by Microsoft, but instead is a product of Frontline Systems Inc. (P.O. Box 4288, Incline Village, NV 89450, www.frontsys.com). Essentially the same code is used in Microsoft Excel Solver, in the Lotus 1-2-3 Solver, and in the Quattro Pro Solver.

USING THE SOLVER FOR LEAST-SQUARES CURVE-FITTING

To use the Solver to perform multiple non-linear least-squares curve fitting, follow the procedure outlined in the accompanying box.

To Use the Solver to Perform
Non-Linear Least-Squares Curve Fitting

1. Start with a worksheet containing the data (independent variable x and dependent variable y_{obsd}) to be fitted.

2. Add a column containing y_{calc} values, calculated by means of an appropriate formula and involving the x values and one or more coefficients to be varied.

3. Add a column to calculate the square of the residual $(y_{obsd} - y_{calc})$ for each data point.

4. Calculate the sum of squares of the residuals.

5. Use **Solver...** to minimize the sum of squares of residuals (the target cell) by changing the coefficients of the function (the changing cells).

The target cell and the changing cells must be on the active sheet. Your model can involve external references to values in other worksheets or workbooks, however.

Since the Solver operates by a search routine, it will find a solution most rapidly and efficiently if the initial estimates that you provide are close to the final values. Conversely, it may not be able to find a solution if the initial estimates are far from the final values. To ensure that the Solver has found a global minimum rather than a local minimum, it's a good idea to obtain a solution using different sets of initial estimates.

The least-squares regression coefficients that it returns may be slightly different, depending on the starting values that you provide.

USING THE SOLVER: AN EXAMPLE

The following example illustrates the ease with which the Solver can be used to perform non-linear least-squares curve fitting. Here we analyze kinetics data (absorbance vs. time) from a biphasic reaction involving two consecutive first-order reactions (A $\rightleftharpoons$ B $\rightleftharpoons$ C) to obtain two rate constants and the molar absorptivity of the intermediate species B.

The equations for the concentrations of the species A, B and C in a reaction sequence of two consecutive first-order reactions can be found in almost any kinetics text. The expression for $[B]_t$ is

$$[B]_t = [A]_0 \frac{k_1}{k_2 - k_1} \{\exp(-k_1 t) - \exp(-k_2 t)\} \tag{12-7}$$

where $[A]_0$ is the initial concentration of the reactant species and k_1 and k_2 are the rate constants. This is an example of an intrinsically non-linear equation.

A stopped-flow spectrophotometer was used to obtain the kinetics data* for the reaction of a nickel(II) complex NiL_2^{2+} of a substituted bidentate diamine ligand (L = tetmeen = 2,3-dimethyl-2,3-diaminobutane) with cyanide ion. The reaction is biphasic; one diamine ligand is replaced by two cyanide ligands, then the second diamine ligand is replaced:

$$NiL_2^{2+} + 2CN^- \rightleftharpoons NiL(CN)_2 + L \qquad (12\text{-}8)$$

$$NiL(CN)_2 + 2CN^- \rightleftharpoons Ni(CN)_4^{2-} + L \qquad (12\text{-}9)$$

The species $NiL(CN)_2$, which is formed and then decays during the reaction, absorbs at 243 nm, where NiL_2^{2+}, $Ni(CN)_4^{2-}$, CN^- or L does not absorb appreciably. The stopped-flow data (measured manually from an oscilloscope trace used to collect the data) are shown in columns A and B of Figure 12-1.

In the expression for $[B]_t$ (equation 12-7), $[A]_0$ is the initial concentration of the reactant species NiL_2^{2+}, and k_1 and k_2 are the rate constants for reactions in equations 12-7 and 12-8. We use $[B]_t$ together with Beer's law to write

$$A_{obsd} = \varepsilon b[B]_t \qquad (12\text{-}10)$$

where A_{obsd} is the measured absorbance, ε is the molar absorptivity of the intermediate species $NiL(CN)_2$, and b is the path length of the stopped-flow cell. The parameters ε, k_1 and k_2 are the regression parameters that we want to obtain by non-linear least-squares curve fitting.

In Figure 12-1, columns C and D contain the concentration of the intermediate species B and the absorbance, calculated using equations 12-9 and 12-10, respectively. The formulas in cells C12 and D12 are (the optical path length was 0.4 cm):

=C_A*k_1*(EXP(-k_2*t)-EXP(-k_1*t))/(k_1-k_2) and

=0.4*E_B*C12

In column E the squares of the residuals are calculated. The sum of squares is obtained in cell E27, using the Σ tool (select cell E27, click twice on the Σ tool).

The changing cells (B8, E5, E6) and the target cell (E27) are shown in bold. Initial values, estimated from the data, were 3000, 1 and 0.2, respectively.

To use the Solver, choose Solver... from the Formula menu. First you may see displayed the message box shown in Figure 12-2. This is only an advisory message, not an error message. Press OK. The Solver Parameters dialog box (Figure 12-3) will then be displayed.

---

*   From J. C. Pleskowicz and E. J. Billo, *Inorg. Chim. Acta* **1985**, *99*, 149.

	A	B	C	D	E
1	Rate of Reaction of Ni(tetmeen)$_2^{2+}$ with Cyanide				
2	Stopped-flow run #173-3 (page 72)				
3	Date:	4/8/96			
4	Operator:	JA Tyrell		*Rate constants:*	
5	Wavelength:	243 nm		$k_1 =$	1.000
6	Conc:	4.00E-05	(C_A)	$k_2 =$	0.200
7	E(A):	0			
8	E(B):	3.00E+03	(E_B)		
9	E(C):	0			
10	t, sec	A(obsd)	[B]	A(calc)	∂^2
12	0.2	0.0047	7.10E-06	0.0085	1.5E-05
13	0.6	0.0129	1.69E-05	0.0203	5.5E-05
14	1.0	0.0163	2.25E-05	0.0271	1.2E-04
15	1.4	0.0188	2.55E-05	0.0306	1.4E-04
16	1.8	0.0201	2.66E-05	0.0319	1.4E-04
17	2.2	0.0208	2.67E-05	0.0320	1.3E-04
18	2.6	0.0208	2.60E-05	0.0312	1.1E-04
19	3.0	0.0205	2.50E-05	0.0299	8.9E-05
20	4.0	0.0178	2.16E-05	0.0259	6.5E-05
21	5.0	0.0149	1.81E-05	0.0217	4.6E-05
22	6.0	0.0118	1.49E-05	0.0179	3.7E-05
23	7.0	0.0090	1.23E-05	0.0147	3.3E-05
24	8.0	0.0070	1.01E-05	0.0121	2.6E-05
25	9.0	0.0052	8.26E-06	0.0099	2.2E-05
26	10.0	0.0038	6.76E-06	0.0081	1.9E-05
27				Σ(∂^2) =	1.03E-03

Figure 12-1. The spreadsheet before optimization of coefficients by the Solver.

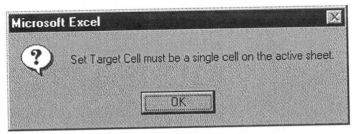

Figure 12-2. The initial Solver message.

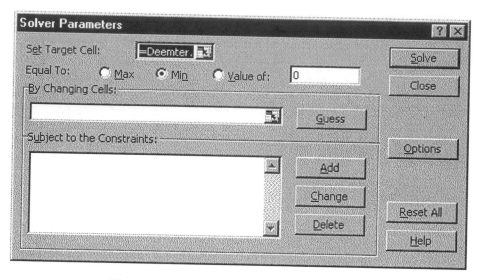

Figure 12-3. The Solver Parameters dialog box.

To solve the problem:

1. In the Set Cell box, type E27, or select cell E27 with the mouse. (If you selected E27 before running the Solver, $E$27 will appear in the Set Cell box.)

2. You want to minimize the sum of squares, so press the Min button.

3. Select cells B8 and E5:E6 so that they appear in the By Changing Cells box.

4. Do not enter any constraints.

5. Press the Solve button.

As the problem is set up and solved, messages will appear in the status bar at the bottom of the screen. The value of the target cell is displayed in the status bar after every iteration. In a few seconds the Solver finds a solution and displays the completion dialog box. You have the options of accepting the Solver's solution or restoring the original values. Press the Accept Solver Solution button. The spreadsheet will be displayed with the final values of the target cells (Figure 12-4).

Excel Tip. Don't introduce constraints (e.g., to force a constant to be greater than or equal to zero) if you're using the Solver to obtain the least-squares best fit. The solution will not be the "global minimum" of the error-square sum, and the regression coefficients may be seriously in error.

Excel Tip. Discontinuous functions in your Solver model can cause problems. They can be either discontinuous mathematical functions such as TAN, which has a discontinuity at π/2, or worksheet functions that are inherently "discontinuous", such as IF, ABS, INT, ROUND, CHOOSE, LOOKUP, HLOOKUP or VLOOKUP.

There are some additional controls in the Solver Parameters dialog box:

- The Add..., Change... and Delete buttons are used to apply constraints to the model. Since the use of constraints is to be avoided, these buttons are not of much interest.
- Pressing the Guess button will enter references to all cells that are precedents of the target cell. In the example above, pressing the Guess button enters the cell references A$12:B$26, $B$6, $B$8, $E$5, $E$6 in the By Changing Cells box.
- The current Solver model is automatically saved with the worksheet. The Reset All button permits you to "erase" the current model and begin again.

	A	B	C	D	E
1	Rate of Reaction of Ni(tetmeen)$_2^{2+}$ with Cyanide				
2	Stopped–flow run #173–3 (page 72)				
3	Date:	4/8/96			
4	Operator:	JA Tyrell		Rate constants:	
5	Wavelength:	243 nm		$k_1 =$	0.639
6	Conc:	4.00E–05	(C_A)	$k_2 =$	0.285
7	E(A):	0			
8	E(B):	2.53E+03	(E_B)		
9	F(C):	0			
10	t, sec	A(obsd)	[B]	A(calc)	∂^2
12	0.2	0.0047	4.66E–06	0.0047	8.5E–11
13	0.6	0.0129	1.16E–05	0.0118	1.3E–06
14	1.0	0.0163	1.62E–05	0.0164	2.6E–09
15	1.4	0.0188	1.89E–05	0.0191	1.0E–07
16	1.8	0.0201	2.04E–05	0.0206	2.2E–07
17	2.2	0.0208	2.09E–05	0.0211	7.6E–08
18	2.6	0.0208	2.07E–05	0.0209	1.2E–08
19	3.0	0.0205	2.01E–05	0.0203	4.6E–08
20	4.0	0.0178	1.75E–05	0.0177	2.1E–08
21	5.0	0.0149	1.44E–05	0.0145	1.3E–07
22	6.0	0.0118	1.15E–05	0.0116	3.9E–08
23	7.0	0.0090	8.98E–06	0.0091	5.7E–09
24	8.0	0.0070	6.94E–06	0.0070	8.6E–11
25	9.0	0.0052	5.31E–06	0.0054	2.8E–08
26	10.0	0.0038	4.04E–06	0.0041	8.2E–08
27				Σ(∂^2) =	2.06E–06

Figure 12-4. Regression coefficients (in bold) returned by the Solver.

Excel Tip.   If, after you've made changes to a worksheet, a Solver model that had previously converged to a reasonable solution refuses to converge, and all attempts to find the problem have failed, use the Reset All button to erase the current model. Then re-enter references to the Target Cell and the Changing Cells. This may solve the problem.

The fit of the curve to the data points is shown in Figure 12-5.

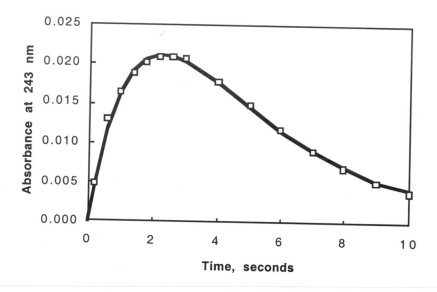

Figure 12-5. Curve calculated from the coefficients obtained by the Solver.

COMPARISON WITH A COMMERCIAL NON-LINEAR LEAST-SQUARES PACKAGE

In Figure 12-6, the Solver results are compared with the results obtained by using NLLSQ, a commercial non-linear regression analysis program (CET Research, Inc., Norman, OK 73070). It is clear that the Solver provides results that are essentially identical to those from the commercial software package. The slight differences (ca. 0.001% or less) arise from the fact that the coefficients are found by a search method; the "final" values will differ depending on the convergence criteria used in each program.

Other software packages that can be used to perform non-linear regression analysis are MathCad (MathSoft, Inc., 201 Broadway, Cambridge, MA 02139) and MatLab (Mathworks Inc., 24 Prime Park Way, Natick, MA 01760).

	A	B	C
33	Comparison of Results from the Solver and from NLLSQ		
34		Solver	NLLSQ
35	k_1	0.63877	0.63878
36	k_2	0.285394	0.285392
37	ε	2526.43	2526.39
38	$\Sigma(\partial^2) =$	2.0620E-06	2.0620E-06

Figure 12-6. Comparison of regression coefficients.

A major disadvantage of the Solver is that it does not provide the standard deviations of the coefficients. This problem is addressed later in this chapter (see "Statistics of Non-linear Regression").

SOLVER OPTIONS

The Options button in the Solver Parameters dialog box activates the Solver Options dialog box (Figure 12-7) and allows you to control the way Solver attempts to reach a solution. The default values of the options are shown in the dialog box.

The Max Time and Iterations parameters determine when the Solver will return a solution or halt. If either Max Time (100 seconds) or Iterations (100) is exceeded before a solution has been reached, the Solver will pause and ask if you want to continue. For most simple problems, these limits will not be exceeded. In any event, you don't need to adjust Max Time or Iterations, since if either parameter is exceeded, the Solver will pause and issue a "Continue anyway?" message.

Both the Precision and Tolerance options apply only to problems with constraints. The Precision parameter determines the amount by which a constraint can be violated. The Tolerance parameter is similar to the Precision parameter, but applies only to problems with integer solutions. Since adding constraints to a model that involves minimization of the error-square sum is not recommended, neither the Precision nor the Tolerance parameter is of use in non-linear regression analysis.

The Convergence parameter was introduced in Excel 97; in earlier versions it was fixed and could not be changed by the user. Unlike the Maximum Change parameter in the Calculations Options dialog box (see Chapter 10, Figure 10-7), which is an absolute convergence limit, the Solver's Convergence parameter is relative; the Solver will stop iterating when the relative change in the target cell value is less than the number in the Convergence box for the last five iterations. Thus you don't have to scale the convergence limit to fit the problem, as is required with the Maximum Change parameter in the Calculations Options dialog box.

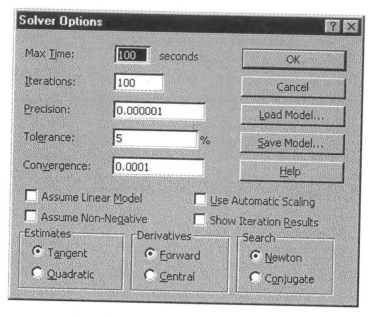

Figure 12-7. The Solver Options dialog box.

If the function is linear, checking the Assume Linear Model box will speed up the solution process.

If the Show Iteration Results box is checked, the Solver will pause and display the result after each iteration.

The Use Automatic Scaling option should be checked if there are large differences in magnitude between the Target Cell and the Changing Cells — for example, if you are varying internuclear bond distance (on the order of 10^{-9} m) to obtain a minimum value of bond energy (on the order of 10^5 J).

The Estimates, Derivatives and Search parameters can be changed to optimize the solution process. The Search parameter specifies which gradient search method to use: the Newton method requires more memory but fewer iterations, the Conjugate method requires less memory but more iterations. The Derivatives parameter specifies how the gradients for the search are calculated: the Central derivatives method requires more calculations but may be helpful if the Solver reports that it is unable to find a solution. The Estimates parameter determines the method by which new estimates of the coefficients are obtained from previous values; the Quadratic method may improve results if the system is highly nonlinear.

The current Solver model is automatically saved with the worksheet. The Save Model... and Load Model... buttons permit you to save multiple Solver models.

THE "USE AUTOMATIC SCALING" OPTION IS IMPORTANT FOR MANY CHEMICAL PROBLEMS

For some chemical models the Solver may refuse to converge satisfactorily. This may occur when there is a large difference in magnitude between changing cells, or between changing cells and the target cell — for example, if you are varying an equilibrium constant with magnitude 1×10^{10} to minimize the sum of squares of residuals. The Solver may "refuse" to vary one or more changing cells or vary them by only an insignificant amount. You can probably overcome this problem by checking the Use Automatic Scaling option in the Solver Options dialog box.

STATISTICS OF NON-LINEAR REGRESSION

As you've seen, the Solver finds the set of least-squares regression coefficients very quickly and efficiently. However, it does not provide the standard deviations of the coefficients. Without these, the Solver's solution is essentially useless. The following illustrates how to obtain the standard deviations of the regression coefficients after obtaining the coefficients by using the Solver.

The standard deviation of the regression coefficient a_i is given by[*]

$$\sigma_i = \sqrt{P_{ii}^{-1}} \ SE(y) \tag{12-11}$$

where P_{ii}^{-1} is the ith diagonal element of the inverse of the P_{ij} matrix.

$$P_{ij} = \sum_{n=1}^{N} \frac{\delta F_n}{\delta a_i} \frac{\delta \Gamma_n}{\delta a_j} \tag{12-12}$$

$\delta F_n/\delta a_i$ is the partial derivative of the function with respect to a_i evaluated at x_n and

$$SE(y) = \sqrt{\frac{SS_{resid}}{N-k}} \tag{12-13}$$

The quantities SS_{resid}, N and k are as defined in Chapter 11.

The $\delta F/\delta a_i$ terms can be calculated for each data point by numerical differentiation. The term a_i is varied by a small amount from its optimized value while the other a_j terms are held constant. The differential $\delta F/\delta a_i \approx \Delta F/\Delta a_i = (F_{new} - F_{opt})/(a_{new} - a_{opt})$ is calculated for each data point. Since Excel carries 15 significant figures, the change in a_i can be made very small, so that $\delta F/\delta a_i =$

---

[*]   K. J. Johnson, *Numerical Methods in Chemistry*, Marcel Dekker, New York, 1980, p. 278.

$\Delta F/\Delta a_i$. This process is repeated for each of the k regression coefficients. Then the cross-products $(\delta F/\delta a_i)(\delta F/\delta a_j)$ are computed for each of the N data points and the $\Sigma(\delta F/\delta a_i)(\delta F/\delta a_j)$ terms obtained. The $\mathbf{P}_{ij}$ matrix of $\Sigma(\delta F/\delta a_i)(\delta F/\delta a_j)$ terms is constructed and inverted. The terms along the main diagonal of the inverse matrix are then used with equation 12-11 to calculate the standard deviations of the coefficients. This method may be applied to either linear or non-linear systems.

To illustrate the procedure, the method is applied to a small data set shown in Figure 12-8. The formula Y = 5*X + 3 + 0.1*(0.5-RAND()) was used to generate Y(data) in column B, with a small amount of experimental "noise" provided by the RAND() function. A linear function was chosen to permit comparison of the standard deviations returned by LINEST with those provided by this method. Because the test equation is a simple linear one, the $\delta F/\delta a_i$ values ($\delta y/\delta_m$ and $\delta y/\delta_b$ in columns H and I of Figure 12-8) are very close to simple integer values. This is not the case when the procedure is applied to more complicated functions.

Compare the standard deviations using the forgoing procedure (cells L24 and M24 in Figure 12-9) with those obtained from LINEST, which are shown in row 32 of Figure 12-10. Once again, it should be made clear that a linear problem was chosen to permit comparison between the standard deviations of the regression coefficients and those from LINEST.

	A	B	C	D	E	F	G	H	I
1	*Method for Obtaining Standard Deviations of Solver Parameters*								
2	1. Regression parms obtained using Solver by minimizing Σd^2 (cell D17).								
3	2. Copy & Paste Y(calc) values to save them as Y(fixed) in column E.								
4	3. Solver parms Copied and Pasted to another location to use as ones to vary (C21:D21).								
5	4. Equations for $\delta Y/\delta m$ and $\delta Y/\delta b$ entered. Cell F12 contains (E12-C12)/($C$19-$C$21).								
6	They should display #DIV/0! initially.								
7	5. ($\delta Y/\delta m$) obtained by multipying parm value by 1.000001.								
8	6. Save column of $\delta Y/\delta m$ values (Copy & Paste Values), then restore original parm value.								
9	7. Repeat process to obtain ($\delta Y/\delta b$)								
10									
11								Saved Values	
11	x	Y(data)	Y(calc)	d^2	Y(fixed)	($\delta Y/\delta m$)	($\delta Y/\delta b$)	($\delta Y/\delta m$)	($\delta Y/\delta b$)
12	0	5.026	5.008	3.5E-04	5.008	#DIV/0!	#DIV/0!	0.0000	1.0000
13	1	7.965	8.006	1.6E-03	8.006	#DIV/0!	#DIV/0!	1.0000	1.0000
14	2	11.012	11.004	6.5E-05	11.004	#DIV/0!	#DIV/0!	2.0000	1.0000
15	3	14.033	14.002	9.4E-04	14.002	#DIV/0!	#DIV/0!	3.0000	1.0000
16	4	16.984	17.001	2.9E-04	17.001	#DIV/0!	#DIV/0!	4.0000	1.0000
17	N = 5		Σd^2 =	3.3E-03					
18			*Values from Solver:*						
19			2.99831	5.00755					
20			*Values to vary:*						
21			2.99831	5.00755					
22									

Figure 12-8. Method for obtaining regression statistics of coefficients obtained by the Solver (part 1).

	K	L	M	N
1	**Standard Deviations of Solver Parameters (continued)**			
2	8. Calculate the $(\partial Y/\partial m)$, $(\partial Y/\partial b)$ products and Σ terms.			
3		$(\partial Y/\partial m)^2$	$(\partial Y/\partial m)(\partial Y/\partial b)$	$(\partial Y/\partial b)^2$
4		0	0	1
5		1	1	1
6		4	2	1
7		9	3	1
8		16	4	1
9	$\Sigma =$	30	10	5
10				
11	9. Construct matrix from the Σ terms & obtain inverse matrix.			
12		**Matrix (P)**		
13		30	10	
14		10	5	
15		**Inverse matrix (P⁻¹)**		
16		0.1	-0.2	
17		-0.2	0.6	
18				
19	10. Calculate standard deviation of the parms using $SE(y)$			
20	& i,i terms from inverse matrix.			
21		**SE(y)**	0.033048936	
22	Formula used: $\sqrt{\{\Sigma d^2/(N-2)\}}$			
23		$\sigma(m)$	$\sigma(b)$	
24		0.01045099	0.025599596	
25	Formula used: $SQRT(P^{-1}ii)*SE(y)$			

Figure 12-9. Method for obtaining regression statistics of coefficients obtained by the Solver (part 2).

A MACRO TO PROVIDE REGRESSION STATISTICS FOR THE SOLVER

The preceding example illustrates the steps in the application of the method. The procedure is cumbersome to apply in practice. The macro SolvStat.xls on the diskette that accompanies this book was written to perform the calculations. It returns the standard deviations of the coefficients, the correlation coefficient and the SE(y) or RMSD; it can be applied to linear or non-linear regression.

	K	L	M
30	*From LINEST:*		
31	*coeff.s*	2.99831	5.00755
32	*std.dev.s*	0.01045099	0.025599596
33	R^2, SE(y)	0.999964	0.033048936

Figure 12-10. Comparison of regression statistics returned by LINEST.

To calculate the standard deviations of regression coefficients obtained by using the Solver, the macro uses the same approach outlined in the preceding example. The $SS_{residuals}$, $SS_{regression}$ and $SE(y)$ are calculated from the known

y's and calculated y's. The partial differentials $\delta y / \delta a_i$ for each of the k regression coefficients are calculated for all N data points by a procedure similar to that used in the Newton–Raphson method of Chapter 10. A table of products of partial differentials is created and used to create a $k \times k$ matrix, and the matrix is inverted. Then equation 12-11 is sued to calculate the standard deviations, using the diagonal elements of the inverted matrix. Finally, the standard deviations, the correlation coefficient and SE(y) are returned to the source worksheet.

USING THE SOLVSTAT MACRO

The SolvStat.xls macro is an Auto_Open macro; when you **Open** the document; it will appear on screen and then **Hide** itself. It installs a new menu command, **Solver Statistics...**, directly under the **Solver...** command in the **Tools** menu. If the Solver Add-In has not been loaded, the **Solver Statistics...** command will be at the top of the menu. The command will remain in the menu until you exit from Excel.

When you choose the **Solver Statistics...** command, a sequence of four dialog boxes will be displayed, and you will be asked to select four cell ranges: (i) the y_{obsd} data, (ii) the y_{calc} data, (iii) the Solver coefficients, (iv) a $3R \times nC$ range of cells to receive the statistical parameters. The Step 1 dialog box is shown in Figure 12-11. The y values can be in row or column format. The Solver coefficients can be in non-adjacent cells.

The array of results returned by the macro is similar to that returned by LINEST, as shown in Figure 12-12.

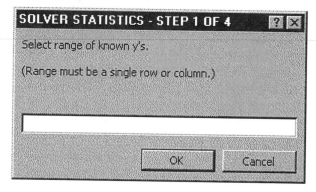

Figure 12-11. SolvStat.xls Step 1 of 4 dialog box.

parm(n)	parm(n-1)	...	parm(2)	parm(1)
std.dev(n)	std.dev(n-1)	...	std.dev(2)	std.dev(1)
R^2	SE(y)			

Figure 12-12. Layout of regression parameters and statistics returned by the SolvStat macro.

	A	B	C
40	Comparison of results from the Solver and the SOLVSTAT macro with a commercial software package		
41		*Solver*	*NLLSQ*
42	k_1	0.63877	0.63878
43	std.dev. of k_1	0.04967	0.04965
44	k_2	0.285394	0.285392
45	std.dev. of k_2	0.01989	0.01987
46	ε	2526.43	2526.39
47	std.dev. of ε	144.18	144.07
48	corr. coeff. R^2	0.9964	0.9964
49	$\Sigma(\partial^2) =$	2.0620E-06	2.0620E-06

Figure 12-13. Regression parameters returned by the SolvStat macro .

The parameters *parm(1)* to *parm(n)* are not calculated by the macro; they are echoed simply to indicate which standard deviation is associated with which coefficient (since the Solver coefficients can be in non-adjacent cells).

If the SolvStat.xls macro is used with the kinetics data of Figure 12-3, the regression parameters shown in Figure 12-13 are returned. The values in rows 42, 44 and 46 can be used to compare the results of the SolvStat macro and a commercial statistical software package.

AN ADDITIONAL BENEFIT FROM USING THE SOLVSTAT MACRO

There is an additional major advantage in using the Solver Add-in and the SolvStat.xls macro, even for functions that are linear or can be rearranged to a linear form. For example, if the function is

$$y = \frac{abx}{1 + bx} \tag{12-14}$$

it can be re-cast as a linear function by taking the reciprocal of each side and rearranging to give

$$\frac{1}{y} = \frac{1}{abx} + \frac{1}{a} \tag{12-15}$$

Plotting $1/y$ vs $1/x$ will yield a straight line with slope $1/ab$ and intercept $1/a$. LINEST can be used to provide the regression coefficients $1/ab$ and $1/a$, and their associated standard deviations. The coefficients a and b can be obtained from the regression coefficients ($a = 1/$intercept, $b =$ intercept/slope). However, relationships dealing with the propagation of error must be used to calculate the standard deviations of a and b from the standard deviations of $1/a$ and $1/ab$. In contrast, when the Solver is used the expression does not need to be rearranged, y_{calc} is calculated directly from equation 12-14, the Solver returns the coefficients a and b, and SolvStat.xls returns the standard deviations of a and b.

USEFUL REFERENCE

James F. Rusling and Thomas F. Kumosinski, *Nonlinear Computer Modeling of Chemical and Biochemical Data*, Academic Press, San Diego, CA, 1996.

PART IV

EXCEL
VISUAL BASIC
MACROS

13

VISUAL BASIC FOR APPLICATIONS: AN INTRODUCTION

In early spreadsheet programs, a macro was simply a string of keystrokes that could be recorded, saved and repeated to automate a simple keyboard operation. In Microsoft Excel, macros are written in a complete programming language that provides the capability to perform iterative calculations, or to take different actions based on the results of logic functions.

Macro programming was introduced in Excel 4.0. A macro written in Excel 4.0 Macro Language (XLM) is a series of statements or commands on an Excel macro sheet, which looks much like an Excel worksheet, with rows and columns. Macro statements in XLM can use any of Excel's worksheet functions, but in addition there are over 400 macro functions that can be used. Macros written in XLM can still be run in Excel 2000.

Beginning with Excel 5.0, Microsoft introduced a new macro language for Excel — Microsoft Visual Basic for Applications, or VBA. The VBA language and programming environment was changed considerably in Excel 97/98, and changed somewhat more in Excel 2000/2001.

VISUAL BASIC PROCEDURES AND MODULES

VBA macros are usually referred to as *procedures*. They are written or recorded on a *module* sheet. A single module sheet can contain many procedures.

THERE ARE TWO KINDS OF MACRO: SUB PROCEDURES AND FUNCTION PROCEDURES

You can create two different kinds of macro: **Sub** procedures, often called command macros, and **Function** procedures, called function macros, custom function macros, or user-defined functions. Although these procedures can use many of the same set of VBA commands, they are distinctly different.

Sub procedures can automate any Excel action. For example, a command macro might be used to create a report by opening a new worksheet, copying selected ranges of cells from other worksheets and pasting them into the new worksheet, formatting the data in the new worksheet, providing headings and printing the new worksheet. Command macros are not associated with a particular cell of a worksheet; command macros are usually "run" by selecting **Macro** from the **Tools** menu. They can also be run by means of an assigned shortcut key, by being called from another macro, or in several other ways.

Function procedures augment Excel's library of built-in functions. A function macro is used in a worksheet in the same way as, for example, the SQRT function. It is entered in a single cell of a worksheet, performs a calculation and returns a single result (or an array result) to the cell in which it is located. For example, a custom function macro named ALPHA can be used to calculate α_j, the fraction of an acid-base species in one of its protonated forms H_jX at a particular pH. The function takes three *arguments*: the pH of the solution, the range of pK_a values of the weak acid and the coefficient j. This function is useful in constructing distribution diagrams, titration curves, and so on.

Both kinds of macro can incorporate decision-making, branching, looping, subroutines and many other aspects of programming languages.

THE STRUCTURE OF A SUB PROCEDURE

The structure of a **Sub** procedure is shown in Figure 13-1. The procedure begins with the keyword **Sub** and ends with **End Sub**. It has a ProcedureName, a unique identifier that you assign to it. (The name can be a long one, since you never have to type it.) A **Sub** procedure has the possibility of using one or more arguments (see "Using Subroutines" in Chapter 14), but for now we will not create **Sub** procedures with arguments. Empty parentheses are required if a **Sub** procedure uses no arguments.

```
Sub ProcedureName(Argument1, ...)
    VBA statements
End Sub
```

Figure 13-1. Structure of a **Sub** procedure.

THE STRUCTURE OF A FUNCTION PROCEDURE

The structure of a **Function** procedure is shown in Figure 13-2. The procedure begins with the keyword **Function** and ends with **End Function**. It has a FunctionName, a unique identifier that you assign to it. (The name should be long enough to indicate the purpose of the function, but not too long, since you will probably be typing it in your worksheet formulas.) A **Function** procedure usually takes one or more arguments; the names of the arguments should also be descriptive. Empty parentheses are required if a **Function** procedure takes no arguments.

```
Function FunctionName(Argument1, ...)
   VBA statements
   FunctionName = result
End Function
```

Figure 13-2. Structure of a user-defined function.

The function's *return statement* directs the procedure to return the result to the caller (usually the cell in which the function was entered). The return statement consists of an assignment statement in which the name of the function is equated to a value, e.g.,

FunctionName = result

THE VISUAL BASIC EDITOR

VBA macros are located on *module sheets*. In Excel 5/95, module sheets were sheets in a workbook, just like worksheets and chart sheets; you could click on a sheet tab and view the VBA code. But beginning with Excel 97, the VBA programming environment became much more sophisticated. It is more professional and has more features, but it's also much more confusing for the beginner.

To access the Visual Basic Editor, choose **Macro** from the **Tools** menu and then **Visual Basic Editor** from the submenu. The Visual Basic Editor screen contains two important windows: the Code window and the Project window. Procedures are viewed or typed in the Code window, which corresponds to a module sheet in Excel 95. Use the Project window to select a particular code module from all the available modules in open workbooks. These are displayed in the Project window, which is usually located on the left side of the screen. If the Project window is not visible, choose **Project Explorer** from the

View menu, or click on the Project Explorer toolbutton [icon] to display it.

In the Project Explorer window you will see a hierarchy "tree" with a node for each open workbook. In the example illustrated in Figure 13-3, a new workbook, Workbook6, has been opened. The node for Workbook6 has a node (a folder icon) labeled Microsoft Excel Objects; click on the folder icon to display the nodes it contains — an icon for each sheet in the workbook and an additional one labeled ThisWorkbook. If you double-click on any one of these nodes you will display the code sheet for it, but these code sheets are for a special type of procedure called an automatic procedure (see Chapter 18).

Sub or **Function** procedures must be created on a module sheet. To insert a module sheet, choose **Module** from the **Insert** menu. A folder icon labeled Modules will be inserted; if you click on this icon the node for Module1 will be displayed.

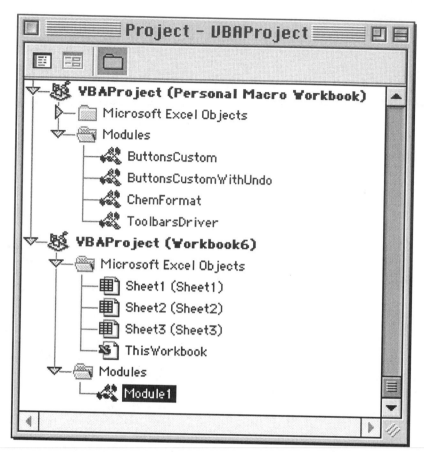

Figure 13-3. The Visual Basic Editor's Project Explorer window.

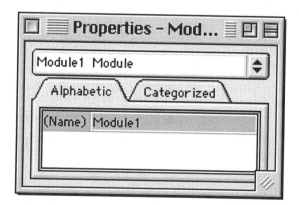

Figure 13-4. The Visual Basic Editor's Properties window.

To change the name of the module from Module1 to a more descriptive one, choose **Properties Window** from the **View** menu, or click on the Properties Window toolbutton. In the Properties Window (Figure 13-4), highlight the name Module1 and type the new name.

GETTING STARTED: USING THE RECORDER TO CREATE A SUB PROCEDURE

Excel provides the Recorder, a useful tool for creating command macros. When you choose **Macro** from the **Tools** menu and **Record New Macro...** from the submenu, all subsequent menu and keyboard actions will be recorded until you press the Stop Macro button or choose **Stop Recording** from the **Macro** submenu. The Recorder is sufficient for creating simple macros, but you can't use it to incorporate logic, branching or looping. After using the Recorder to create some simple macros, you'll view it as simply a tool to create fragments of macro code for incorporation into more complex macros. Macros that involve only the use of menu or keyboard commands can be created using the Recorder.

The Recorder creates Visual Basic commands. You don't have to know anything about Visual Basic to record a command macro in Visual Basic. This provides a good way to gain some initial familiarity with Visual Basic.

To illustrate the use of the Recorder, we'll record the action of applying scientific number formatting to a number in a cell. First, select a cell in a worksheet and enter a number. Now choose **Macro** from the Tools menu, then **Record New Macro...** from the submenu. The Record Macro dialog box (Figure 13-5) will be displayed.

The Record Macro dialog box displays the default name that Excel has assigned to this macro: Macro1, Macro2, etc. Change the name in the Macro Name box to ScientificFormat (no spaces are allowed in a name). The "Store

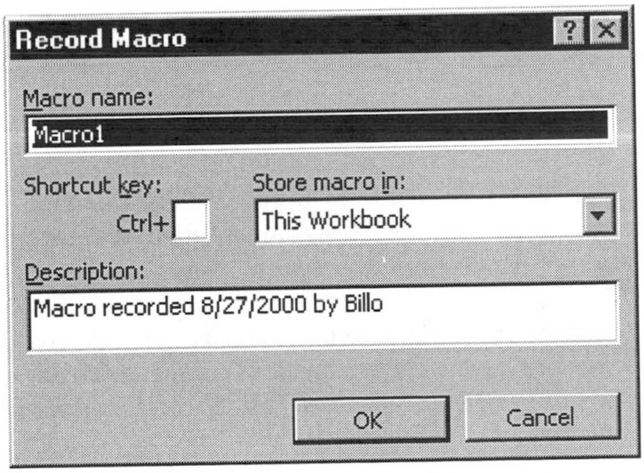

Figure 13-5. The Record Macro dialog box.

Macro In" box should display This Workbook; if not, choose This Workbook. Enter "e" in the box for the shortcut key, then press OK. The Macro Stop toolbar will appear (Figure 13-6), indicating that a macro is being recorded. If the Macro Stop toolbar doesn't appear, you can always stop recording by using the **Tools** menu: in the Macro submenu the **Record New Macro...** command will be replaced by **Stop Recording**.

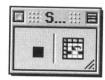

Figure 13-6. The Macro Stop toolbar.

Now choose **Cells...** from the **Format** menu, choose the Number tab and choose Scientific number format, then press OK. Finally, press the Macro Stop button.

To examine the macro code that you have just recorded, choose **Macro** from the **Tools** menu and **Visual Basic Editor** from the submenu. Click on the node for the module in the active workbook. This will display the code module sheet containing the Visual Basic code. The macro should look like the example shown in Figure 13-7.

```
Sub  ScientificFormat()
'
' Macro1  Macro
' Macro recorded 7/12/2000 by E. J. Billo

    Selection.NumberFormat = "0.00E+00"

End Sub
```

Figure 13-7. Macro for scientific number-formatting, recorded in VBA.

This macro consists of a single line of code. You'll learn about Visual Basic code in the chapters that follow.

To run the macro, enter a number in a cell, select the cell, then press the shortcut key combination that you designated when you recorded the macro. The number should be displayed in the cell in scientific format.

*Excel Tip. If you have trouble locating the code module containing your macro, here's what to do "when all else fails": choose **Macro** from the **Tools** menu and **Macros...** from the submenu. Highlight the name of the macro in the Macro Name list box, and press the Edit button. This will display the code module sheet containing the Visual Basic code.*

THE PERSONAL MACRO WORKBOOK

The Record Macro dialog box allows you to choose where the recorded macro will be stored. There are three possibilities in the "Store Macro In" list box: This Workbook, New Workbook and Personal Macro Workbook.   The Personal Macro Workbook (PERSONAL.XLS in Excel for Windows, or Personal Macro Workbook in Excel for the Macintosh) is a workbook that is automatically opened when you start Excel.   Since only macros in open workbooks are available for use, the Personal Macro Workbook is the ideal location for macros that you want to have available all the time.

Normally the Personal Macro Workbook is hidden (choose **Unhide...** from the **Window** menu to view it).   If you don't yet have a Personal Macro Workbook, you can create one by recording a macro as described earlier, choosing Personal Macro Workbook from the "Store Macro In" list box.

RUNNING A SUB PROCEDURE

In the preceding example, the macro was run by using a shortcut key. There are a number of other ways to run a macro. One of the most common is to use the Macro dialog box. Again, enter a number in a cell, select the cell, then choose **Macro** from the **Tools** menu and **Macros...** from the submenu. The Macro dialog box will be displayed (Figure 13-8). This dialog box lists all macros in open workbooks. To run the macro, select it from the list, then press the Run button.

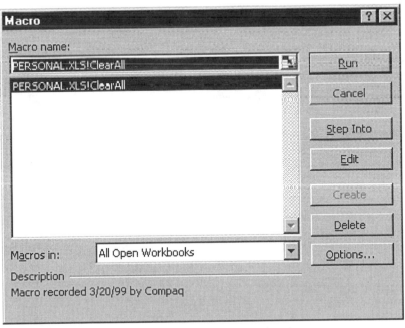

Figure 13-8. The Macro dialog box.

Here are some (but not all) of the ways to run a command macro:

- from the Macro dialog box, as described in this chapter
- by means of a shortcut key, as described in this chapter
- by means of a custom menu command, as described in Chapter 18
- by means of a custom button on a toolbar, as described in Chapter 19
- by means of a button on a worksheet, as described in Chapter 8
- as a subroutine called by another macro, as described in Chapter 15
- by means of an automatic procedure (Excel 5/95) or event-handler procedure (Excel 97) , as described in Chapter 18

ASSIGNING A SHORTCUT KEY
TO A SUB PROCEDURE

If you didn't assign a shortcut key to the macro when you recorded it, but would like to do so "after the fact", choose **Macro** from the **Tools** menu and **Macros...** from the submenu. Highlight the name of the macro in the Macro Name list box, and press the Options... button. You can now enter a letter for the shortcut key: CONTROL+(key) or SHIFT+CONTROL+(key) in Excel for Windows, OPTION+COMMAND+(key) or SHIFT+OPTION+COMMAND+(key) in Excel for the Macintosh.

GETTING STARTED:
CREATING A SIMPLE CUSTOM FUNCTION

As a simple first example of a **Function** procedure, we'll create a custom function to convert temperatures in degrees Fahrenheit to degrees Celsius.

Function procedures can't be recorded; you must type them on a module sheet. You can have several macros on the same module sheet, so if you recorded the ScientificFormat macro earlier in this chapter, you can type this custom function procedure on the same module sheet. If you do not have a module sheet available, insert one by choosing **Module** from the **Insert** menu.

Type the macro as shown in Figure 13-9. DegF is the argument passed by the function from the worksheet to the module (the Fahrenheit temperature); the single line of VBA code evaluates the Celsius temperature and returns the result to the *caller* (in this case, the worksheet cell in which the function is entered).

```
Function FtoC(DegF)
    FtoC = (DegF – 32) * 5 / 9
End  Function
```

Figure 13-9. Fahrenheit to Celsius custom function.

USING A FUNCTION MACRO

A custom function is used in a worksheet formula in exactly the same way as any of Excel's built-in functions. You can enter it in a formula by using Paste Function, or by typing it. The workbook containing the custom function must be open.

To use Paste Function, follow the procedure described in Chapter 3. Select the worksheet cell or the point in a worksheet formula where you want to enter the function (e.g., cell B2 in Figure 13-10). Choose **Function...** from the **Insert** menu or press the Paste Function toolbutton f_x to display the Paste Function dialog box. Scroll through the Function Category list and select the User Defined category. The FtoC function will appear in the Paste Function list box. When you press OK, the Formula Palette (the Paste Function Step 2 of 2 dialog box) will be displayed. Enter the argument, or click on the cell containing the argument to enter the reference (cell A2 in Figure 13-10), then press the OK button.

	A	B
1	°F	°C
2	212	100

Figure 13-10. A custom function.

You can also type the function name, with or without the opening parenthesis, and then press CONTROL+A or CONTROL+SHIFT+A, as described in Chapter 3. The function placeholder argument will be displayed, highlighted so that you can enter a value or reference (Figure 13-11).

	A	B
1	°F	°C
2	212	100
3	32	=FtoC(DegF)

Figure 13-11. Entering a custom function by using CONTROL+SHIFT+A.

Unfortunately, if you're entering the custom function in a different workbook than the one that contains the custom function, the function name must be entered as an external reference, e.g. PERSONAL.XLS!FtoC. This can make typing the function rather cumbersome, and means that you'll probably enter the function by using Excel's Paste Function. But, see "Creating Add-In Function Macros" in Chapter 17 for a solution to the problem.

RENAMING A MACRO

To rename a **Sub** procedure, access the Visual Basic Editor and click on the module containing the macro. The name of the macro is in the first line of code, immediately following the **Sub** keyword. Simply edit the name.

HOW DO I SAVE A MACRO?

A macro is part of a workbook, just like a worksheet or a chart. To save the macro, you simply **Save** the workbook.

14

PROGRAMMING WITH VBA

This chapter provides an overview of macro programming using VBA. If you are familiar with programming in other versions of BASIC or in FORTRAN, many of the programming techniques described in this chapter will be familiar.

CREATING VISUAL BASIC CODE

VBA has a wide range of commands, functions and methods that can be used to create custom applications, and this tends to make VBA confusing for the beginner (or even the non-beginner). This chapter covers the basics that you will need to get started.

ENTERING VBA CODE

When you type VBA code in a module, it's good programming practice to use TAB to indent related lines for easier reading.

As you type your VBA code, Excel checks each line for errors. A line that contains one or more errors will be displayed in red, or another color if previously specified. Variables usually appear in black. Other colors are also used: comments (see later) are usually green and VBA keywords (**Function**, **Range**, etc.) usually appear in blue.

You can enter numbers in E format but they will automatically be converted to floating point. You can't enter numbers as percentages; the percent symbol has another meaning in VBA (see "VBA Data Formats" later in this chapter).

If you type a long VBA expression, it will not wrap to the next line but will simply disappear off the screen. You will have to insert a *line-continuation character* (a space followed by the underscore character followed by RETURN) to cause a line break in a line of VBA code, as in the following example:

ReturnValu = **InputBox**("Enter validation code number", _
"Validation of this copy of SOLVER.STATS")

The line-continuation character can't be used within a string.

Several VBA statements can be combined in one line by separating them with colons. For example, the procedure in Figure 14-1 can be replaced by the more compact one in Figure 14-2 or even by the one in Figure 14-3.

251

```
Sub MultiBeeper()
For x = 1 To 10
Beep
Next
End Sub
```

Figure 14-1. A simple VBA **Sub** procedure.

```
Sub MultiBeeper()
For x = 1 To 10:  Beep:  Next
End Sub
```

Figure 14-2. A **Sub** procedure with several statements combined.

```
Sub MultiBeeper() : For x = 1 To 10: Beep: Next : End Sub
```

Figure 14-3. A **Sub** procedure in one line.

MAKING A REFERENCE TO A CELL OR RANGE OF CELLS

One of the most important things you'll need to master is the technique of making a reference to a cell or range of cells. In brief, you'll need to be able to send values from a worksheet to a module sheet so that you can perform operations on the worksheet data, and you'll need to be able to send the results back from the module sheet to the worksheet.

You can reference a particular cell or range within a worksheet in a number of ways: by selecting a cell or range by using the **Select** keyword, by using the **Selection** or **ActiveCell** keywords, or by using the **Range** method or the **Cells** method.

Don't use the **Select** keyword unless you actually need to select cells in a worksheet. For example, to copy a range of cells from one worksheet to another, you could use the statements shown in Figure 14-4, and in fact this is exactly the code you would generate using the Recorder. But you can do the same thing much more efficiently, and without switching from one worksheet to another, by using the code shown in Figure 14-5.

```
Range("D1:D20").Select
Selection.Copy
Sheets("Sheet15").Select
Range("A1").Select
ActiveSheet.Paste
```

Figure 14-4. VBA code fragment by the Recorder.

```
Range("D1:D20").Copy (Sheets("Sheet15").Range("A1"))
```

Figure 14-5. A more efficient way to accomplish the same thing, without selecting cells.

MAKING A REFERENCE TO THE ACTIVE CELL
OR A SELECTED RANGE OF CELLS

You may wish to create a macro that operates on the contents of a user-selected cell or range. To do this you can use either the **ActiveCell** keyword or the **Selection** keyword.

Note the difference between the active cell and the selection: **Selection** can be a range of cells or a single cell; **ActiveCell** (always a single cell) refers to the cell in the upper-left corner of a range if applied to a range of cells.

MAKING A REFERENCE TO A CELL OTHER THAN THE ACTIVE CELL

Instead of a macro that gets the contents of the active cell or a selection, you may want to create a macro that copies values from specified rows and columns in a worksheet, independent of where the cursor has been "parked" by the user. To accomplish this, you can use either the **Range** method or the **Cells** method to make a reference to a cell or range. The syntax of the **Cells** method is **Cells**(*row, column*). The first two of the following refer to cell B3, the next two to the range B3:E27.

    Range("B3")

    Cells(3,2)

    Range("B3:E27")

    Range(Cells(3, 2), Cells(27, 5))

The preceding examples are "absolute" references, since they always refer to B3 or B3:E27. You can also use what could be called a "computed" reference, in which the reference depends on the value of a variable. The **Cells** method is conveniently used in this way. For example, the expression

    Cells(x,2)

allows you to select any cell in column B, depending on the value assigned to the variable x. The **Range** method can be used in a similar way by using the concatenation operator, e.g.,

    Range("B" & x)

REFERENCES USING THE UNION OR INTERSECT METHOD

VBA can create references by using methods that are the equivalents of the union operator or intersection operator described in Chapter 3. The **Union** method creates a reference that includes multiple selections, e.g., A1,B5 or G3:L3,G5:L5. The syntax of the **Union** method is **Union**(*range1, range2*). The **Intersect** method creates a reference that is common to two references (e.g.,

F4:F6 E5:I5). The syntax of the **Intersect** method is **Intersect**(*range1*, *range2*). Both *range1* and *range2* must be Range objects.

GETTING VALUES FROM A WORKSHEET

To transfer values from worksheet cells to a procedure, use a reference to a worksheet range in an assignment statement, like the following.

> variable1 = **ActiveCell.Value**

> QZ = **Worksheets**("Sheet1").**Range**("A9").**Value**

The **Value** keyword can usually be omitted:

> MyVal = **Range**("A" & x)

> CellVal = **Cells**(StartRow+x,StartCol)

The corresponding **Formula** property is used to obtain the formula in a cell, rather than its value.

SENDING VALUES TO A WORKSHEET

To send values from a module sheet back to a worksheet, simply use an assignment statement like the following. Numbers, text, the contents of a VBA variable, even a worksheet formula can be entered into worksheet cells or ranges.

> **Cells**(1, 2).**Value** = 5

> **Range**("E1").**Value** = "Jan.-Mar."

> **Worksheets**("Sheet1").**Range**("A1") = variable2

> **Cells**(1, 3) = "=sum(F1:F10)"

CREATING VISUAL BASIC CODE: COMPONENTS OF VISUAL BASIC STATEMENTS

VBA macro code consists of *statements*. Statements are constructed by using VBA commands, operators, variables, functions, objects, properties or methods. (VBA Help refers to keywords such as **Beep**, **Do** or **Exit** as statements, but here they'll be referred to as commands, and we'll use "statement" in a general way to refer to a line of VBA code.)

OPERATORS

VBA operators include the arithmetic operators (+, -, *, /, ^), the text concatenation operator (&), the comparison operators (=, <, >, <=, >=, <>) and the Boolean or logical operators, which are discussed later.

> *Excel Tip.*   *Be sure to leave a space on either side of the concatenation operator; otherwise it will be mistaken for a type-declaration character (see "VBA Data Types" later in this chapter).*

VARIABLES AND ARGUMENTS

Variables are the names you create to indicate the storage locations of values or references. Arguments are variables that are passed from a worksheet to a procedure, for example.

The value of a variable is determined by an *assignment statement*. An assignment statement assigns the result of an expression to a variable or object; the form of an assignment statement is:

> *variable = expression*

In an Excel worksheet, you have to use Define Name or Create Name to assign a name to a variable, but variable names in VBA are automatically assigned as you type the VBA code in a module. There are just a few rules for naming variables or arguments:

- You can't use any of the VBA reserved words, such as Function, Range or Value.

- The first character must be a letter.

- A name cannot contain a space or a period.

- The type-declaration characters (%, $, #, !, &) cannot be embedded in a name.

You can use either upper- or lowercase letters. If you change the case of a variable name in the line you are currently typing (e.g., from variable1 to Variable1), VBA will change all instances of the old form of the variable to the new form.

You should make variable names as descriptive as possible, but avoid overly long names which are tedious to type. You can use the underscore character to indicate a space between words (e.g., formula_string). Don't use a period to indicate a space, since VBA reserves the period character for use with objects. The most popular form for variable names uses upper- and lowercase letters (e.g., FormulaString).

OBJECTS, PROPERTIES AND METHODS

VBA is an *object-oriented* programming language. *Objects* in Microsoft Excel are the familiar components of Excel, such as a worksheet, a chart, a toolbar or a range. Objects have *properties* and *methods* associated with them. Objects are the nouns of the language, properties are the adjectives, and methods are the verbs.

OBJECTS

Some examples of VBA objects are the **Workbook** object, the **Worksheet** object, the **Chart** object and the **Range** object. A complete list of objects in Microsoft Excel is listed in Excel's On-line Help. You can also use the Object Browser to see the complete list of objects. To display the Object Browser dialog box, choose **Object Browser** from the **View** menu.

There is a hierarchy of objects. You specify the object by specifying its location in a hierarchy, separated by periods, e.g.,

Workbooks("Book1").**Worksheets**("Sheet3").**Range**("E5")

Figure 14-6 shows an abbreviated hierarchical list of the most useful objects.

You can also refer to *collections* of objects. A collection is a group of objects of the same kind. A collection has the plural form of the object's name (e.g., **Worksheets**). **Worksheets** refers to all worksheets in a particular workbook. To reference a particular worksheet, you use the reference **Worksheets**(*NameText*); for example, **Worksheets**("Spectrum1").

To distinguish a particular collection of worksheets in a particular workbook, use the period operator to connect object names. For example, to refer to the Worksheet Spectrum1 in the Workbook Deconvolution, use the reference **Workbooks**("Deconvolution").**Worksheets**("Spectrum1"). To refer to a particular cell within the worksheet, use

Workbooks("Deconvolution").**Worksheets**("Spectrum1").**Range**("E5")

If Deconvolution is the only open workbook, and Spectrum1 is the active sheet, you can omit references to them and simply use the reference **Range**("E5").

Application	MenuBar	Menu	MenuItem	
	ToolBar	ToolbarButton		
	Workbook	Chart	Axis	AxisTitle
				GridLines
				TickLabels
			etc.	
		Worksheet	PageSetup	
			Range	Areas
				etc.

Figure 14-6. Partial list of objects, arranged in hierarchical order.

SOME USEFUL OBJECTS

The objects you probably will use most often in the beginning are the **Workbook** object, the **Worksheet** object and the **Range** object. (For reasons that will become clear in a while, there is no Cell object.) Later you may use the **Dialog, MenuBar, ToolBar** or **Chart** objects and the objects they in turn contain.

"OBJECTS" THAT ARE REALLY PROPERTIES

Although **ActiveCell** and **Selection** are properties, not objects, you can treat them like objects. (**ActiveCell** is a property of the **Application** object, or the **ActiveWindow** property of the **Application** object.) The **Application** object has the following properties that you can treat just as though they were objects: the **ActiveWindow, ActiveWorkbook, ActiveSheet, ActiveCell, Selection** and **ThisWorkbook** properties.   Since there is only one **Application** object, you can omit the reference to **Application** and simply use **ActiveCell**.

YOU CAN DEFINE YOUR OWN OBJECTS

VBA allows you to equate a variable to an object, but the variable does not automatically become an object. If you then attempt to use the variable in an expression that requires an object, you'll get an "Object required" error message. The **Set** command lets you define a variable or property as an object. The following example makes the **InputBox** method return an Object (so that you can use its **Address** property in addition to its **Value** property, for example):

Set known_Ys = Application.InputBox _
("Select the range of Y values", "STEP 1 OF 2", , , , , , 8)

PROPERTIES

Objects have *properties* that can be set or read. Some properties of the **Range** object are the **ColumnWidth** property, the **NumberFormat** property, the **Font** property, the **Value** property. A property is connected to the object it modifies by a period, e.g.,

CelFmt = Range("E5").NumberFormat

returns the number formatting of cell E5 and assigns it to the variable CelFmt, and

Range("E5"). NumberFormat = "0.000"

sets the number formatting of cell E5. (Some properties, such as **Column** or **Count**, are read-only.)

Properties can also modify properties.

SOME USEFUL PROPERTIES

There is a large and confusing number of properties: VBA Help lists a total of 486 property names. The list of properties belonging to the **Range** object alone contains more than 40 entries. Some of the most useful properties of the **Range** object are listed in Table 14-1.

Table 14-1. Some Useful VBA Properties

Column	Returns a number corresponding to the first column in the range.
ColumnWidth	Returns or sets the width of all columns in the range.
Count	Returns the number of items in the range.
Font	Returns or sets the font of the range.
Formula	Returns or sets the formula.
Name	Returns or sets the name of the range.
NumberFormat	Returns or sets the format code for the range.
Row	Returns a number corresponding to the first row in the range.
RowHeight	Returns or sets the height of all rows in the range.
Text	Returns or sets the text displayed by the cell.
Value	Returns or sets the contents of the cell or range.

USING PROPERTIES

In a VBA macro, many times you'll need to determine the value or state of an object's property, or change it. There are two kinds of property: read-only and read-write. Properties can have values that are numeric, string or logical.

To return an object's property, use the following syntax:

VariableName = ObjectName.PropertyName

To set an object's property (only if it is a read-write property, of course), use the following syntax:

ObjectName.PropertyName = expression

For example, to change the background color of a range of cells to blue, use the statement

Range("A5:E5").Interior.ColorIndex = 8

METHODS

Objects also have *methods*. Methods can operate on an object or on a property of an object. Some methods that can be applied to the **Range** object are the

Copy method, the **Cut** method, the **FillDown** method or the **Sort** method. For example,

>**Range**("A1:E1").**Clear**

clears the formulas and formatting in the range A1:E1.

SOME USEFUL METHODS

VBA Help lists 267 methods. Many of them correspond to familiar menu commands. For example, **Copy**, **Cut**, **Clear** and **Sort** can be performed on a range of cells. Some useful VBA methods are listed in Table 14-2.

Table 14-2. Some Useful VBA Methods

Activate	Activates an object (sheet, etc.).
Clear	Clears an entire range.
Close	Closes an object.
Copy	Copies an object to a specified range or to the Clipboard.
Cut	Cuts an object to a specified range or to the Clipboard.
FillDown	Copies the cell(s) in the top row into the rest of the range.
Select	Selects an object.

TWO WAYS TO SPECIFY ARGUMENTS

VBA methods usually take one or more arguments. The **Sort** method, for example, takes 10 arguments. The syntax of the **Sort** method is

>*object.***Sort***(key1, order1, key2, order2, key3, order3, header, orderCustom, matchCase, orientation)*

The object argument is required; all other arguments are optional.

You can specify the arguments of a method in two ways. One way is to list the arguments in order as they are specified in the preceding syntax, i.e.,

>**Selection.Sort   Range**("A2"), 1

(In the preceding example, only the arguments key1 and order1 were specified; the remaining arguments are optional and were not required.)

The second way is to use the name of the argument as it appears in the preceding syntax, with the := operator, to specify the value of the argument. The arguments can appear in any order, as in the following (all in one line of code, of course):

>**Selection.Sort** Key1:=**Range**("A2"), Order1:=xlAscending, Key2:=**Range**("B2"), Order2:=xlAscending, Key3:=**Range**("C2"), Order3:=xlDescending, Header :=xlGuess, OrderCustom:=1, MatchCase:=**False**, Orientation:=xlTopToBottom

ARGUMENTS WITH OR WITHOUT PARENTHESES

As well as performing an action, methods create a return value. The return value can be either **True** or **False**. Even the **ChartWizard** method creates a return value: **True** if the chart was created successfully, **False** if the method failed. Usually you aren't interested in these return values.

An example of a method that creates a useful return value is the **CheckSpelling** method. The **CheckSpelling** method has the following syntax:

Application.CheckSpelling(word)

If you use this method, you'll need the return value (either **True** or **False**) to determine whether the word is spelled correctly.

If you want to use the return value of a method, you must enclose the arguments of the method in parentheses. If the arguments are not enclosed in parentheses, then the return value will not be available for use. Put another way, the expression

result = **Application.CheckSpelling(ActiveCell.Value)**

does not produce a syntax error, while the expression

result = **Application.CheckSpelling ActiveCell.Value**

does give a syntax error.

SOME USEFUL FUNCTIONS

The functions available in VBA are similar to the functions available in Excel itself. There are 108 VBA functions listed in VBA Help. Table 14-3 lists some of the more useful ones for numerical calculations.

USING WORKSHEET FUNCTIONS WITH VBA

Many useful worksheet functions do not have a VBA equivalent. To use them, use the syntax

Application. WorksheetFunction

and supply arguments for the function just as you would in a worksheet. For example, to use the FIND function in VBA, use the code

Application.Find(find_text,within_text,start_num)

e.g.,

Application.Find("]",**Range**("A1"),1)

Table 14-3. Some Useful VBA Functions

Abs	Returns the absolute value of a number.
Asc	Returns the character code of a character.
Chr	Returns the character corresponding to a code.
Exp	Returns e raised to a power.
Fix	Returns the integer part of a number (truncates).
Int	Returns the integer part of a number (rounds down).
IsArray	Returns **True** if the variable is an array.
IsNull	Returns **True** if the expression is null (i.e., contains no valid data).
IsNumeric	Returns **True** if the expression can be evaluated to a number.
LBound	Returns the lower limit of an array dimension.
LCase	Converts a string into lowercase letters.
Left	Returns the leftmost characters of a string.
Len	Returns the length (number of characters) in a string.
Log	Returns the natural (base-e) logarithm of a number.
Mid	Returns a specified number of characters from a string.
Right	Returns the rightmost characters of a string.
RTrim	Returns a string without trailing spaces.
Sqr	Returns the square root of a number.
Str	Converts a number to a string.
UBound	Returns the upper limit of an array dimension.
UCase	Converts a string into uppercase letters.

SOME USEFUL VBA COMMANDS

Commands in VBA are similar to commands in BASIC or FORTRAN. Table 14-4 lists some of the most useful VBA commands.

VBA DATA TYPES

VBA uses a range of different data types. Table 14-5 lists the built-in data types.

Unless you declare a variable's type, VBA will use the **Variant** type. You can save memory space if your procedure deals only with integers, for example, by declaring the variable type.

You can also declare a variable's type by appending a type-declaration character to the variable name, a technique from older versions of BASIC. The type-declaration characters include % for integer variables and $ for string variables.

Table 14-4. Some Useful VBA Commands

Beep	Makes a "beep" sound.
Dim	Declares an array (see Chapter 15) and allocates storage for it.
Do...Loop	Delineates a block of statements to be repeated.
Else	Optional part of **If...Then** structure.
ElseIf	Optional part of **If...Then** structure.
End	Terminates a procedure.
End If	Terminates block of statements begun by **If**.
Exit	Exits a **Do...**, **For...**, **Function...** or **Sub...** structure.
For Each...Next	Delineates a block of statements to be repeated.
For...Next	Delineates a block of statements to be repeated.
Function	Marks the beginning of a **Function** procedure.
GoSub...Return	Delineates a subroutine.
GoTo	Unconditional branch.
If...Then...Else	Delineates a block of conditional statements.
On...GoSub	Branch to one of several specified subroutines.
On...GoTo	Branch to one of several specified lines.
Select Case	Executes one of several blocks of statements.
Set	Assigns an object reference to a variable or property.
Stop	Stops execution.
Sub	Marks the beginning of a **Sub** procedure.
Until	Optional part of **Do...Loop** structure.
While	Optional part of **Do...Loop** structure.
With...EndWith	Delineates a block of statements to be executed on a single object.

THE VARIANT DATA TYPE

The **Variant** data type is the default data type in VBA. Like Excel itself, the **Variant** data type handles and interconverts between many different kinds of data: integer, floating point, string, etc. The **Variant** data type automatically chooses the most compact representation. But if your procedure deals with only one kind of data, it will be more efficient and usually faster to declare the variables as, for example, **Integer**.

Table 14-5. VBA's Built-in Data Types

Data type	Storage required	Range of values
Logical	2 bytes	**True** or **False**
Integer	2 bytes	–32,768 to 32,767
Long integer	4 bytes	–2,147,483,648 to 2,147,483,647
Single precision	4 bytes	–3.402823E38 to –1.401298E–45 for negative values; 1.401298E–45 to 3.402823E38 for positive values
Double precision	8 bytes	–1.79769313486232E308 to –4.94065645841247E–324 for negative values; 4.94065645841247E–324 to 1.79769313486232E308 for positive values
Currency	8 bytes	–922,337,203,685,477.5808 to 922,337,203,685,477.5807
Date	8 bytes	
Object	4 bytes	Any Object reference
String	1 byte/character	
Variant	16 bytes + 1 byte/character	Any numeric value up to the range of a **Double** or any text

STRING DATA TYPES

Strings can be stored either as *variable-length strings* (the default data type) or as *fixed-length strings*. To declare a string variable as fixed-length, use the statement in a **Dim** statement (more about the **Dim** statement in Chapter 15)

String * *length*

For example, the following statement sets aside storage for a two-dimensional array of names and addresses, containing fixed-length strings of 32 characters:

Dim AddressList(4,500) **As String** * 32

If a string of length less than 32 characters is assigned to the array AddressList, trailing spaces are added to fill out the string length. If a string of more than 32 characters is assigned to AddressList, the string will be truncated.

THE BOOLEAN (LOGICAL) DATA TYPE

A logical variable can have only the value **True** or **False**. The keywords **True** and **False** are often implied; thus the expressions

If (j > N) = **True Then** ALPHA = "n/a"

and

If (j > N) **Then** ALPHA = "n/a"

are equivalent.

You can use other data types as Boolean variables. When a variable is used in a logical expression, zero is converted to **False** while any non-zero value is converted to **True**. Thus the expression

If j **Then** *expression*

tests for a non-zero value of the variable j.

When Boolean variables are converted to other data types, **False** becomes zero but **True** is converted to –1.

DECLARING VARIABLES OR ARGUMENTS IN ADVANCE

VBA uses the **Variant** data type as the default data type for variables and arguments. The **Variant** data type permits Excel to switch between floating-point, integer and string variables as required.

You can force a particular variable or argument to take a specified data type. For variables, use the **Dim** statement, e.g.,

Dim ChemFormula **As String**

SPECIFYING THE DATA TYPE OF AN ARGUMENT

You can specify the data type of an argument passed to a **Function** procedure by using the **As** keyword in the **Function** statement. For example, if the **Function** procedure MolWt has two arguments, formula (a string) and decimals (an integer), then the statement

Function MolWt (formula **As String**, decimals **As Integer**)

declares the type of each variable. If an argument of an incorrect type is supplied to the function, a #VALUE! error message will be displayed.

SPECIFYING THE DATA TYPE RETURNED BY A FUNCTION PROCEDURE

You can also specify the data type of the return value. If none is specified, the **Variant** data type will be returned. In the example of the preceding section, MolWt returns a floating-point result. The **Variant** data type is satisfactory;

however, if you wanted to specify double precision floating-point, use an additional **As Type** expression in the statement, e.g.,

Function MolWt (formula **As String**, decimals **As Integer**) **As Double**

PROGRAM CONTROL

If you are familiar with computer languages such as BASIC or FORTRAN, you will find yourself quite comfortable with most of the material in this section.

DECISION-MAKING (BRANCHING)

VBA supports **If...Then** and **If...Then...Else** or **ElseIf** structures, very similar to the Excel worksheet function IF. In addition, VBA provides the **Select Case** decision structure, similar to the ON *value* GOTO statement in BASIC.

The **If...Then** statement can be on a single line:

If (x = j) **Then** numerator = 10 ^ (logbeta - pH * x)

or it can be followed by multiple statement lines as in Figure 14-7, (the Block If statement).

```
If (pKa_logical = False) Then
    logbeta = logbeta + pKs_or_logKs(x)
    denom = denom + 10 ^ (logbeta - pH * x)
    etc.
EndIf
```

Figure 14-7. Example of VBA **If...EndIf** structure.

If...Then...Else or **ElseIf** structures are also possible. For example:

If LogicalExpression **Then** statement **Else** statement

or, as illustrated in Figure 14-8, you can employ several logical expressions.

```
If LogicalExpression1 Then
    statements
ElseIf LogicalExpression2
    statements
ElseIf LogicalExpression3
    statements
    etc.
EndIf
```

Figure 14-8. The VBA **If...ElseIf...EndIf** structure.

The **Select Case** statement provides an efficient alternative to the series of **ElseIf** conditionN statements when conditionN is a single expression that can take various values. The syntax of the **Select Case** statement is illustrated in Figure 14-9.

TestExpression is evaluated and used to direct program flow to the appropriate **Case**. ExpressionListN can be a single value, a list of values separated by commas or a range of values. The optional **Case Else** statement is executed if TestExpression doesn't match any of the values in any of ExpressionListN.

The example shown in Figure 14-10 illustrates the use of **Select Case** to calculate the pK_a value of a polyprotic acid. Since data at or near the equivalence points cause large calculation errors, the pK_a is calculated only for n-bar values in the range 0.2 – 0.8, 1.2 – 1.8, 2.2 – 2.8 or 3.2 – 3.8. The expression used to calculate the pK_a from the n-bar parameter depends on the number of protons bound, i.e., on the value of n-bar. The **Select Case** statement is used to direct program flow to the appropriate expression.

Note that a range of values is indicated by using the **To** keyword.

```
Select Case TestExpression
    Case ExpressionList1
    statements
    Case ExpressionList2
    statements
    Case ExpressionList3
    statements
    Case Else
    statements
End Select
```

Figure 14-9. The VBA **Select Case** structure.

LOGICAL OPERATORS

You are already familiar with the **And** and **Or** operators, but VBA provides in addition the **Xor** (exclusive or) operator. The operators have the following syntax:

expression1 **And** expression2   **True** if both expressions are **True**.

expression1 **Or** expression2   **True** if either expression is **True**.

expression1 **Xor** expression2   **True** if one expression is **True**, the other **False**.

```
NBar = (ZP * CR + CA + COH - CH - CNa) / CR

Select Case NBar
Case 3.2 To 3.8
    pK = pH + Application.Log((NBar - 3) / (4 - NBar))
Case 2.2 To 2.8
    pK = pH + Application.Log((NBar - 2) / (3 - NBar))
Case 1.2 To 1.8
    pK = pH + Application.Log((NBar - 1) / (2 - NBar))
Case 0.2 To 0.8
    pK = pH + Application.Log((NBar) / (1 - NBar))
Case Else
    pK = ""
End Select
    :
    :
End Sub
```

Figure 14-10. An example of the **Select Case** structure.

The preceding expressions must evaluate to **True** or **False**; that is, they must be logical expressions. The logical operators are almost always used in combination with **If** statements.

More than one **And** or **Or** can be combined in a single statement. For example

> **If** Char = " " **Or** Char = "^" **Or** Char = "," **Or** Char = "(" **Or** Char = "/" **Then**...

evaluates to **True** if any one of the logical expressions is **True**.

Parentheses are often necessary to control the logic of the expression. For example, each of the expressions

> **If** (expression1 **And** expression2) **Or** expression3 **Then**...

> **If** expression1 **And** (expression2 **Or** expression3) **Then**...

has eight different possible combinations of expression1, expression2 and expression3; two of them give different outcomes depending on which expression is used.

LOOPING

The loop structures in VBA are similar to those available in other programming languages.

FOR...NEXT LOOPS

The syntax of the **For...Next** loop is given in Figure 14-11.

```
For Counter = Start To End Step Increment
    statements
Next Counter
```

Figure 14-11. The VBA **For...Next** structure.

Both **Step** Increment in the **For** statement, and Counter following the **Next** are optional. If Increment is omitted, it is set equal to 1. Increment can be negative.

FOR EACH...NEXT LOOPS

The **For Each...Next** loop structure is similar to the **For...Next** loop structure, except that it executes the statements within the loop for each object within a group of objects. Figure 14-12 illustrates the syntax of the statement.

```
For Each Element In Group
    statements
Next Element
```

Figure 14-12. The VBA **For Each...In...Next** structure.

An important point: the **For..Each...Next** loop returns an object variable in each pass through the loop. You can access or use all of the properties or methods that apply to Element. For example, in a loop such as

For Each cel **In Selection**

the variable cel is an object that has all the properties of a cell (a **Range** object): **Value**, **Formula**, **NumberFormat**, etc.

DO WHILE... LOOP

The **Do While...Loop** is used when you don't know beforehand how many times the loop will need to be executed. The syntax is shown in Figure 14-13.

```
Do While LogicalExpression
    statements
Loop
```

Figure 14-13. The VBA **Do While...Loop** structure.

```
Do
    statements
Loop While LogicalExpression
```

Figure 14-14. Alternate form of **Do While...Loop** structure.

An alternate format of this type of loop places **While** *LogicalExpression* at the end of the loop, as exemplified in Figure 14-14.

Note that this form of the **Do While** structure executes the loop at least once.

EXITING FROM A LOOP OR FROM A PROCEDURE

Often you use a loop structure to search through an array or collection of objects, looking for a certain value or property. Once you find a match, you don't need to cycle through the rest of the loops. You can exit from the loop using the **Exit For** (from a **For...Next** loop or **For Each...Next** loop) or **Exit Do** (from a **Do While**... loop). The **Exit** statement will normally be located within an **If** statement. For example,

> **If** CellContents.**Value** <= 0 **Then Exit For**

Use the **Exit Sub** or **Exit Function** to exit from a procedure. Again, the **Exit** statement will normally be located within an **If** statement.

Exit statements can appear as many times as needed within a procedure.

SUBROUTINES

Although all **Sub** procedures are subroutines, by "subroutine" we mean a subprogram that is called by another VBA program. It's good programming practice to break up a complicated task into simpler tasks and write subroutines to do each task. The separate subroutines are called by a main program.

There are several ways to execute a subroutine within a main program. The two most common are by using the **Call** command, or by using the name of the subroutine. These are illustrated in Figure 14-15. MainProgram calls subroutines Task1 and Task2, each of which requires arguments.

```
Sub MainProgram()
etc.
Call Task1(argument1,argument2)
etc
Task2 argument3,argument4
etc
End Sub

Sub Task1(ArgName1,ArgName2)
etc
End Sub

Sub Task2(ArgName3,ArgName4)
etc
End Sub
```

Figure 14-15. A main program illustrating the different syntax of subroutine calls.

The two methods use different syntax if the subroutine requires arguments. If the **Call** command is used, the arguments must be enclosed in parentheses. If only the subroutine name is used, the parentheses must be omitted. Note that the variable names of the arguments in the calling statement and in the subroutine do not have to be the same.

SCOPING A SUBROUTINE

A subroutine can be **Public** or **Private**. Public subroutines can be called by any subroutine in any module. The default for any **Sub** procedure is **Public**. A **Private** subroutine can be called only by other subroutines in the same module. To declare the subroutine Task3 as a private subroutine, use the statement

> **Private Sub** Task3()

INTERACTIVE MACROS

VBA provides two built-in dialog boxes for display of messages or for input, **MsgBox** and **InputBox**.

MSGBOX

The **MsgBox** dialog box allows you to display a message, such as "Please wait..." or "Access denied". The box can display one of three message icons, and there are many possibilities in the number and function of buttons that can be displayed.

The syntax of **MsgBox** is

> **MsgBox** (*prompt_text, buttons, title_text, helpfile, context*)

where *prompt_text* is the message displayed within the box, *buttons* specifies the buttons to be displayed and *title_text* is the title to be displayed in the Title Bar of the box. For information about *helpfile* and *context*, refer to *Microsoft Excel Visual Basic Reference*.

For example, the VBA expression

> **MsgBox** "You entered " & incr & "." & **Chr**(13) & **Chr**(13) & _
> "That value is too large." & **Chr**(13) & **Chr**(13) & "Please try again.", 64

produces the message box shown in Figure 14-16.

The value of *buttons* determines the type of message icon and the number and type of response buttons; it also determines which button is the default button. The possible values are listed in Table 14-6. The values of *buttons* are built-in constants — for example, the value 64 for *buttons* can be replaced by the

Figure 14-16. A **Msgbox** display.

variable name vbInformation. (For a complete list of names of these built-in constants, refer to *Microsoft Excel Visual Basic Reference*.) The values 0 – 5 specify the number and type of buttons, values 16 – 64 specify the type of message icon, and values 0, 256, 512 specify which button is the default button. You can add together one number from each group to form a value for *buttons*. For example, to specify a dialog box with a Warning Query icon, with Yes, No and Cancel buttons, and with the No button as default, the values 32 + 3 + 256 = 291. The same result can be obtained by using the expression

buttons = vbInformation + vbYesNoCancel + vbDefaultButton2

and using the variable buttons in the **MsgBox** function.

Table 14-6. Values for the buttons Parameter of **MsgBox**

buttons	Description
0	Display OK button only.
1	Display OK and Cancel buttons.
2	Display Abort, Retry and Ignore buttons.
3	Display Yes, No and Cancel buttons.
4	Display Yes and No buttons.
5	Display Retry and Cancel buttons.
16	Display Critical Message icon.
32	Display Warning Query icon.
48	Display Warning Message icon.
64	Display Information Message icon.
0	First button is default.
256	Second button is default.
512	Third button is default.

MSGBOX RETURN VALUES

MsgBox returns a value that indicates which button was pressed. This allows you to take different actions depending on whether the user pressed the Yes, No or Cancel buttons, for example. To get the return value of the Message Box, use an expression like

>   RtnValu **= MsgBox** (*prompt_text, buttons, title_text, helpfile, context*)

The return values of the buttons are as follows: OK, 1; Cancel, 2; Abort, 3; Retry, 4; Ignore, 5; Yes, 6; No, 7.

INPUTBOX

The **InputBox** allows you to pause a macro and request input from the user. There are both an **InputBox** Function and an **InputBox** method.

The syntax of the **InputBox** Function is

>   **InputBox***(prompt_text, title_text, default, x_position, y_position, helpfile, context)*

where *prompt_text* and *title_text* are as in **MsgBox**. *Default* is the expression displayed in the input box, as a string. The horizontal distance of the left edge of the box from the left edge of the screen, and the vertical distance of the top edge from the top of the screen are specified by *x_position* and *y_position*, respectively. For information about *helpfile* and *context*, refer to *Microsoft Excel Visual Basic Reference*.

If the user presses the OK button or the RETURN key, the **InputBox** function returns as a value whatever is in the text box. If the Cancel button is pressed, the function returns a null string. The following example produces the input box shown in Figure 14-17.

>   ReturnValu **= InputBox**("Enter validation code number", _
>   "Validation of this copy of SOLVER.STATS")

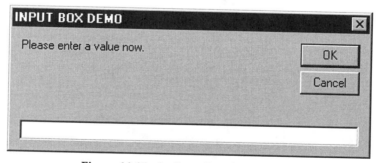

Figure 14-17. An InputBox display.

The syntax of the **InputBox** method is

Object.InputBox*(prompt_text, title_text, default, x_position, y_position, helpfile, context, type_num)*

The differences between the **InputBox** function and the **InputBox** method are the following: (i) *default* can be any data type and (ii) the additional argument *type_num* specifies the data type of the return value. The values of *type_num* and the corresponding data types are listed in Table 14-7. Values of *type_num* can be added together. For example, to specify an input dialog box that would accept number or string values as input, use the value 1 + 2 = 3 for *type_num*.

Table 14-7. InputBox Data Type Values

type_num	Data type
0	Formula
1	Number
2	String
4	Logical
8	Reference (as a Range object)
16	Error value
64	Array

TESTING AND DEBUGGING

When an error occurs during execution of a procedure, VBA will stop execution and display a run-time error message. There are a large number (over 50) of informative run-time error messages. These error messages are for the most part self-explanatory. Here are some examples:

Subscript out of range

Attempted to access an element of an array outside its specified dimensions.

Property or method not found

Object does not have the specified property or method.

Argument not optional

A required argument was not provided.

The line of code in which the error occurred will be highlighted, usually in yellow (see Figure 14-18). As a rule, after you have corrected the error in your VBA code, the line will still be highlighted, and you won't be able to run the macro. Press F5 to force the procedure to run.

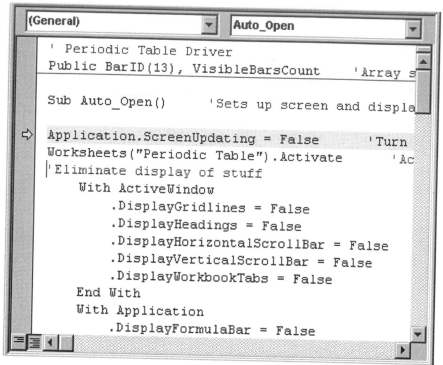

Figure 14-18. VBA code with a highlighted line.

TRACING EXECUTION

When your program produces an error during execution, or executes but doesn't produce the correct answer, it is often helpful to execute the code one statement at a time and examine the values of selected variables during execution. If your procedure contains logical constructions (**If** or **Select Case**, for example), simply stepping through code will allow you to verify the logic.

STEPPING THROUGH CODE

To step through the code of a **Sub** procedure, follow the steps in the following box. There are two ways you can begin the process.

As you step through the code, the next statement to be executed is highlighted, as shown in Figure 14-18. Use the Step Into toolbutton ⬚ or press F8 to step through the procedure. Press F5 to run the macro from the current line.

To Step Through VBA Code

1. Activate a worksheet, choose **Macro** from the **Tools** menu and **Macros...** from the submenu.

2. Select the macro in the Macro Name list box and press the Step Into button. This will display the code module containing the procedure. The first line of the procedure will be highlighted, usually in yellow (see Figure 14-18).

3. Choose **Toolbars** from the **View** menu and **Debug** from the submenu to display the Debug toolbar (Figure 14-19).

or...

1. Activate the Visual Basic Editor. The VBA code module must be visible.

2. Choose **Toolbars** from the **View** menu and **Debug** from the submenu to display the Debug toolbar (Figure 14-19).

Figure 14-19. The VBA Debug toolbar.

ADDING A BREAKPOINT

A breakpoint allows you to halt execution at a specified line of code, rather than having to step through the code from the beginning. There are several ways to add a breakpoint:

- Opposite the line of code where you want to set the breakpoint, click in the gray bar on the left side of the VBA module sheet. The line of code will be highlighted (usually in red-brown) and a breakpoint indicator, a large dot of the same color, will be placed in the margin (see Figure 14-20).

- Place the cursor in the line of code where you want to set a breakpoint. Press the Toggle Breakpoint button 🖐 on the Debug toolbar.

- Insert a **Stop** statement in the VBA code. See Appendix D for details.

- Enter a break expression in the Add Watch dialog box (see "Examining the Values of Variables" later in this chapter).

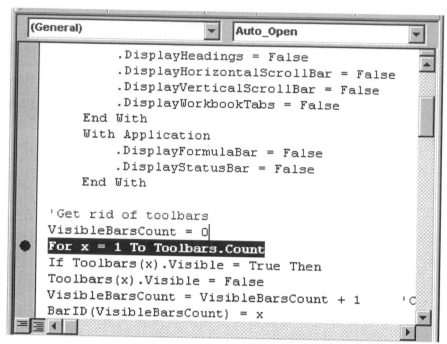

Figure 14-20. VBA code with a breakpoint.

When you run the macro, the code will execute until the breakpoint is reached, at which point execution will stop. You can now step through the code one statement at a time or examine the values of selected variables.

Since you can't "run" a **Function** procedure, the *only* way to step through a **Function** procedure is to add a breakpoint, then recalculate a formula containing the custom function.

To remove a breakpoint, click on the breakpoint indicator, or place the cursor on the highlighted line and press the Toggle Breakpoint button, or delete a **Stop** statement.

EXAMINING THE VALUES OF VARIABLES

You can also display the values of selected variables as the code is executed. There are several ways to select variables or expressions to be displayed:

- Highlight the variable or expression and then choose **Quick Watch...** or press the Quick Watch button 66° on the Debug toolbar, to display the Quick Watch dialog box (Figure 14-21).

- Highlight the variable or expression and then choose **Add Watch...** from the **Debug** menu to display the Add Watch dialog box (Figure 14-22).

To see the values of the selected variables or expressions, you must be in Step mode. The variables will be listed in the Watches pane (Figure 14-23), which is usually located below the Code window. The current values of the variables will be displayed as you step through the code.

Figure 14-21. The VBA Quick Watch dialog box.

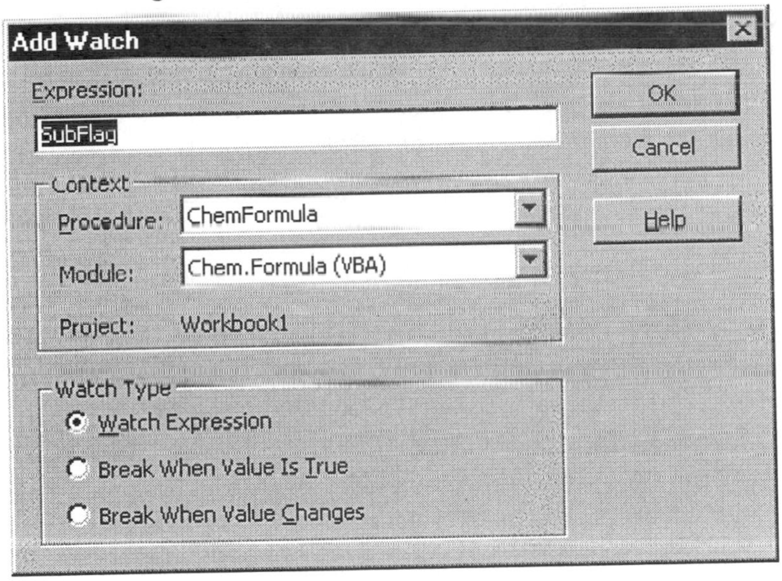

Figure 14-22. The VBA Add Watch dialog box.

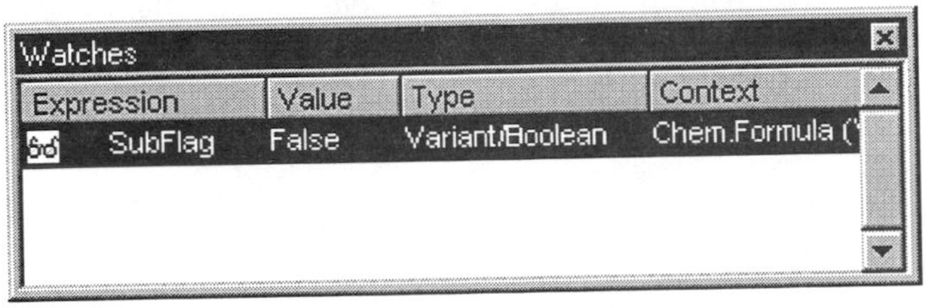

Figure 14-23. The VBA Watches pane.

To remove a variable or expression from the Watches pane, select it in the Watches pane, choose **Edit Watch** from the **Debug** menu, then press DELETE.

Watch expressions are not saved with your code.

USING CONDITIONAL WATCH

A conditional watch expression causes VBA to enter break mode only when a variable changes in value or when an expression evaluates to **True**.

To establish conditional watch expressions, choose **Add Watch** from the **Debug** menu and press the appropriate Watch Type button (see Figure 14-22). There are three possibilities, which are indicated by different icons in the Watches pane:

👀   Watch expression (current value is displayed in the Watches pane when VBA enters break mode)

✋   Conditional break expression (break occurs when expression is True)

✋   Conditional break expression (break occurs when expression changes)

Excel Tip. *You can adjust the widths of the Expression, Value and Context columns in the Watches pane by placing the mouse pointer on the separator bar to the left of the Value or Context header; the pointer will change to the*  *pointer shape, and you can drag the separator bar to adjust the column width.*

USEFUL REFERENCES

Microsoft Excel Visual Basic Reference, 2nd edition, Microsoft Press, 1995. A useful reference — the same information as On-line Help.

Reed Jacobson, *Microsoft Excel 2000 Visual Basic for Applications : Fundamentals*, Microsoft Press, Redmond, WA, 1999.

John Walkenbach, *Excel 2000 Programming for Windows for Dummies*, IDG Books Worldwide, San Mateo, CA, 1999. A good book for beginners in VBA programming.

John Walkenbach, *Microsoft® Excel 2000 Power Programming with VBA*, IDG Books Worldwide, San Mateo, CA, 1999.

15

WORKING WITH ARRAYS IN VBA

Many scientists make extensive use of arrays in their calculations. Because arrays in VBA can be quite confusing, this chapter provides detailed coverage of this important topic.

VISUAL BASIC ARRAYS

If you're familiar with other programming languages you are probably familiar with the concept of an array. An array is a collection of related variables denoted by a single name, such as Sample. You can then specify any element in the array by using an index number: Sample(1), Sample(7), etc.

DIMENSIONING AN ARRAY

The **Dim** statement is used to declare the size of an array. Unless specified otherwise, VBA arrays begin with a index of 0. The statement

> **Dim** Sample(10)

establishes array storage for 11 elements, Sample(0) through Sample(10).

You can specify the lower bound of an array in the **Dim** statement. For example,

> **Dim** Sample (1 **To** 10)

allows you to specify any value, even negative values, for the lower or upper bounds of an array. Alternatively, you can use the **Option Base 1** statement, which specifies that all arrays begin with a lower index of 1. The **Option Base 1** statement is used at the module level: that is, it must appear in a module sheet ahead of any procedures.

> *Excel Tip.* *Since worksheet ranges, worksheet functions and worksheet arrays use (or assume) a lower array index of 1, you should always include the* **Option Base 1** *statement at the beginning of each code module that uses arrays.*

Several variables can be dimensioned in a single **Dim** statement.

It's considered good programming practice to put the **Dim** statements at the beginning of the procedure.

USE THE NAME OF THE ARRAY VARIABLE
TO SPECIFY THE WHOLE ARRAY

You can refer to the complete array by using the array variable name in your code. Sometimes the array name can be used with parentheses, other times without parentheses.

MULTIDIMENSIONAL ARRAYS

Arrays can be multidimensional. Excel permits arrays with up to 60 dimensions; one-, two- and three-dimensional arrays are common. To create a 2-D array with dimensions 2 × 500, use the statement

Dim Spectrum (2,500)

Since multidimensional arrays such as the one above can use up significant amounts of memory, it's important to define the data type of the variable. The complete syntax of the **Dim** statement is

Dim VariableName(Lower **To** Upper) **As** Type

The optional Lower **To** can be omitted. Type can be **Integer**, **Single**, **Double**, **Variant**, etc. See the complete list of data types in "VBA Data Types" in chapter 13.) A **Variant** array can hold values of different data types (e.g., 1.173 and "Billo, E. Joseph").

Several variables can be dimensioned in a single **Dim** statement, but there must be a separate **As Type** for each variable.

RETURNING THE DIMENSIONS OF AN ARRAY

Use the **LBound** and **UBound** functions to obtain the dimensions of an array during execution of your procedure. The **LBound** function returns the lower index of an array. For example, for the array Sample described previously, **LBound**(Sample) returns 1 and **UBound**(Sample) returns 10.

The complete syntax of **LBound** and **UBound** is **LBound**(arrayname, dimension). For the array Spectrum dimensioned thus:

Dim Spectrum (2,500)

the statement **UBound**(Spectrum,1) returns 2 and **UBound**(Spectrum,2) returns 500.

DYNAMIC ARRAYS

If you don't know what array size you will need to handle a particular problem, you can create a *dynamic array*. This will allow you to declare a variable as an array but set its size later. Dimension the array using the **Dim** command, using empty parentheses; use the **ReDim** command later to specify the array size, e.g.,

```
Dim MeanX(), MeanY()
    :
'Get number of cells to use in calculation
Ncells = XValues.Count
ReDim MeanX(Ncells), MeanY(Ncells)
```

You can also use the **ReDim** command to change the number of dimensions of an array.

The **ReDim** command can appear more than once in a procedure.

PRESERVING VALUES IN DYNAMIC ARRAYS

You can also use the **ReDim** statement to re-size an existing array during execution of the code. For example, your procedure initially may need an array of 5×2000 elements, but later, during execution on a particular data set, a much smaller number of array elements may be needed. You can free up memory by re-sizing the array during execution.   However, the **ReDim** statement re-initializes the array (numbers are set to zero, strings to null, objects to **Nothing**, for example). You can preserve the values in an existing array by using the **Preserve** keyword, e.g.,

```
Dim MeanX(), MeanY()
    :
ReDim Preserve MeanX(Ncells / 2), MeanY(Ncells / 2)
```

But, there's a limitation. Only the upper bound of the last dimension of a multidimensional array can be changed. Thus, the following code is valid:

```
Dim BothMeans(2, 1000)
    :
ReDim Preserve BothMeans(2,Ncells / 2)
```

but the following code will generate a run-time error:

```
Dim BothMeans(1000, 2)
    :
ReDim Preserve BothMeans(Ncells / 2, 2)
```

If you use **Preserve**, you can't use the **ReDim** command to change the number of dimensions of an array.

Working With Arrays
in Sub Procedures: Passing Values
From Worksheet to VBA Module

There are at least two ways to get values from a worksheet into a VBA array. You can either set up a loop to read the value of each worksheet cell and write the value to the appropriate element of an array, or you can assign the VBA array to a worksheet range. These two methods are described next. There is a definite difference with respect to speed. (See "Speed Differences in Reading or Writing Arrays Created by Two Different Methods" later in this chapter.)

USING A LOOP TO TRANSFER VALUES
FROM A WORKSHEET TO A VBA ARRAY

The **Sub** procedure shown in Figure 15-1 reads values from worksheet cells and writes them to elements of a VBA array by means of a loop. Either the **Cells** method or the **Range** method can be used; in the example that follows, the **Range** method is used, with the concatenation operator, to access the appropriate cell in the range. In this example the values to be copied are in column A, beginning in row 2. To illustrate an additional point, the values extend down to a row that must be determined by the VBA code.

The code to obtain the row number of the last-used row in the block of cells in column A was obtained by using the Recorder to record the VBA code corresponding to CONTROL+SHIFT+(down arrow).

```
Sub ArrayDemo1()
'Reads values from column A, beginning in row 2,
'into a VBA array.
'++++++++++++++++++++++++++++++++++++++++++++++++
Dim TestArray()
LastRow = Range("A2").End(xlDown).Row
NCells = LastRow - 1
ReDim TestArray(NCells)
For x = 2 To LastRow
TestArray(x - 1) = Range("A" & x)
Next x
MsgBox "First array element = " & TestArray(1) & Chr(13) & _
"Last array element = " & TestArray(NCells)
End Sub
```

Figure 15-1. Reading values into a VBA array.

A RANGE SPECIFIED IN A SUB PROCEDURE
CAN BE SET EQUAL TO AN ARRAY VARIABLE

If a variable in a VBA **Sub** procedure is set equal to a range of cells in a worksheet, that variable becomes an array. No **Dim** statement is necessary. Thus the following expression in Figure 15-2 creates a VBA array called TestArray:

TestArray = **Range**("A2:A10")

The worksheet array can be a range reference or a name that refers to a reference. Thus, if the name XRange had been assigned to the range "A2:A10", then the following expression would also create a VBA array called TestArray:

TestArray = **Range**("XRange")

A one-row or one-column reference becomes a one-dimensional array; a rectangular range becomes a two-dimensional array of dimensions *array(rows, columns)*.

The lower index of these arrays is always 1. Although arrays created *within* VBA have a lower array index of zero unless specified otherwise (by means of the **Option Base 1** statement, for example), when you transfer a range of cells *from a worksheet to VBA*, an array is created with lower array index of 1.

```
Sub ArrayDemo2()
'Reads values from column A, beginning in row 2,
'into a VBA array.
'++++++++++++++++++++++++++++++++++++++++++++++
TestArray = Range("A2:A10")
MsgBox "Fifth array element = " & TestArray(5, 1)
MsgBox "Now changing first array element to zero."
TestArray(5, 1) = 0
MsgBox "Confirming that first array element = " & TestArray(5, 1)
MsgBox "Please examine first array element on worksheet."
End Sub
```

Figure 15-2. Another way to create a VBA array.

SOME WORKSHEET FUNCTIONS USED WITHIN VBA
CREATE AN ARRAY

If you use a worksheet function within VBA that returns an array, the lower array index will be 1. Such worksheet functions include: LINEST, TRANSPOSE, MINVERSE, MMULT. Other functions that return arrays include the VBA function **Caller** when used with a menu command or toolbutton.

Note that the values in the range of cells have not been transferred to a VBA array; the VBA variable simply "points" to the range on the worksheet.

AN ARRAY OF OBJECT VARIABLES

There is an important difference between equating a range of cells in a worksheet to a simple variable in VBA, thus:

ar = **Range**("A2:B9")

or equating a range of cells in a worksheet an object variable by using the **Set** command, thus:

Set ar = **Range**("A2:B9")

Equating a variable in VBA to a worksheet range automatically creates an array in VBA in which each array element contains the value stored in the cell.   Using the **Set** command to equate an object variable in VBA to a worksheet range automatically creates an array of "cell" objects in VBA. If you change the value of one of the array elements by means of your VBA code, the value in that cell in the worksheet will change, as illustrated by the **Sub** procedure in Figure 15-3.

For an array of object variables, you must use a different approach to obtain the upper or lower bounds of the array indices, thus:

ar.**Rows.Count**

or

ar.**Columns.Count**.

```
Sub ArrayDemo2a()
'Reads values from column A, beginning in row 2,
'into a VBA array of object variables.
'+++++++++++++++++++++++++++++++++++++++++++++
Set TestArray = Range("A2:A10")
MsgBox "Fifth array element = " & TestArray(5)
MsgBox "Now changing fifth array element to zero."
TestArray(5) = 0
MsgBox "Confirming that fifth array element = " & TestArray(5)
MsgBox "Please examine fifth array element on worksheet."
End Sub
```

Figure 15-3. Creating an array of object variables.

WORKING WITH ARRAYS IN SUB PROCEDURES: PASSING VALUES FROM A VBA MODULE TO A WORKSHEET

There are at least two ways to send values from a VBA array to a worksheet. You can either set up a loop and write the value of each array element to a worksheet cell, or you can assign the value of the VBA array to the value of a worksheet range.

USING A LOOP TO TRANSFER VALUES FROM A VBA ARRAY TO A WORKSHEET

To write the value of each array element to a worksheet cell, you can use either the **Cells** method or the **Range** method. In the example shown in Figure 15-4, the **Cells** method is used.

```
Sub ArrayDemo3()
'Demo to illustrate writing array values to a sheet
'by means of a loop.
'+++++++++++++++++++++++++++++++++++++++++++++
Dim TestArray(10)
'Puts the numbers 11, 12, 13... in a VBA array.
For x = 1 To 10: TestArray(x) = 10 + x: Next
'Then writes the array elements to cells D1:D10.
For x = 1 To 10
Cells(x, 4).Value = TestArray(x)
Next x
End Sub
```

Figure 15-4. Passing values from a VBA array to a worksheet.

This method is straightforward, although sometimes not as convenient as the method that follows.

EQUATING A WORKSHEET RANGE TO AN ARRAY VARIABLE

In the example shown in Figure 15-5, a 2-D range on a worksheet becomes an array in VBA. Then the array elements are written back to the worksheet with a single line of code.

This is much more convenient than the loop method, where two nested loops would be required to read or write a 2-D range. However, a problem arises when you use this method with a 1-D range, as described next.

```
Sub ArrayDemo4()
'Demo to illustrate writing array values to a sheet
'by writing the array.
'++++++++++++++++++++++++++++++++++++++++++++++
TestArray = ActiveSheet.Range("A1:B5000")
Range("D1:E5000") = TestArray
End Sub
```

Figure 15-5. Another way to write values from a VBA array to a worksheet.

A 1-DIMENSIONAL ARRAY ASSIGNED TO A WORKSHEET RANGE CAN CAUSE PROBLEMS

Arrays can cause some confusion when you write the array back to a worksheet by assigning the value of the array to a worksheet range.

VBA considers a one-dimensional array to have the elements of the array in a row. This can cause problems when you select a range of cells in a column and assign an array to it, as in the **Sub** procedure shown in Figure 15-6.

```
Sub ArrayDemo5()
'Illustrates that 1-D VBA arrays have elements in a row,
'not a column.
'Was supposed to put the numbers 11, 12, 13... in E1:E10.
'But instead writes element(1) in all cells.
'++++++++++++++++++++++++++++++++++++++++++++++
Dim TestArray(10)
'Puts the numbers 11, 12, 13... in a VBA array.
For x = 1 To 10: TestArray(x) = 10 + x: Next
'Then writes the array elements to cells E1:E10.
Range("E1:E10").Value = TestArray
End Sub
```

Figure 15-6. From a VBA array to a worksheet: **the row-column problem.**

If you run the preceding **Sub** procedure, you will find that cells E1 through E10 will all contain 11, the first element of the array. However, if you write the array to a row of cells instead of a column, thus:

Range("E1:N1").**Value** = TestArray

each cell of the range will receive the correct array value.

There are at least three ways to "work around" this problem caused by "horizontal" and "vertical" arrays. One way is to use a loop to write the elements of the array to individual worksheet cells in a column (Figure 15-7), instead of using the more convenient procedure of assigning the array variable to a range of cells in the worksheet.

```
Sub ArrayDemo6()
'One method to "work around" the row-column problem:
'write the array elements using a loop.
'++++++++++++++++++++++++++++++++++++++++++++++++++++
Dim TestArray(10)
'Puts the numbers 11, 12, 13... in a VBA array.
For x = 1 To 10: TestArray(x) = 10 + x: Next
'Then writes the array elements to cells E1:E10.
DestRow = 0 : DestCol = 5
For x = 1 To 10
Cells(DestRow + x, DestCol) = TestArray(x)
Next
End Sub
```

Figure 15-7. One method to "work around" the row–column problem.

A second way is to specify both the row and the column dimensions of the array, so as to make it an array in a column, as illustrated in the **Sub** procedure shown in Figure 15-8.

```
Sub ArrayDemo7()
'Second method to "work around" the row-column problem:
'specify the row and column dimensions.
'++++++++++++++++++++++++++++++++++++++++++++++++++++
'Puts the numbers 11, 12, 13... in a VBA array.
Dim TestArray(10, 1)
For x = 1 To 10: TestArray(x, 1) = 10 + x: Next
'Then writes the array elements to cells E1:E10.
Range("E1:E10").Value = TestArray
End Sub
```

Figure 15-8. A second way to "work around" the row–column problem.

A third way is to use the TRANSPOSE worksheet function (Figure 15-9):

```
Sub ArrayDemo8()
'Another method to "work around" the row-column problem: use
Transpose. Note that Transpose creates a 1-base array.
'++++++++++++++++++++++++++++++++++++++++++++++++++++
Dim TestArray(10)
'Puts the numbers 11, 12, 13... in a VBA array.
For x = 1 To 10: TestArray(x) = 10 + x: Next
NewArray = Application.Transpose(TestArray)
Range("E1:E10").Value = NewArray
End Sub
```

Figure 15-9. A third way to "work around" the row–column problem.

SPEED DIFFERENCES IN READING OR WRITING ARRAYS
CREATED BY TWO DIFFERENT METHODS

There is a marked difference in the speed of execution of the two methods of reading or writing arrays just described. When you are writing values from a VBA array to a range of worksheet cells, assigning the array name to the range reference, as illustrated by **Sub** procedure ArrayDemo3, is much faster than using a loop, as in ArrayDemo4. The same is true when you are reading worksheet cells into a VBA array. However, VBA code that involves accessing the individual array elements (such as summing or comparing all the elements of an array) executes much faster when the cell values have been read into elements of a VBA array that was declared by means of a **Dim** statement. Thus if your arrays are large, the method you use to "create" the array should be guided by what you intend to do with the array.

WORKING WITH ARRAYS
IN FUNCTION PROCEDURES:
FROM WORKSHEET TO MODULE

You can create **Function** procedures that use arrays as arguments, or return an array as a result.

A RANGE PASSED TO A FUNCTION PROCEDURE
AUTOMATICALLY BECOMES AN ARRAY

If a range argument is passed in a function macro, the range automatically becomes an array in the VBA procedure. No **Dim** statement is necessary. Thus the expression

Function Deming(XValues, Yvalues)

passes the worksheet ranges XValues and Yvalues to the VBA procedure where they become arrays.

A one-row or one-column reference becomes a one-dimensional array; a rectangular range becomes a two-dimensional array of dimensions *array(rows, columns)*.

The **Function** procedure in Figure 15-10 is identical to Excel's **INDEX** worksheet function: it passes a range and a number as arguments and returns the xth element of the array.

The name of the array in the procedure must be the name of the placeholder argument of the function.

```
Function ArrayDemo9(ArrayRef, x)
'Shows how to pass an array as an argument.
'+++++++++++++++++++++++++++++++++++++++++++++++++++
ArrayDemo9 = ArrayRef(x)
End Function
```

Figure 15-10. Passing an array as an argument in a **Function** procedure.

PASSING AN INDEFINITE NUMBER OF ARGUMENTS USING THE PARAMARRAY KEYWORD

Occasionally a **Function** procedure needs to accept an indefinite number of arguments. The SUM worksheet function is an example of such a function; its syntax is =SUM(number1,number2,...). To allow a **Function** procedure to accept an indefinite number of arguments, use the **ParamArray** keyword in the argument list of the function, as in the following expression

> **Function** ConcatenateSpecial(**ParamArray** String1())

Only one argument can follow the **ParamArray** keyword, and it must be the last one in the function's list of arguments. The argument declared by the **ParamArray** keyword is an array of **Variant** elements. Empty parentheses are required.

Interestingly, even if you use the **Option Base 1** statement, the lower bound of the array is zero.

Elements in the array of arguments passed using the **ParamArray** keyword can themselves be arrays.

RETURNING AN ARRAY OF VALUES AS A RESULT

There are several ways to enable a **Function** procedure to return an array of values. The most obvious is to assemble the values in an array and return the array. The procedure shown in Figure 15-11 illustrates a function that returns an array of values: the letters a, b, c and d in a 2 × 2 array. The function requires no arguments. The user must select a 2 × 2 range of cells, enter the function and press CONTROL+SHIFT+ENTER.

A second approach is to use the **Array** keyword. The **Array** function returns a variant that contains an array. In the **Sub** procedure shown in Figure 15-12, the variable MyArray contains an array of text values, but an array can contain text, number values, the value of a VBA variable, even a **Range** reference.

To use this function the user must select a horizontal range of cells, enter the function, and press CONTROL+SHIFT+ENTER.

```
Function ArrayDemo11()
'Shows one way to return an array of values:
'Create an actual array by using the Dim statement.
'++++++++++++++++++++++++++++++++++++++++++++++++++++
Dim ReturnValu(2, 2)
ReturnValu(1, 1) = "a"
ReturnValu(1, 2) = "b"
ReturnValu(2, 1) = "c"
ReturnValu(2, 2) = "d"
ArrayDemo11 = ReturnValu()
End Function
```

Figure 15-11. A **Function** procedure that returns an array.

```
Function ArrayDemo12()
'Shows another way to return an array of values: the Array keyword
'Array returns a single variable of Variant type
'that contains an array.
'Only 1-dimensional array, though.
'++++++++++++++++++++++++++++++++++++++++++++++++++++
MyArray = Array("Jan", "Feb", "Mar", "Apr", "May", "Jun", _
"Jul", "Aug", "Sep", "Oct", "Nov", "Dec")
ArrayDemo11 = MyArray
End Function
```

Figure 15-12. Using the **Array** keyword in a **Function** procedure.

The **Array** keyword can accommodate only a one-dimensional array. To use this approach to return a two-dimensional array, you can create an array of arrays, as illustrated in Figure 15-13. This function returns the letters a, b, c and d in a 2 × 2 array.

```
Function ArrayDemo13()
'Shows another way to return a 2-D array of values.
'Use the Array function in a nested fashion.
ArrayDemo13 = Array(Array("a", "b"), Array("c", "d"))
End Function
```

Figure 15-13. Using the **Array** keyword to return a 2-D array.

In each of the preceding examples, you must select an appropriate range of cells to contain the result, enter the function, and press CONTROL+SHIFT+ENTER (Excel for Windows) or CONTROL+SHIFT+RETURN (Excel for the Macintosh).

16

CREATING COMMAND MACROS

A command macro (a **Sub** procedure) can automate any sequence of actions that can be performed by the use of menu commands or keystrokes. Many simple but useful command macros can be created entirely by using the Recorder, as described in Chapter 13. But command macros can also carry out much more complicated actions. In this chapter we'll look at some examples of creating more advanced **Sub** procedures.

CREATING ADVANCED MACROS IN VBA

There are literally hundreds of VBA objects, properties and methods, and this can be confusing for the beginner. Chapter 14 focused on the basic tools needed to create macros to automate chemical worksheet calculations: how to transfer values from a sheet to a VBA module, how to perform calculations within a VBA module, how to perform logical branching and iterative looping, and how to send values back from a VBA module to a worksheet. In this chapter we'll use these tools to create some useful **Sub** procedures.

CREATING A SUB PROCEDURE
TO FORMAT TEXT AS A CHEMICAL FORMULA

As a first example, we'll create the ChemicalFormat macro to format text in a selected cell as a chemical formula. The macro will be a simple one. It will examine each character of a text string; if the character is a number, the character will be subscripted. You'll find this macro useful if you label rows or columns in your worksheets with chemical formulas, such as "$CH_3CH=CH_2$" or "moles of H_3PO_4"; the macro will be a real timesaver.

We need to accomplish four steps in this procedure: (i) obtain the contents of the active cell, (ii) set up a loop so that we can examine each character in turn, (iii) test to see if the character is a number and (iv) subscript the character.

The code required for the first three steps was described in Chapter 14. We'll (i) use the **ActiveCell** keyword to obtain the string, (ii) use **For... Next** to create a loop and the **Mid** function to access each character within the loop and (iii) use an **If** statement to test each character. The only bit of code we'll need that wasn't described in Chapter 14 is the code to subscript a character. You could spend some time looking in the VBA On-line Help, but a much easier way to obtain the correct code is to use the Recorder. Simply turn on the Recorder, perform the action of subscripting a character, then go to the Visual Basic Editor and examine the recorded code. You'll find that the expression to subscript specified characters in a text string in the active cell uses the **Characters**(*start*, *length*) method to specify the characters:

object.**Characters**(*start*, *length*).**Font.Subscript** = **True**

Simply **Copy** the code fragment and **Paste** it into your macro.

The complete macro is shown in Figure 16-1. Right now you'll have to use the Macro Run dialog box to run the macro, which isn't very convenient. Of course you could assign a shortcut key to the macro, as described in Chapter 13. In Chapter 19 you'll learn how to create a custom toolbutton and assign the macro to it.

```
Sub ChemicalFormat1()
For x = 1 To Len(ActiveCell)
char = Mid(ActiveCell, x, 1)
If Asc(char) >= 48 And Asc(char) <= 57 Then ActiveCell. _
Characters(Start:=x, Length:=1).Font.Subscript = True
Next x
End Sub
```

Figure 16-1. A simple **Sub** procedure to format text as a chemical formula.

ADDING ENHANCEMENTS TO THE CHEMICALFORMAT MACRO

The simple macro in Figure 16-1 formats the text in a single cell. We'd like our macro to be able to format text in a single cell or in a range of cells. To do this, we simply need to add an outer **For..Each...Next** loop. We'll add the lines of code

For Each cel **In** **Selection**

and

Next cel

at the beginning and the end, respectively, of the original macro, and change three instances of **ActiveCell** to cel. Note that in the **For Each** cel **In Selection** loop shown in Figure 16-2, the variable cel is an object variable; thus we can use the code

cel.**Characters**(Start:=x, Length:=1).**Font.Subscript** = **True**

In addition, the macro subscripts every number character in a string. We'd like our macro to be able to handle formulas of hydrates (e.g., $CuSO_4 \cdot 5H_2O$) or other text strings containing numbers that should not be subscripted, such as "$CaSO_4 \cdot 1/2H_2O$" or "H_3PO_4 (85%)". The **Sub** procedure in Figure 16-2 includes code that handles these situations, by using a logical variable FirstFlag. Number characters are subscripted only when FirstFlag = **True**, and FirstFlag is set to **False** when any of the special characters are encountered while looping through the string.

```
Sub ChemicalFormat2()
For Each cel In Selection
FirstFlag = True
  For x = 1 To Len(cel)
  char = Mid(cel, x, 1)
  If IsNumeric(char) Or char = "." Then
    If FirstFlag = True Then GoTo EndLoop
    If char = "." Then GoTo EndLoop
    cel.Characters(Start:=x, Length:=1).Font.Subscript = True
  Else
    FirstFlag = False
    If char = " " Or char = "*" Or Asc(char) = 165 Or char = "," _
    Or char = "(" Or char = "/" Then FirstFlag = True
  End If
EndLoop:     Next x
Next cel
End Sub
```

Figure 16-2. The ChemicalFormat macro with some additional features added.

Some examples of text formatted by using the ChemicalFormat2 macro are shown in Figure 16-3.

A
1 $CuSO_4 \cdot 5H_2O$
2 $CaSO_4 \cdot 1/2H_2O$
3 H_3PO_4 (85%)
4 $H_2SO_4 + 2NaOH = Na_2SO_4 + 2H_2O$

Figure 16-3. Some examples of formatting with the ChemicalFormat macro.

ADDING MORE ENHANCEMENTS

Finally, we'd like our macro to be able to format text in a worksheet cell or range, in a chart title, or in a textbox. As well, we don't want the macro to crash

```
Sub ChemicalFormat3()
'Formats text, e.g., H2SO4, as a chemical formula (subscripts numbers).
'Operates on a cell, a range of cells, text in a chart, or a textbox.
'Last modified 9/22/00
'MAKE SURE WE ARE ON A WORKSHEET OR CHART SHEET
If TypeName(ActiveSheet) <> "Worksheet" And _
TypeName(ActiveSheet) <> "Chart" Then Beep: Exit Sub
'GO TO CORRECT CODE FOR SELECTION TO BE FORMATTED
Select Case TypeName(Selection)
Case "Range"   'FORMAT TEXT IN A CELL OR RANGE
  For Each cel In Selection
  Call DoFormat(cel.Value, cel)
  Next cel
  Exit Sub
Case "AxisTitle"    'FORMAT TEXT IN A CHART
   Call DoFormat(Selection.Characters.Text, Selection)
   Exit Sub
Case "ChartTitle"  'FORMAT TEXT IN A CHART
   Call DoFormat(Selection.Characters.Text, Selection)
   Exit Sub
Case "TextBox" 'FORMAT TEXT IN A TEXT BOX
   Call DoFormat(Selection.Characters.Text, Selection)
   Exit Sub
End Select
End Sub


Sub DoFormat(FormulaString, FormulaObject)
FirstFlag = True
  For x = 1 To Len(FormulaString)
  char = Mid(FormulaString, x, 1)
   If IsNumeric(char) Or char = "." Then
    If FirstFlag = True Then GoTo EndLoop
    If char = "." Then GoTo EndLoop
      FormulaObject.Characters(Start:=x, Length:=1).Font.Subscript
= True
  Else
     FirstFlag = False
     If char = " " Or char = "*" Or Asc(char) = 165 Or char = "," _
     Or char = "(" Or char = "/" Then FirstFlag = True
  End If
EndLoop:     Next x
End Sub
```

Figure 16-4. The ChemicalFormat macro with some additional features added.

if we attempt to run it when no sheet is active, for example. The **Sub** procedure shown in Figure 16-4 uses the VBA keywords **TypeName(ActiveSheet)** and **TypeName(Selection)** to determine that the correct kind of sheet is active and that an appropriate selection has been made, before attempting to format the selection.

To develop the code to handle text in a chart or in a text box, a one-line **Sub** procedure containing either

> **Msgbox   TypeName(ActiveSheet)**

or

> **Msgbox   TypeName(Selection)**

was used to display the keywords associated with selected chart elements or a text box. Again, this was faster and more convenient than looking in reference books or using the On-line Help. Once the correct keywords had been found, i t was a relatively simple matter to modify ChemicalFormat2 to handle text in different environments.

The inner-loop code in Figure 16-2 was used as a general subroutine (now called DoFormat) to examine the text, find the number characters and format them. The main program simply ensures that a worksheet or chart is active, determines the kind of text to be formatted, then calls the subroutine. The subroutine call passes two arguments, a simple variable containing the text to be examined (FormulaString) and an object variable containing the text to be subscripted (FormulaObject).

Since several different kinds of text were to be formatted, a **Select Case** construction was used. As it turned out, the same syntax is used to subscript text in a chart or in a text box. Nevertheless, the original **Select Case** structure was left in place.

CREATING A SUB PROCEDURE
TO APPLY DATA LABELS IN A CHART

As described in Chapter 5, the Data Labels tab in the Format Data Series dialog box allows you to add data labels to a chart (to each data point in an XY chart, or to each column in a column chart, for example). But only the x values or the y values can be used as labels. You'd probably like to use some other text as data labels. You can do this manually, by adding the data labels, then manually editing each one to enter the text you want. A macro will make the task much easier.

In the following section we will develop a **Sub** procedure to assign text values in a range of cells as the data labels for a chart. Here's how we'll proceed: we'll use the Recorder to obtain the code for adding data labels, selecting a data label and entering text. We'll use this code as a framework on which to construct our macro.

The code shown in Figure 16-5 was obtained when the Recorder was used to record the following actions: selecting a chart, selecting a data series, choosing **Selected Data Series** from the **Format** menu and applying data labels, selecting a data label, and typing the text "abc" in place of the original label. In this **Sub** procedure we have almost all the code we need to create the initial version of our macro.

```
Sub Macro1()
' Macro recorded 9/25/00 by Dr. Billo
ActiveSheet.ChartObjects("Chart 1").Activate
ActiveChart.SeriesCollection(1).Select
ActiveChart.SeriesCollection(1).ApplyDataLabels _
Type:=xlDataLabelsShowValue, AutoText:=True, LegendKey:=False
ActiveChart.SeriesCollection(1).DataLabels.Select
ActiveChart.SeriesCollection(1).Points(4).DataLabel.Select
Selection.Characters.Text = "abc"
Selection.AutoScaleFont = False
With Selection.Characters(Start:=1, Length:=3).Font
        .Name = "Geneva"
        .FontStyle = "Regular"
        .Size = 10
        .Strikethrough = False
        .Superscript = False
        .Subscript = False
        .OutlineFont = False
        .Shadow = False
        .Underline = xlUnderlineStyleNone
        .ColorIndex = xlAutomatic
    End With
    ActiveChart.ChartArea.Select
End Sub
```

Figure 16-5. Macro1 recorded.

In the first version of our macro, the chart must be the active document, and the user must have selected the data series to which the data labels are to be attached before running the macro. In a later version we'll allow the user to select the data series while the macro is running.

The four lines of code from **Macro1** that we'll use (after modifying them) are as follows:

```
ActiveChart.SeriesCollection(1).ApplyDataLabels _
Type:=xlDataLabelsShowValue, AutoText:=True, LegendKey:=False

ActiveChart.SeriesCollection(1).DataLabels.Select
```

ActiveChart.SeriesCollection(1).Points(4).DataLabel.Select

Selection.Characters.Text = "abc"

To complete our macro, we'll need to add some code (i) so that the user can specify a range of cells that contain the labels, (ii) to set up a loop so that we can loop through all the data points and (iii) to change **SeriesCollection(1)** to the general case.

To specify the range of cells to use as data labels, we'll use the **InputBox** method. We'll define the range as an object variable, so that we can use a **For...Each...Next** loop. For a code example, see "You Can Define Your Own Objects" in Chapter 14. To convert **SeriesCollection(1)** into the general case, remember that a specific item in a collection can be referred to by its index number or by its name. We'll use the code

SeriesName = Selection.Name

and

ActiveChart.SeriesCollection(SeriesName)

to specify the selected data series.

After the macro was completed, some testing revealed that in most cases, the code containing the **Select** keyword could be eliminated. The completed macro is shown in Figure 16-6.

```
Sub Labeler1()
' Get the name of the chart data series.
SeriesName = Selection.Name

' Input the range of cells to be used as labels.
Set LabelRange = Application.InputBox("Select the range of cells
containing the labels", "DATA LABELER STEP 1 OF 1", , , , , , 8)

' Apply the standard data labels.
ActiveChart.SeriesCollection(SeriesName).ApplyDataLabels _
Type:=xlDataLabelsShowValue, AutoText:=True, LegendKey:=False

' Set up a loop to replace data labels with new ones.
x = 1
For Each cel In LabelRange
    ActiveChart.SeriesCollection(SeriesName).Points(x). _
DataLabel. Characters.Text = cel.Value
    x = x + 1
Next cel
End Sub
```

Figure 16-6. The completed DataLabeler macro.

A more sophisticated version of the macro can be found on the CD-ROM that acompanies this book.

17

CREATING CUSTOM FUNCTIONS

Chapter 13 provided an introduction to **Sub** procedures and **Function** procedures. By now it should be clear that a **Sub** procedure (a command macro) is a computer program that you "run"; it can perform actions such as formatting, opening or closing documents, etc. A **Function** procedure (a user-defined function) is a computer program that calculates a value and returns it to the cell in which it is typed. A **Function** procedure cannot change the worksheet environment (e.g., it can't make a cell Bold).

The following sections provide some examples of more advanced custom functions.

A CUSTOM STATISTICAL FUNCTION

One of the underlying assumptions of the least squares method is that there is no error in the measurement of the independent variable (the y values). This assumption is often not valid, and one of the most obvious cases of this is found in method-comparison analysis. A typical example of method-comparison analysis involves the comparison of two different instruments, a current production instrument and an improved model. Measurements made on a series of samples with the two instruments, and plotted by using current instrument readings as the x values and new instrument readings as the y values, should ideally result in a straight line of unit slope and zero intercept. The actual slope and intercept of the line can provide estimates of the proportional and constant error between the two methods.

It is obvious that there is error in both the x and the y values. Calculation of the least-squares slope and intercept by "standard" methods is clearly not valid. In the custom function that follows, the method of Deming is used to calculate the regression parameters for the straight line $y = mx + b$. The Deming regression calculation assumes Gaussian distribution of errors in both x and y values and uses duplicate measurements of x values (and of y values) to estimate the standard errors. A portion of a data table is shown in Figure 17-1.

	H	I	J
1	Sample		
2	#	X	Y
3	1	25.10	24.95
4		26.50	26.55
5	2	20.60	21.70
6		23.55	22.20
7	3	45.30	44.60
8		44.85	44.80
9	4	28.45	29.60
10		29.45	30.00
11	5	46.70	47.45
12		48.05	47.20
13	6	27.65	26.45
14		27.30	27.25
15	7	27.05	26.70
16		26.90	26.90

Figure 17-1. Portion of data table to calculate Deming regression parameters.

The equations[*] for the Deming slope and intercept are

$$m = U + \sqrt{U^2 + (1/\lambda)}$$

and

$$b = \bar{y} - m\,\bar{x}$$

where

$$U = \frac{S_y^2 - (1/\lambda)\,S_x^2}{2\,r\,S_x\,S_y}$$

$$\lambda = \frac{S_{ex}^2}{S_{ey}^2}$$

$\bar{x}$ and $\bar{y}$ are the means and S_x^2, S_y^2 are the variances of the x_{mean} and y_{mean} values, respectively, r is the Pearson correlation coefficient and S_{ex}^2 and S_{ey}^2 are the error variances of the x values and y values, respectively, calculated from the equation

$$S_e = \sqrt{\frac{\Sigma(\text{difference between duplicates})^2}{2\,N}}$$

---

[*]   See, for example, P. J. Cornbleet and N. Gochman, *Clin. Chem.* **1979**, *25*, 432.

```
' Deming Regression
' Calculates Deming regression parameters for Y = mX + b
' Equations from Cornbleet & Gochman, Clin. Chem. 1979, 25, 432.
' Copyright 1997 by E. J. Billo
' Begun 5/10/97.  Last modified 10/3/00

Function Deming(XValues, Yvalues)
Dim MeanX(), MeanY()
'Get number of cells to use in calculation loop
Ncells = XValues.Count
ReDim MeanX(Ncells / 2), MeanY(Ncells / 2)

' Step thru pairs of cells, calculating sums for statistics calcs
For x = 2 To Ncells Step 2
    MeanX(x / 2) = (XValues(x - 1) + XValues(x)) / 2
    MeanY(x / 2) = (Yvalues(x - 1) + Yvalues(x)) / 2
    SumX = SumX + MeanX(x / 2): SumY = SumY + MeanY(x / 2)
    SumX2 = SumX2 + (MeanX(x / 2)) ^ 2
    SumY2 = SumY2 + (MeanY(x / 2)) ^ 2
    SumXY = SumXY + MeanX(x / 2) * MeanY(x / 2)
    SumDeltaX2 = SumDeltaX2 + (XValues(x - 1) - XValues(x)) ^ 2
    SumDeltaY2 = SumDeltaY2 + (Yvalues(x - 1) - Yvalues(x)) ^ 2
Next

' Calculate some intermediate statistical quantities
XBar = SumX / N: YBar = SumY / N
Sx2 = (N * SumX2 - SumX ^ 2) / (N * (N - 1))
Sy2 = (N * SumY2 - SumY ^ 2) / (N * (N - 1))
Sdx2 = SumDeltaX2 / (2 * N)
Sdy2 = SumDeltaY2 / (2 * N)
rPearson = (N * SumXY - SumX * SumY) / _
Sqr((N * SumX2 - SumX ^ 2) * (N * SumY2 - SumY ^ 2))

' Calculate quantities that are specific to the Deming calculation
lambda = Sdx2 / Sdy2
U = (Sy2 - Sx2 / lambda) / (2 * rPearson * Sqr(Sx2) * Sqr(Sy2))
Slope = U + Sqr(U ^ 2 + 1 / lambda)
Intercept = YBar - Slope * XBar

Deming = Array(Slope, Intercept)
End Function
```

Figure 17-2. A **Function** procedure to calculate Deming regression parameters.

The macro listing is shown in Figure 17-2. The two arguments used by the macro are a range of cells containing pairs of x measurements on a given sample and a corresponding range of pairs of y measurements.

Within the procedure, the major part of the code is virtually identical to the code that would be used to obtain the usual least-squares regression parameters for $y = mx + b$, namely, obtaining the sum of x values, the sum of squares of x values, etc. The difference is that pairs of values are used; the mean value of each pair of x values or of y values is used as the independent variable or dependent variable, respectively, and the estimate of the standard errors for the two sets of data is obtained from the differences between pairs.

Near the end of the code some calculations that are specific to the Deming equations are performed. The function returns an array of two values, the slope and intercept of the Deming regression. Since an array is returned, the user must press CONTROL+SHIFT+ENTER (Windows) or CONTROL+SHIFT+RETURN (Macintosh).

This function could be located in the Statistical category, rather than in the User Defined category. See "Assigning a Custom Function to a Function Category" later in this chapter.

A CUSTOM FUNCTION THAT TAKES AN OPTIONAL ARGUMENT

A custom function can have optional arguments. Use the **Optional** keyword in the list of arguments to declare an optional argument. The optional argument or arguments must be last in the list.

You will need to determine the presence or absence of optional arguments in a procedure by using the **IsMissing** keyword. As well, you may need to provide a default value if an argument is omitted.

In the example shown in Figure 17-3, the custom function MolWt calculates the formula weight of a compound from formula (a text string in a cell). The formula weight is calculated in the subroutine Parse (not shown), which returns FW, the formula weight, and NumDecimals, the number of decimal places allowed in the result, based on the atomic weights used. The user can specify, by the optional argument decimals, the number of decimal places to be displayed in the returned value. If the optional argument decimals is not supplied, the function returns the value of FW, but if decimals is supplied, the worksheet function FIXED is used to set the number of decimal places in the result.

The MolWt function can be found on the CD-ROM that acompanies this book.

```
Function MolWt(formula As String, Optional decimals)
' If decimals is omitted, returns value as a number.
' If decimals is specified, returns value as text,
' formatted with specified number of decimals.
' Maximum number of decimal places allowed is determined in subroutine
' by decimal places of atomic weights.

Call Parse(formula, FW, NumDecimals)
If IsMissing(decimals) Then MolWt = FW: Exit Function
If NumDecimals < decimals Then MolWt = Application.Fixed(FW, _
NumDecimals): Exit Function
MolWt = Application.Fixed(FW, decimals)
End Function
```

Figure 17-3. A **Function** procedure with an optional argument.

A CUSTOM FUNCTION THAT TAKES
AN INDEFINITE NUMBER OF ARGUMENTS

Many of Excel's worksheet functions can accept an indefinite number of arguments. The SUM function is an example of such a function; its syntax is =SUM(number1,number2,...). To create a custom function with an indefinite number of arguments, use the **ParamArray** keyword. An argument declared with **ParamArray** is an array of **Variants**, and each element of the array can itself be an array.

The next example shows a custom function that accepts a number of worksheet ranges as arguments and combines them into a single array. This custom function is designed to be used with Excel's LINEST function (see Chapter 11). LINEST requires that the known_x's argument be a contiguous range of cells; you can't use a non-adjacent selection as the argument, even if the ranges are enclosed in parentheses. But when you perform multiple linear regression using several columns of x values, oftentimes those columns of x values are not contiguous, as illustrated in Figure 17-4. The expression =LINEST(E4:E13,(A4:A13,C4:C13),1,1) returns the #REF! error message. You could rearrange the worksheet to have all the columns of x values adjacent, but with the custom function shown in Figure 17-5 you can combine the separated ranges into a single array and use it as the known_x's argument of LINEST.

Since LINEST requires that there be the same number of values for each independent variable, we first check to make sure that each range in the array of arguments has the same number of rows. An interesting point: although we specified **Option Base** 1, the **ParamArray** keyword produces an array whose lower array index is zero. Thus we must loop from 0 to **UBound**(rng). Note that **UBound**(rng) is the number of elements in the array, while rng(z).**Columns.Count** returns the number of columns in an individual element in the array.

	A	B	C	D	E
2		Intermediate		Intermediate	
3	X1	Calculation	X2	Calculation	Y
4	0.386	#&:	0.149	#&:	19.84
5	0.636	#&:	0.405	#&:	37.07
6	2.428	#&:	5.893	#&:	65.641
7	1.438	#&:	2.067	#&:	103.47
8	1.452	#&:	2.108	#&:	151.04
9	4.049	#&:	16.392	#&:	209.41
10	6.523	#&:	42.556	#&:	277.28
11	5.407	#&:	29.235	#&:	355.57
12	7.328	#&:	53.699	#&:	443.44
13	5.562	#&:	30.938	#&:	541.07

Figure 17-4. LINEST fails when used with a non-adjacent selection of known_x's

If an array element is not the correct size, we need to return an error value. Use the **CVErr** keyword to return a worksheet error value that Excel can handle appropriately. The error values are listed in Appendix D.

After checking all the elements in the array, we assemble them into the final array of dimensions XSize rows by YMax columns, and return the array to the worksheet.

Now the LINEST expression

=LINEST(E4:E13,MakeArray(A4:A13,C4:C13),1,1)

returns the correct result. You can use names for cell ranges in the MakeArray function, e.g.,

=LINEST(YValues,MakeArray(Xvalues1,Xvalues2),1,1.

```
Option Base 1

Function MakeArray(ParamArray rng())
' Combines individual 1-D or 2-D arrays into a final 2-D array.
' In this version all individual arrays must be vertical.
' All individual arrays must have same number of rows.

Dim Result() As Variant
Dim XSize As Integer, YSize As Integer, YMax As Integer

' Z index is used to select within array of arrays.
' x and y are column & row indices of individual arrays.

' First, calculate size of final array.
' And check to make sure all arrays contain same number of rows.
' Despite use of Option Base 1, LBound = 0
XSize = rng(0).Rows.Count        'returns "Base 1" value
YMax = 0
For z = 0 To UBound(rng)
YSize = rng(z).Columns.Count      'returns "Base 1" value
If XSize <> rng(z).Rows.Count Then MakeArray = CVErr(xlErrRef): _
 Exit Function
YMax = YMax + YSize
Next z
ReDim Result(XSize, YMax)

' Now combine each individual array into final array.
YStart = 0
For z = 0 To UBound(rng)
YSize = rng(z).Columns.Count
For y = 1 To YSize
For x = 1 To XSize
Result(x, YStart + y) = rng(z)(x, y)
Next x, y
YStart = YStart + YSize
Next z
MakeArray = Result
End Function
```

Figure 17-5. A **Function** procedure to combine arrays.

PROVIDING A DESCRIPTION FOR A CUSTOM FUNCTION IN THE PASTE FUNCTION DIALOG BOX

When you use Paste Function to enter a function, the function's description appears at the bottom of the dialog box. For a custom function, the default description is "Choose the Help button for Help on this function and its arguments", but you can provide a custom description for your custom function. There are two ways to do this: you can use the Macro dialog box (normally used only for **Sub** procedures), or you can write and run a simple one-line VBA **Sub** procedure. Either way, the description becomes part of Excel and does not need to be entered each time you start Excel.

To use the Macro dialog box to enter a description, carry out the steps in the following box.

To Provide a Description for a Custom Function

1. Choose **Macro** from the **Tools** menu and **Macros...** from the submenu.
2. Type the name of the custom function in the Macro Name box.
3. Press the Options... button to display the Macro Options dialog box.
4. Type the description in the Description box, then press OK.

To enter a description by means of VBA code, type and run a **Sub** procedure similar to the one shown in Figure 17-6.

```
Sub CustomFunctionDescription()
Application.MacroOptions macro:="FtoC", Description:=_
"Converts Fahrenheit temperature to Celsius"
End Sub
```

Figure 17-6. A **Sub** procedure to add a function description.

For Excel's built-in functions, the Formula Palette (Step 2 of Paste Function) provides help information about each argument as you begin to enter it. There's no way to provide similar information about arguments in the Formula Palette for custom functions. But since the same description text appears in the Step 1 and the Step 2 dialog boxes, you can provide information about the arguments in the description. Unfortunately the limit for Description is 255 characters. You can provide line breaks in the text by using **Chr**(10) or **Chr**(13), but the Formula Palette can only display two lines of text.

ASSIGNING A CUSTOM FUNCTION
TO A FUNCTION CATEGORY

Initially, your custom function will be located in the User Defined category in the Paste Function dialog box. You can specify the category in which the function will be located, by typing and running a **Sub** procedure similar to the one shown in Figure 17-7. The category values are listed in Appendix D.

```
Sub CustomFunctionDescription()
Application.MacroOptions macro:="FtoC", Category:=3
End Sub
```

Figure 17-7. A **Sub** procedure to change function category.

CREATING ADD-IN FUNCTION MACROS

Saving a custom function as an Add-In is by far the most convenient way to use it. Here are some of the advantages:

- An Add-In custom function is listed in the Paste Function list box without the workbook name preceding the name of the function, making it virtually indistinguishable from Excel's built-in functions.
- If the Add-In workbook is placed in the Excel Startup folder, the Add-In will be available every time you start Excel.
- An Add-In workbook can be hidden so it can't be viewed by the user.

HOW TO CREATE AN ADD-IN MACRO

To save a workbook as an Add-In, choose **Save As...** from the **File** menu. Choose Microsoft Excel Add In from the Save File As Type drop-down list box, then press OK. In Excel for Windows, Add-In macros are automatically given the filename extension .xla.

Your custom functions will now be indistinguishable from Excel's built-in functions. By giving them names using lower-case letters, you can distinguish them from Excel's built-in functions.

Command macros can also be saved as Add-Ins.

HOW TO PROTECT AN ADD-IN WORKBOOK

To protect an Add-In workbook, activate the Visual Basic Editor. Choose **Properties...** from the **Tools** menu and choose the Protection tab. Check the Lock Project For Viewing box, enter a password, and **Save** the workbook.

ADVANTAGES AND DISADVANTAGES
OF USING FUNCTION MACROS

Some function macros perform calculations that could be performed by using a worksheet formula or formulas, while in other cases implementing the function on the worksheet is impossible (when looping or branching is required, for example).

The main advantage of using a custom function instead of worksheet formulas is minimization of errors occurring when formulas are entered. The main disadvantage is that custom functions are slower than worksheet formulas. For example, in a sheet that uses worksheet formulas to calculate the four alpha values for phosphoric acid in 0.1 pH increments from pH 0 to 14 (calculation of 564 cells) using worksheet formulas (see Chapter 20), there was no detectable delay when the sheet was recalculated. On the same PC, when the ALPHA function was used, it took three seconds to recalculate the sheet.

18

CREATING CUSTOM MENUS AND MENU BARS

In this chapter you'll learn how to add a new command to a menu, how to add a new menu to a menu bar and how to create a whole new menu bar with separate menus and commands. In this way you can add new capabilities to Excel or even create a complete custom application.

In earlier versions of Microsoft Excel, menu bars contained only menus, and toolbars only toolbuttons. Beginning with Excel 97, toolbuttons and menu items are both members of the class of CommandBarButton objects. *Command bars* can contain buttons or menus or both. In general, however, mixing buttons and menu commands on the same toolbar doesn't seem to be a good idea, so this chapter deals exclusively with menus and Chapter 19 with toolbars.

MODIFYING MENUS OR MENU BARS

You can modify existing menus or menu bars by using the **Customize** command in the **Tools** menu. Each version of Excel has used a different method to customize menus: Excel 5/95 used the *Menu Editor* to modify menus, Excel 97 displayed a *customizable copy* of the menu bar in which to make changes, but in Excel 2000 menus are customized in exactly the same way as toolbars — by dragging the menu command to the desired place on the menu bar.

ADDING OR REMOVING A MENU COMMAND

You can add any of Excel's built-in menu commands to a menu, or remove menu commands that you don't use. The procedures in the following boxes cover these possibilities.

Some commands in the list of commands in the Commands tab have an icon beside them, others don't. Commands with an icon are toolbuttons, and those without an icon are menu commands.

To Add a Built-in Menu Command to a Menu

1. The toolbar that contains the menu to be customized must be visible.

2. Choose **Customize...** from the **Tools** menu, or choose **Toolbars...** from the **View** menu and choose **Customize...**, or right-click on any toolbar and choose **Customize...** from the shortcut menu.

3. (This step necessary only in Excel 97/98.) Choose the Toolbars tab, then check the box for, or double-click on, the Worksheet Menu Bar or Chart Menu Bar to display a customizable copy of the menu bar.

4. Choose the Commands tab.

5. In the Categories box, choose the desired category, then choose the new command from the list in the Commands box and drag it to the desired location in the copy of the menu. The mouse pointer will display a large "X" beside the mouse pointer image if the mouse pointer is over a location at which the command cannot be installed. If the command can be installed, there will be a small "+".

To Remove a Menu Command from a Menu

1. The toolbar that contains the menu to be customized must be visible.

2. Choose **Customize...** from the **Tools** menu, or choose **Toolbars...** from the **View** menu and choose **Customize...**, or right-click on any toolbar and choose **Customize...** from the shortcut menu.

3. (This step necessary only in Excel 97/98.) Choose the Toolbars tab, then check the box for, or double-click on, the Worksheet Menu Bar or Chart Menu Bar to display a customizable copy of the menu bar.

4. Drag the menu command off the menu (when you click on it, the menu command is highlighted) and release the mouse button. The menu command will disappear. You can remove a complete menu from a menu bar in the same way.

CREATING A NEW MENU BAR

In previous chapters you learned how to create useful **Sub** procedures. You may find that you use a particular macro so often that it would be more convenient to have it on one of Excel's drop-down menus, rather than having to "run" it by means of the **Run** command from the **Macro** menu. Running a command macro by means of an Excel menu command makes the custom command accessible to Excel users who are not familiar with the use of macros.

If you have written a number of related macros, you may even decide to create a custom menu bar with several custom menus, each containing custom commands. While both "Custom Menu Command" and "New Menu" can be found in Customize, there's no "New Menu Bar". That's because menu bars are really toolbars. To create a new menu bar, you simply create a new toolbar and place menus on it instead of toolbuttons.

You can install either custom menus or any of Excel's built-in menus on your custom menu bar. To install custom menus, follow the procedure in the following box; to install any of Excel's built-in menus, follow the same procedure but choose "Built-in Menus" from the Categories box. You can apply the same methods used for custom menus to add commands to, remove commands from, or rename the built-in menus. Menus installed on a custom menu bar are displayed with a drop-down arrow next to the menu name. Figure 18-1 shows such a custom menu bar.

Figure 18.1   A custom menu bar with three menus: Excel's built-in **File** and **Edit** menus and a custom menu.

ADDING A CUSTOM MENU TO A MENU BAR

To add a custom menu to a menu bar, use the procedure in the following box.

To Add a Custom Menu to a Menu Bar

1. The toolbar that contains the menu to be customized must be visible.

2. Choose **Customize...** from the **Tools** menu, or choose **Toolbars...** from the **View** menu, or right-click on any toolbar and choose **Customize...** from the shortcut menu.

3. (This step necessary only in Excel 97/98.) Choose the Toolbars tab, then check the box for, or double-click on, the Worksheet Menu Bar or Chart Menu Bar to display a customizable copy of the menu bar.

4. Choose the Commands tab.

5. Scroll through the Categories box and choose "New Menu".

6. Select New Menu and drag it to the desired location on the toolbar.

7. Right-click on the custom menu item to display the shortcut menu. Type the name of the new menu in the Name input box (the default text is "New menu").

Excel Tip: *If you add a built-in menu to a toolbar and make changes to the menu, the changes will also appear in the built-in menu on the menu bar. Thus if you want a modified File menu on a custom menu bar, you should add a custom menu, rename it File, and then add the desired menu commands.*

ADDING A CUSTOM MENU COMMAND TO A MENU

To add a custom menu command to a menu, use the procedure in the following box.

When you exit from Excel, any changes you made to the menu bars or toolbars are saved, and this menu bar/toolbar configuration will be displayed the next time you start Excel.

**To Add a Custom Menu Command to a Menu Bar
and Assign a Macro to It**

1. The toolbar that contains the menu to be customized must be visible.

2. Choose **Customize...** from the **Tools** menu, or choose **Toolbars...** from the **View** menu and choose **Customize...**, or right-click on any toolbar and choose **Customize...** from the shortcut menu.

3. (This step necessary only in Excel 97/98.) Choose the Toolbars tab, then check the box for, or double-click on, the Worksheet Menu Bar or Chart Menu Bar to display a customizable copy of the menu bar.

4. (This step necessary only in Excel 97/98.) In the copy of the menu bar, click on the menu to which you want to add the command, to display a customizable copy of the menu.

5. In the Customize dialog box, choose the Commands tab.

6. Scroll through the Categories box and choose Macros.

7. Select Custom Menu Item and drag it to the desired location on the menu (a black bar will show where the command will be placed).

7. Right-click on the custom menu item to display the shortcut menu and choose Assign Macro...

8. Excel displays the Assign Macro dialog box. Choose the name of the macro to be assigned to the custom menu item (best located in the Personal Macro Workbook). Click the OK button.

9. In the shortcut menu, type the text for the new menu command in the Name input box (the default text is "&Custom Menu Command").

10. To insert a separator bar above the new menu command, choose Begin a Group from the shortcut menu; to remove a separator bar, uncheck Begin a Group from the shortcut menu.

MODIFYING MENUS OR MENU BARS BY USING VBA

You can also modify existing menus or menu bars by using Visual Basic.

If you create a macro to be used by other people, you can make it easy for them to use the macro by installing a custom menu command. As well, you can install the new menu command by means of an automatic procedure, so that all the users have to do is open the macro document. There are two ways you create an automatic procedure: by creating an Auto_Open macro in a module sheet, or by creating a **Workbook_Open** event procedure. These are described in the sections that follow.

In contrast to changes to menus made by using the **Customize...** command, menus and menu commands installed by using VBA are not saved when you quit Microsoft Excel.

ADDING A MENU COMMAND BY MEANS OF AN AUTO_OPEN MACRO

In Excel 5/95, any macro with the name Auto_Open will be run automatically whenever the workbook containing that macro is opened. You can still employ Auto_Open procedures in Excel 97 or Excel 2000.

To install a new menu command whenever Excel is started, create an Auto_Open procedure in the Personal Macro Workbook. To install a new menu command when a particular workbook is opened, create an Auto_Open procedure in that workbook. Use the **MenuBars** and **Menus** VBA keywords to specify where the new **MenuItem** is to be added, with the **Add** keyword. The syntax of the **Add** method is:

MenuItems.Add(caption, onAction, shortcutKey, before)

The text of the menu item that will appear in the menu is Caption. The name of the macro that runs when the new menu item is selected is OnAction. ShortcutKey is available only for Macintosh. If Before, which specifies the position in the menu, is omitted, the command is added at the end of the menu.

Figure 18-2 illustrates an Auto_Open procedure that installs a new menu command in the Tools menu and assigns the **Sub** procedure named CommandHandler to it. To install your own custom menu command in a menu, replace Tools, New Menu Item and CommandHandler with the name of the menu, the text of your menu command and the name of your macro.

Menu commands installed by this method remain in the menu until you exit from Excel.

```
Sub AUTO_OPEN()
MenuBars("Worksheet").Menus("Tools").MenuItems.Add _
Caption:="New Menu Item", OnAction:="CommandHandler", Before:=1
End Sub
```

Figure 18.2   An Auto_Open **Sub** procedure to install a custom menu command.

ADDING A MENU COMMAND
BY MEANS OF AN EVENT-HANDLER PROCEDURE

Event-handler procedures were introduced in Excel 97; they can be used in Excel 97 or Excel 2000. Event-handler procedures provide more versatility than the automatic procedures provided with Excel 5/95.

Event-handler procedures are not written in the usual VBA modules, but in the code modules associated with a worksheet or workbook. For example, to install a new menu command when a particular workbook is opened, put the event-handler procedure in the code window for **ThisWorkbook** (use the Project Explorer to locate the node for **ThisWorkbook**, then click on it to display the Code window).

Use the **CommandBars** VBA keyword to specify where the new Control is to be added, with the **Add** keyword. The syntax of the **Add** method is:

Controls.Add(Type, Id, Parameter, Before, Temporary)

Type is msoControlButton. Id and Parameter can be omitted; see VBA On-line Help for an explanation of the use of these arguments. If Before, which specifies the position in the menu, is omitted, the command is added at the end of the menu. Set Temporary to **True** to make the custom menu command temporary; temporary menu commands remain in the menu until you exit from Excel.

```
Private Sub Workbook_Open()
'This procedure installs a new menu command
'at the top (the default position) of the Tools menu.

Set NMC = Application.CommandBars("Tools").Controls.Add _
(Type:=msoControlButton, temporary:=True)
   NMC.Caption = "New Menu Command..."
   NMC.OnAction = "MyProcedure"
End Sub
```

Figure 18.3. An event-handler procedure to install a custom menu command.

Figure 18-3 illustrates a Workbook_Open procedure that installs a new menu command in the **Tools** menu and assigns the **Sub** procedure named MyProcedure to it. In the preceding code, replace New Menu Command... and MyProcedure with the text of your menu command and macro name. In this example, MyProcedure, which simply beeps ten times, is a **Sub** procedure located in a module sheet (Figure 18-4).

```
Sub MyProcedure()
For x = 1 To 10: Beep: Next
End Sub
```

Figure 18.4. The **Sub** procedure that runs when the custom menu command is chosen from the **Tools** menu.

```
Private Sub Workbook_Open()
'This procedure installs a new menu command
'immediately below the Goal Seek command in the Tools menu
'with a separator bar below it.

posn = Application.CommandBars("Tools").Controls _
("Goal Seek...").Index

Set NMC = Application.CommandBars("Tools").Controls.Add _
(Type:=msoControlButton, before:=posn + 1, temporary:=True)
   NMC.Caption = "New Menu Command..."
   NMC.OnAction = "MyProcedure"

Set NMC = Application.CommandBars("Tools").Controls.Add _
(Type:=msoControlButton, before:=posn + 2, temporary:=True)
   NMC.Caption = "-"

End Sub
```

Figure 18.5.   An event-handler procedure to install a custom menu command.

You may want to install a new menu command immediately above or below a menu command that is already present in the menu. Since menus can be customized, you shouldn't use a numerical value for the position; instead, determine the position of the command programmatically. Use the **Index** property as shown in Figure 18 5.

Use Caption = "-" (the hyphen) to insert a separator bar, with OnAction:="" or omitted.

From the preceding examples it should be clear that you can create automatic procedures that install a menu command when a particular workbook is opened and remove the menu command when that workbook is closed (using the **Workbook_BeforeClose** event procedure). Or you can install a command when a particular worksheet is activated and remove it when that worksheet is deactivated.

19

CREATING CUSTOM TOOLBUTTONS
AND TOOLBARS

In this chapter you'll learn how to customize Excel's built-in toolbars and how to create new toolbuttons to simplify some of the operations that you perform often.

CUSTOMIZING TOOLBARS

Some of the toolbuttons on Excel's toolbars are not very useful for scientists. You can remove the toolbuttons you don't use, giving a less cluttered workspace and providing room for other, more useful toolbuttons.

MOVING AND CHANGING THE SHAPE OF TOOLBARS

Excel's toolbars are usually located along the edges of the screen, most commonly along the top edge. Excel initially displays the Standard and Formatting toolbars, which provide tools for common Excel actions. There are a number of other toolbars provided with Excel, including the Chart, Drawing and Forms toolbars. You can also create custom toolbars, containing the tools you use most.

A toolbar can be moved from its position at the top of the screen and placed anywhere on the screen. Simply drag the *move handle* (located at the left side of a docked toolbar) to another location. A toolbar that is moved from a position along the edge of the screen is called a *floating toolbar*. It appears in its own window, with a title bar and a Close box (a title bar, Close box and a Size box in the Macintosh version of Excel). If you change the height or width of a floating toolbar, the tools are automatically rearranged to fit the new shape. If a floating toolbar (Figure 19-1) is dragged near the edge of the screen, Excel places it in a *toolbar dock.* There are four toolbar docks, one along each of the edges of the Excel application window.

If a toolbar does not fill the whole window (from left to right or vertically, depending on its orientation) the toolbar dock is visible as the extra blank space at either side of the toolbar, separated from the toolbar by the toolbar border.

Figure 19-1. The Excel 2000 Standard toolbar.

Excel Tip.   A toolbar that contains a drop-down list box, such as the Style box, cannot be placed in a vertical position in the left or right toolbar docks.

ACTIVATING OTHER TOOLBARS

You can display a toolbar by choosing **Toolbars** from the **View** menu, then checking the box for the desired toolbar in the submenu. You can also display a toolbar by choosing **Customize...** from the **Tools** menu, choosing the Toolbars tab (Figure 19-2) and checking the box for the desired toolbar.

Remove a toolbar by choosing **Toolbars** from the **View** menu and unchecking the box for the toolbar, or by dragging the toolbar from the toolbar dock and then pressing the Close box. You can have several toolbars in the Excel window at once.

*Excel Tip.   Right-click on any toolbar (Windows) or hold down the CONTROL key and click on any toolbar (Macintosh) to display the **Toolbar** shortcut menu.*

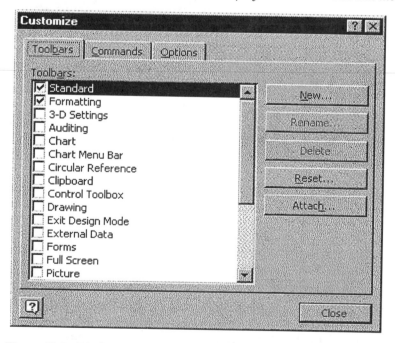

Figure 19-2. Displaying a built-in toolbar with the Toolbars dialog box.

ADDING OR REMOVING TOOLBUTTONS FROM TOOLBARS

Removing seldom-used toolbuttons from a toolbar provides a less cluttered working environment and makes room for other, more useful tools. To delete a toolbutton from a toolbar or to add a built-in toolbutton to a toolbar, follow the procedures in the following boxes.

To Delete a Toolbutton from a Toolbar

1. Choose **Toolbars** from the **View** menu and choose **Customize...**, or choose **Customize...** from the **Tools** menu, or right-click on a toolbar and choose **Customize...** from the shortcut menu.

2. Drag the toolbutton off the toolbar (when you click on it, the toolbutton outline is highlighted) and release the mouse button. The toolbutton will disappear.

To Add a Built-in Toolbutton to a Toolbar

1. Choose **Toolbars** from the **View** menu and choose **Customize...**, or choose **Customize...** from the **Tools** menu, or right-click on a toolbar and choose **Customize...** from the shortcut menu.

2. Choose the Commands tab.

3. Select the desired toolbutton from the various categories (Figure 19-3). Drag the toolbutton to the desired place on the toolbar. Spaces between toolbuttons may be added or removed by using the procedure in the following box.

The built-in Microsoft Excel toolbars can be restored to their "factory-installed" condition by using the Reset button in the Toolbars tab of the Toolbars dialog box (Figure 19-2). Individual toolbuttons can be restored by following the procedure described earlier. Custom toolbars cannot be reset.

Toolbuttons can be organized into logical groups by grouping them together using *separator bars*.

To Insert or Remove a Separator Bar Between Toolbuttons

1. Choose **Toolbars** from the **View** menu and choose **Customize...**, or choose **Customize...** from the **Tools** menu, or right-click on a toolbar and choose **Customize...** from the shortcut menu.

2. Right-click on the button to the left of which you want to insert or remove a separator bar, to display the shortcut menu.

3. Choose Begin a Group.

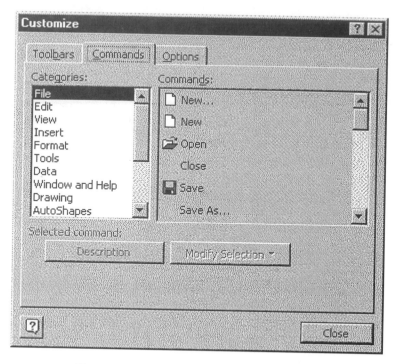

Figure 19-3. Using the Customize dialog box.

CREATING A NEW TOOLBAR

There are two ways to create a custom toolbar. One way is to modify an existing toolbar (such as the Standard toolbar). The other way is to create a new toolbar and then proceed to add built-in tools, as described earlier, or custom toolbuttons, as described later in this chapter. This way you can leave the Standard and Formatting toolbars (for example) unmodified and display your own custom toolbar. To create a new toolbar, use the procedure in the following box.

To Create a Custom Toolbar

1. Choose **Toolbars** from the **View** menu and choose **Customize...**, or choose **Customize...** from the **Tools** menu, or right-click on a toolbar and choose **Customize...** from the shortcut menu.

2. Choose the Toolbars tab.

3. Press the New... button. Enter a name for the custom toolbar (see Figure 19-4), then press OK. The (empty) custom toolbar will appear.

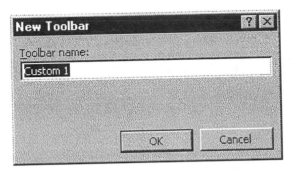

Figure 19-4. Naming a custom toolbar with the New Toolbars dialog box.

CREATING CUSTOM TOOLBUTTONS

Some of your macros will be written for a very specific purpose, such as to prepare a specialized report. The workbook containing the macro will be opened only when you want to assemble the report. Other macros automate tasks that you perform often, and you'll want to have them available whenever you're using Excel. These macros should be saved in the Personal Macro Workbook. To make a command macro even easier to use, you can add a custom toolbutton to a toolbar and assign the macro to it. The three macros described in this section — the NumberFormatConvert macro, the FullPage macro and the ChemicalFormat macro — are particularly convenient to use when they are assigned to a button.

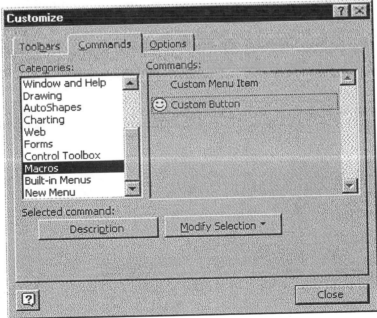

Figure 19-5. A custom toolbutton that can be placed on a toolbar.

The NumberFormatConvert macro toggles a number between floating-point and scientific formats. The FullPage macro sets margins to zero and deletes the header and footer, to provide maximum space on a page for printing the spreadsheet. The ChemicalFormat macro converts text in a cell to a chemical formula (e.g., H2SO4 becomes H_2SO_4).

The Macros category in the Commands tab (Figure 19-5) contains a custom toolbutton that can be assigned to a macro by following the procedure in the following box.

**To Add a Custom Toolbutton to a Toolbar
and Assign a Macro to It**

1. Choose **Toolbars** from the **View** menu and choose **Customize...**, or choose **Customize...** from the **Tools** menu, or right-click on a toolbar and choose **Customize...** from the shortcut menu.

3. Scroll through the Categories box and choose Macros.

4. Select the "happyface" toolbutton (you'll create a new image for it later) and drag it to the desired location on the toolbar.

5. Now right-click on the custom button to display the shortcut menu, and choose **Assign Macro....** In the Assign Macro dialog box (Figure 19-6), select the name of the macro to be assigned to the toolbutton. Click the OK button.

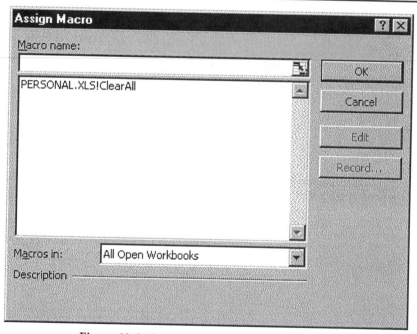

Figure 19-6. Assigning a macro to a toolbutton.

Excel Tip.   When you place a custom toolbutton on a toolbar, you don't have to assign a macro to it right away. Later, if you click a custom toolbutton that doesn't have a macro assigned to it, the Assign to Tool dialog box will appear.

THE NUMBERFORMATCONVERT MACRO

In scientific spreadsheet computing, it's common to convert numbers from floating-point format to scientific, and vice versa. Although Excel provides toolbuttons in the Formatting toolbar to format numbers with Currency, Percent or Comma format, there isn't a toolbutton for Scientific format. The macro listing in Figure 19-7 toggles a number between Floating Point or General format and Scientific format.

THE MACRO

After getting the number format of the active cell (or top left cell of a range selection), the macro examines each character of the format. If characters corresponding to percent, currency, date or time formats are found, nothing is done. If an "E" is found, the range is formatted "General". If none of these characters is found, the number format is changed to the default Scientific format, 0.00E+00.

```
Sub NumberFormatConvert()
'   Toggles between floating point and scientific number formats
'     Begun 3/21/97.  Last modified 6/30/00

'Check for a document
If TypeName(ActiveSheet) <> "Worksheet" Then Beep: Exit Sub
'If not worksheet, exit.
If TypeName(Selection) <> "Range" Then Exit Sub

'Get number format of cell
fmt = ActiveCell.NumberFormat

'Examine each character of number format
For x = 1 To Len(fmt)
c = Mid(fmt, x, 1)
  If c = "%" Or c = "$" Then Exit  Sub
  If c = "m" Or c = "d" Or c = "y" Or c = "h" Or c = "s" _
Then Exit Sub
  If c = "E" Then Selection.NumberFormat = "General": Exit  Sub
Next

'All others get scientific number format
Selection.NumberFormat = "0.00E+00"
End Sub
```

Figure 19-7. Simple number-formatting macro to assign to a toolbutton.

Writing a macro that performs Floating Point/Scientific number formatting and returns a number with the same number of significant figures as in the original number is much more difficult. You may wish to try to write one.

THE FULLPAGE MACRO

The FullPage macro maximizes the space on a page that is available for printing a worksheet, by eliminating margins, header and footer. To do this by using menu commands requires choosing **Page Setup** from the **File** menu, setting Left Margin, Right Margin, Top Margin and Bottom Margin to zero, choosing Header and deleting the header text, then choosing Footer and deleting the footer text. The FullPage macro was written to do this at the click of a toolbutton; the listing is shown in Figure 19-8.

THE MACRO

Two custom toolbuttons, a FullPage(Portrait) toolbutton and a FullPage(Landscape) toolbutton, were created and positioned on the left side of the Standard toolbar. The FullPage macro was assigned to both buttons, as described earlier.

When either button is pressed, the macro uses **Application.Caller** to determine which button was pressed. **Application.Caller** returns a two-element array, the button position and the toolbar name. These are used to obtain the button caption text.

```
Sub FullPage ()

'Get info about which button pressed
WhichButton = Application.Caller
ButtonPosition = WhichButton(1)
BarName = WhichButton(2)
ButtonCaption = CommandBars(BarName).
Controls(ButtonPosition).Caption

'Use Excel4Macro command since it executes MUCH faster than the VBA
equivalent
If ButtonCaption = "Full Page (Portrait)" Then
ExecuteExcel4Macro   ("PAGE.SETUP("""","""",0,0,0,0,,,1,1,1)")
'Double quotes needed
Else
ExecuteExcel4Macro   ("PAGE.SETUP("""","""",0,0,0,0,,,1,1,2)")
End If

End Sub
```

Figure 19-8. FullPage Landscape and Portrait macros to assign to toolbar buttons.

This macro uses the **ExecuteExcel4Macro** method. The Excel 4 Macro PAGE.SETUP executes much faster than its VBA equivalent.

The button caption text is used to set the orientation argument (the eleventh argument) in the Excel 4 Macro to 1 (for Portrait) or 2 (for Landscape). The first six arguments set the header and footer to null, and the margins to zero.

CREATING A CUSTOM TOOLBUTTON IMAGE

Use the Button Editor to edit an existing toolbutton image or create a new one. First, choose **Toolbars** from the **View** menu and choose **Customize...** from the shortcut menu. Right-click on the toolbutton image you wish to edit to display the **Toolbars** shortcut menu. Choose **Edit Button Image...** from the shortcut menu to display the Button Editor (Figure 19-9).

The button image will be displayed in the Picture area, 16 pixels wide by 15 pixels high. You add or remove pixels from the image. Click on any pixel in the Picture area to add a pixel of a selected color. Click a second time if you want to remove that color. Make the background color "Erase". This will provide a gray background identical to the rest of the button.

To edit large toolbuttons, choose **Toolbars** from the **View** menu and check the Large Buttons check box. Then display the button image to be edited. The pixel area displayed is 24 pixels wide by 23 high.

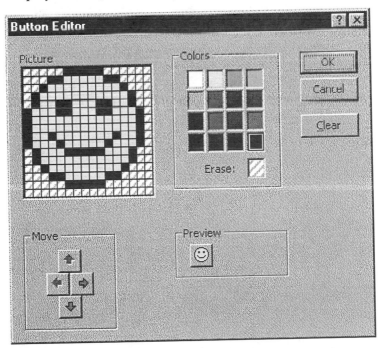

Figure 19-9. The Button Editor dialog box.

The Move buttons allow you to shift the image up, down, left or right, but only if the background is "Erase". If you want to create a new picture, press the Clear button before you begin. You will start with a complete "Erase" background. The Preview window shows the appearance of the toolbutton image.

Custom toolbuttons were created for the Full Page Portrait ▦, Full Page Landscape ▦, Chemical Format H₂0 and Toggle Between Floating Point and Scientific ⋮ macros. The Chemical Format macro is described in Chapter 16. Module sheets containing these macros should be saved in the Personal Macro Workbook.

HOW TO ADD A TOOLTIP TO A CUSTOM BUTTON

If you create a custom toolbutton by dragging the Custom button onto a toolbar, the ToolTip message is simply "Custom". You can add your own ToolTip text. Figure 19-10 shows a custom ToolTip message displayed with the Full Page Landscape toolbutton. To add or change ToolTip text, use the procedure in the following box

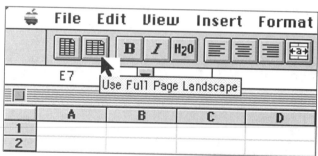

Figure 19-10. A ToolTip for a custom toolbutton.

To Add a ToolTip to a Custom Toolbutton

1. Choose **Toolbars** from the **View** menu and choose **Customize...**, or choose **Customize...** from the **Tools** menu, or right-click on a toolbar and choose **Customize...** from the shortcut menu.
2. Right-click on the custom button to display the shortcut menu.
3. In the Name box, type the ToolTip text.

CREATING TOOLBUTTONS OR TOOLBARS BY MEANS OF A MACRO

You can use VBA to create and manipulate toolbuttons or toolbars. Refer to Excel's On-line Help for more information.

PART V

SOME APPLICATIONS

20

ANALYSIS OF
SOLUTION EQUILIBRIA

In this chapter you'll learn how to calculate species distributions of polyprotic weak acid species, how to apply Gran's method for the estimation of end-points in titrations and a general method for the calculation of titration curves.

SPECIES DISTRIBUTION DIAGRAMS

To illustrate the variation in composition of an aqueous solution of a polyprotic weak acid species, it is useful to plot a species distribution curve such as the one for citric acid shown in Figure 20-1 . The parameter plotted versus pH for each species is α, the fraction of the total citric acid concentration in the form of a particular species.

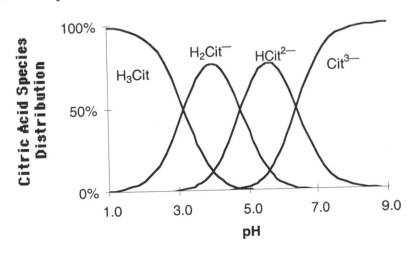

Figure 20-1. Species distribution diagram for citric acid.

For a weak acid H_KA, the fraction of the acid in the form containing j protons is:

$$\alpha_j = \frac{[H_jA]}{[H_KA]_T} \tag{20-1}$$

For example, the expression for α_0 for the Cit^{3-} species is:

$$\alpha_0 = \frac{[Cit^{3-}]}{[H_3Cit]_T} \tag{20-2}$$

The equations that follow use protonation constants rather than dissociation constants. The stepwise protonation constant K^H of a weak base species B is the equilibrium constant for the following reaction (charges are omitted for clarity):

$$B + H^+ \rightleftharpoons HB$$

$$K_1^H = \frac{[HB]}{[B][H^+]}$$

and in general:

$$K_j^H = \frac{[H_jB]}{[H_{j-1}B][H^+]}$$

KH values are typically much greater than 1 and are often reported as $\log KH$ values. K_a values are typically very small, and are often reported as pK_a values ($-\log K_a$). Since KH is the reciprocal of K_a, it follows that numerically, $\log K^H$ is identical to pK_a for a given conjugate acid/conjugate base equilibrium.

It is also convenient to use the cumulative protonation constant β_j for the overall reaction

$$B + jH^+ \rightleftharpoons H_jB$$

$$\beta_j = \frac{[H_jB]}{[B][H^+]^j}$$

The relationship between K's and β_j is:

$$\beta_j = K_1^H K_2^H \cdots K_j^H$$

The expression for the α_j value for a particular protonated species is easily calculated from the expression

$$\alpha_j = \frac{\beta_j [H^+]^j}{\sum\limits_{j=0}^{K} \beta_j [H^+]^j} \tag{20-3}$$

where $\beta_0 = 1$. Thus the expression for α_2 for citric acid is:

$$\alpha_2 = \frac{K_1^H K_2^H [H^+]^2}{1 + K_1^H[H^+] + K_1^H K_2^H[H^+]^2 + K_1^H K_2^H K_3^H[H^+]^3} \qquad (20\text{-}4)$$

The custom function ALPHA provides a convenient way to calculate alpha values for polyprotic species, and thus to construct distribution diagrams. The macro is shown in Figure 20-2. The syntax of the function is ALPHA(*j*, *pH*, *pKs_or_logKs*, *pKa_logical*). Either pK_a values or $\log K^H$ values can be used. If pK_a values are used, pKa_logical is TRUE; if $\log K^H$ values are used, pKa_logical is FALSE or omitted.

```
'ALPHA Function
'Written by        E. J. Billo
'Begun             1/25/96
'Last Modified     2/7/96


Function ALPHA(j, pH, pKs_or_logKs, pKa_logical)
'pKa_logical:      TRUE if table of pKs_or_logKs is pKa values
'                  FALSE if table of pKs_or_logKs is logK values
N = pKs_or_logKs.Count
If (j < 0) Or (j > N) Or (pH = "") Or (N = 0) Then ALPHA = "n/a":
GoTo 1000
logbeta = 0
denom = 1
numerator = 1
If (pKa_logical = False) Then          'Calculation using protonation
constants
    For x = 1 To N
    logbeta = logbeta + pKs_or_logKs(x)
    denom = denom + 10 ^ (logbeta - pH * x)
    If (x = j) Then numerator = 10 ^ (logbeta - pH * x)
    Next
Else          'Calculation using dissociation constants
    For x = N To 1 Step -1
    logbeta = logbeta + pKs_or_logKs(x)
    denom = denom + 10 ^ (logbeta - pH * (N - x + 1))
    If (N - x + 1 = j) Then numerator = 10 ^ (logbeta - pH * (N - x + 1))
    Next
End If
ALPHA = numerator / denom
1000  End Function
```

Figure 20-2. The ALPHA **custom function**

ANALYSIS OF TITRATION DATA

The location of the end-point of a titration by using either the first or second derivative of the titration data was discussed in Chapter 9. These methods use only the data points near the end-point. Another approach, Gran's method, makes use of the complete data set. It is useful when either (i) the inflection at the end-point is poorly defined or (ii) data at the end-point is missing.

Consider an acid–base titration. At any point before the end-point, the concentration of unreacted H^+ is given by 10^{-pH}. Thus, to estimate the volume required to reach the end-point, it is merely necessary to plot 10^{-pH} versus titrant volume V and extrapolate to $10^{-pH} = 0$. If dilution by the titrant is important, then the function 10^{-pH} should be multiplied by $(V_0 + V)/V_0$, where V_0 is the initial volume.

A similar approach can be used with other electrodes. In the following example, the titration of a chloride sample with standard silver nitrate, the potential of a silver electrode in combination with a saturated calomel reference electrode was used to follow the course of the titration. The potential of the electrode pair is a direct measure of the free chloride ion concentration: as the chloride ion concentration decreases, the potential increases. The titration results are shown in Figure 20-3.

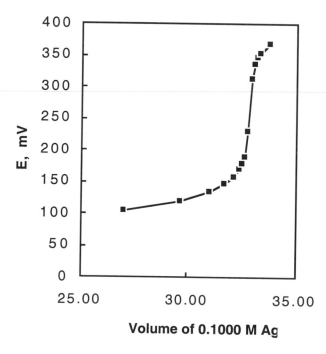

Figure 20-3. Potentiometric titration curve of chloride titrated with silver ion.

The chloride ion concentration at any point on the titration curve can be calculated from equation 20-5, which can be derived from the Nernst equation.

$$[Cl^-] = antilog \left[\frac{-(E - offset)}{slope} \right] \qquad (20\text{-}5)$$

Offset is used to scale the E values into a more convenient range and can be any potential value; *slope* is the Nernst slope, theoretically 59.2 mV at 25°C.

Figure 20-4 shows the titration data V and E; column C contains the formula =10^(-(E-offset)/slope). Figure 20-5 shows the linear relationship between the Gran function (column C in Figure 20-4) and the volume.

The end-point is considered to be the volume where the straight-line portion of the Gran plot crosses the x axis. You can make an expanded chart of the titration data near the end-point and estimate the end-point reading visually. From Figure 20-6 the end-point can be estimated to be approximately 32.82 mL.

	A	B	C
1	**Gran's Method** Used to Determine the		
2	Endpoint of a Titration of Chloride with		
3	0.1000 M silver nitrate.		
4	Unreacted chloride calculated from:		
5	=10^-((E-offset)/slope))		
6	(A rearranged form of the Nernst equation.)		
7			
8	offset =	-20	
9	slope =	59	
10			
11	V	E	[Cl-], mM
12	27.00	104	7.91
13	29.70	120	4.24
14	31.1	135	2.36
15	31.73	147	1.48
16	32.13	159	0.92
17	32.40	171	0.58
18	32.54	180	0.41
19	32.67	191	0.27
20	32.81	230	0.06
21	32.94	313	0.00
22	33.08	337	0.00
23	33.21	348	0.00
24	33.35	355	0.00
25	33.75	370	0.00

Figure 20-4. Gran's method calculations.

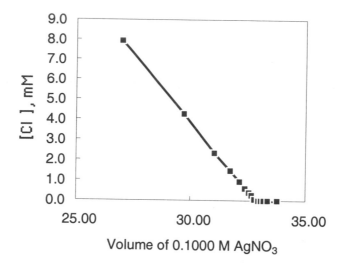

Figure 20-5. Gran plot.

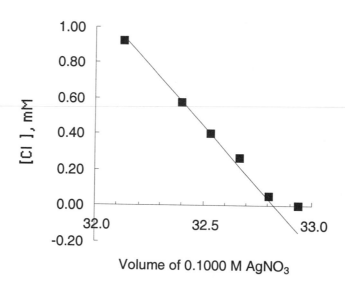

Figure 20-6. Using a Gran plot for graphical estimation of end-point.

Alternatively the end-point can be obtained algebraically. **LINEST** was used to obtain the slope and intercept of the straight-line portion of the data, shown in Figure 20-7 (the last five rows of data points were not included). The intercept is 44.44 and the slope is –1.35.

	E	F
17	*From LINEST:*	
18	-1.3534	44.4350
19	0.01	0.18
20	0.99987	0.03

Figure 20-7. Slope and intercept of Gran plot.

To obtain the end-point volume, the value of V where $[Cl^-] = 0$, you need to calculate the *x intercept*. The x intercept = –intercept/slope = 32.83 mL, the end-point volume for the titration.

A second example illustrates how to obtain an end-point that was missed. To determine the concentration of a weak base compound, a known excess of HCl was added to a solution of the compound, and a micrometer syringe buret. was used to back-titrate the excess acid with standard base. Unfortunately the student doing the titration did not take small increments near the inflection point (which was not very pronounced, in any case), and the result shown in Figure 20-8 was obtained.

Nonetheless, the end-point can be estimated by using Gran's method. From the pH measurements, the concentration of free $[H^+]$ was calculated using the relationship $[H^+] = 10^{-pH}$, as illustrated in Figure 20-9. The data, when plotted (Figure 20-10), gave an excellent straight line with R^2 (omitting the last four data points) of 0.99993. The end-point volume was calculated to be 0.722 mL.

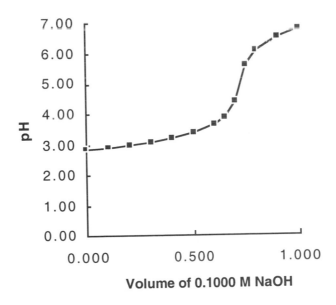

Figure 20-8. Acid-base titration with "missed" end-point.

	A	B	C
3	V	pH	[H+]
4	0.000	2.85	0.0014
5	0.100	2.91	0.0012
6	0.200	2.99	0.0010
7	0.300	3.08	0.0008
8	0.400	3.20	0.0006
9	0.500	3.36	0.0004
10	0.600	3.62	0.0002
11	0.650	3.86	0.0001
12	0.700	4.36	0.0000
13	0.750	5.60	0.0000
14	0.800	6.08	0.0000
15	0.900	6.50	0.0000
16	1.000	6.75	0.0000

Figure 20-9. Gran's method calculations.

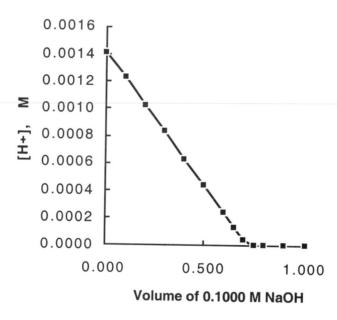

Figure 20-10. Gran plot to obtain "missed" end-point.

SIMULATION OF TITRATION CURVES USING A SINGLE MASTER EQUATION

In calculating titration curves, separate equations for different regions of the curve ("before the equivalence point", "at the equivalence point", "after the equivalence point", etc.) are often employed. This section illustrates how to use a single "master" equation to calculate points on a titration curve. Instead of calculating pH as a function of the independent variable V, it is convenient to use pH as the independent variable and V as the dependent variable. The species distribution at a particular pH value is calculated from the $[H^+]$, and the volume of titrant required to produce that amount of each species is calculated. For example, in the titration of a weak monoprotic acid HA, we can calculate the concentration of A^- at a particular pH and then calculate the number of moles of base required to produce that amount of A^-. In general $(J - j)$ moles of base per mole of acid are required to produce the species H_jA from the original acid species H_JA.

We start with the relationship:

(initial moles of available protons) – (moles of titrant base added) =

(moles of bound protons) + (moles of $[H^+]$)   (20-6)

Then in the general case, for C_H moles of strong acid + C_{HA} moles of weak acid H_JA in an initial volume V_0, titrated with V milliliters of standard base of concentration C_{OH}, the relationship is:

$$C_H + J\,C_{HA} + [OH^-]\,(V_0 + V) - C_{OH}V - j\Sigma[H_jA] + [H^+]\,(V_0 + V)\quad(20\text{-}7)$$

From equation 20-7, after rearrangement, we obtain equation 20-8. (The $[OH^-]$ is an indicator of one of the sources of protons, namely H_2O.) The $[H_jA]$ values are calculated employing the usual α factors, from K's and $[H^+]$.

$$V_{calc} = \frac{C_H + J\,C_{HA} + [OH^-]\,(V_0 + V) - [H^+](V_0 + V) - j\Sigma[H_jA]}{C_{OH}}\quad(20\text{-}8)$$

Equation 20-8 permits the calculation of all points on a titration curve by means of a single equation. As written, it handles strong acids, weak acids or mixtures, and it is readily expanded to handle mixtures of polyprotic acids.

Figure 20-11 illustrates a portion of a spreadsheet for the calculation of the titration curve of 2.500 mmol of a weak acid ($pK_a = 5$) with 0.1000 M strong base. The volume required to obtain a given pH value was calculated for pH values from 3 to 12 in increments of 0.20. The formula used to calculate V in cell C9 is

=(CHA+10^(pKw-A9)*(V_0)-10^(-A9)*(V_0)-B9*CHA)/COH.

The terms in the formula are in the same order as those in equation 20-8.

The titration curve is shown in Figure 20-12.

	A	B	C	D
1	**Weak Acid–Strong Base Titration Curve**			
2	weak acid, mmol		2.500	(CHA)
3	titrant base, M		0.1000	(COH)
4	total volume, mL		50.0	(V_0)
5	pKa		5.00	(logK)
6	pcKw		14.00	(pcKw)
7				
8	**pH**	**alpha1**	**V(calc)**	
9	3.00	9.9E–01	0	
10	3.50	9.7E–01	0.61	
11	4.00	9.1E–01	2.22	
12	4.50	7.6E–01	5.99	
13	5.00	5.0E–01	12.50	
14	5.50	2.4E–01	18.99	
15	6.00	9.1E–02	22.73	
16	6.50	3.1E–02	24.23	
17	7.00	9.9E–03	24.75	
18	8.00	1.0E–03	24.98	
19	9.00	1.0E–04	25.00	
20	10.00	1.0E–05	25.05	
21	10.50	3.2E–06	25.16	
22	11.00	1.0E–06	25.50	
23	11.50	3.2E–07	26.58	
24	12.00	1.0E–07	30.00	

Figure 20-11. Spreadsheet for weak acid-strong base titration curve.

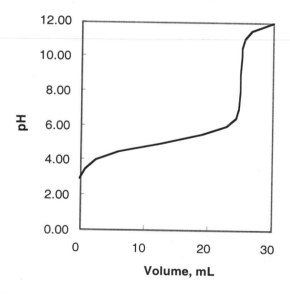

Figure 20-12. Weak acid–strong base titration curve.

21

ANALYSIS OF
SPECTROPHOTOMETRIC DATA

In this chapter you'll learn how to handle calibration curves that are not straight lines, how to analyze the spectra of mixtures of components and how to deconvolute a spectrum into its individual absorption bands.

CALIBRATION CURVES FOR SPECTROPHOTOMETRY

Linear calibration lines can be handled quite easily by using LINEST. However, when a calibration curve is not linear, the problem is a little more difficult. The calibration curve in Figure 21-1 shows readings on a series of sodium standards, made using a CIBA-Corning Model 410 flame photometer. The calibration line is noticeably curved.

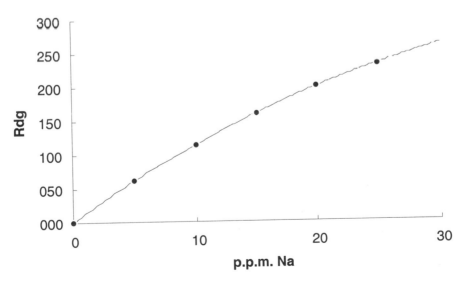

Figure 21-1. Flame photometry calibration curve.

339

	D	E	F	G
4		Power Series in X		
5	ppm Na	ppm^2	ppm^3	Rdg
6	0	000	000	000
7	5	025	125	062
8	10	100	1000	115
9	15	225	3375	160
10	20	400	8000	200
11	25	625	15625	233
12				
13		From LINEST:		
14	d	c	b	a
15	0.00141	-0.193	13.280	0.079
16	0.00050	0.019	0.193	0.498
17	0.999986	0.50787	#N/A	#N/A

Figure 21-2. Fitting a calibration curve to a cubic equation.

One way to handle a curved calibration line is to fit the line to a power series. A cubic equation ($y = a + bx + cx^2 + dx^3$) is usually sufficient to fit a case such as Figure 21-1. (In any event, since there are only six known points, you couldn't use a polynomial with more than five adjustable parameters.) You can use either LINEST or the Solver to obtain the coefficients of the power series. Figure 21-2 shows a spreadsheet in which LINEST is used to find the regression coefficients for the equation $Rdg = a + b \times ppm + c \times (ppm)^2 + d \times (ppm)^3$.

In this case either

$$Rdg = a + b \times ppm + c \times (ppm)^2 + d \times (ppm)^3$$

with $a = 0.079$, $b = 13.28$, $c = -0.193$ and $d = 0.00141$ or

$$Rdg = b \times ppm + c \times (ppm)^2 + d \times (ppm)^3$$

with $b = 13.30$, $c = -0.195$ and $d = 0.00144$ give excellent fits to the data points.

To obtain concentration information (x) from a flame photometer reading (y), it is necessary to find the value of x that gives the observed value of y. The **Goal Seek** command in **Tools** menu performs this task very conveniently (see "Solving a Problem Using Goal Seek..." in Chapter 10).

An even simpler way to obtain concentration values from flame photometer readings is use LINEST or the Solver to fit the data to a power series, but using the concentration values as the y (dependent) variables and the readings as the independent variables. In this way you will obtain a polynomial such as

$$ppm = a + b \times Rdg + c \times (Rdg)^2 + d \times (Rdg)^3$$

	M	N	O	P
		Power Series in Y		
3				
4	Rdg	Rdg^2	Rdg^3	ppm Na
5	000	000	000	0
6	062	3844	238328	5
7	115	13225	1520875	10
8	160	25600	4096000	15
9	200	40000	8000000	20
10	233	54289	12649337	25
11				
12		From LINEST:		
13	d	c	b	a
14	3.75653E-07	4.09564E-05	0.077250495	-0.00811603
15	1.02243E-07	3.62575E-05	0.003375014	0.080465303
16	0.999969982	0.081032976	#N/A	#N/A

Figure 21-3. Alternate approach.

from which the concentration can be calculated directly. Figure 21-3 illustrates this approach.

Thus the equation

$$\text{ppm} = -0.00812 + 0.07725 \times \text{Rdg} + (4.096 \times 10^{-5}) \times \text{Rdg}^2 + (3.7565 \times 10^{-7}) \times \text{Rdg}^3$$

yields the concentration directly.

ANALYSIS OF SPECTRA OF MIXTURES

A common analytical problem in spectrophotometry is the analysis of a mixture of components. If the spectra of the pure components are available, the spectrum of a mixture can be analyzed to determine the concentrations of the individual components. If the mixture contains N components, then absorbance measurements at N suitable wavelengths are necessary to solve the set of N linear equations in N unknowns.

APPLYING CRAMER'S RULE TO A SPECTROPHOTOMETRIC PROBLEM

As a simple example of the analysis of mixtures, consider an aqueous solution containing a mixture of Co^{2+}, Ni^{2+} and Cu^{2+}, to be analyzed by spectrophotometric measurements at three different wavelengths. The spectra of the individual ions and of a mixture are shown in Figures 21-4 and 21-5. The most suitable wavelengths for analysis are 394, 510 and 808 nm (determined from an examination of Figure 21-4 and the data table). The molar absorptivities of the three species at these wavelengths are shown in Figure 21-6, together with absorbance readings for a mixture of the three ions, measured in a 1.00-cm cell.

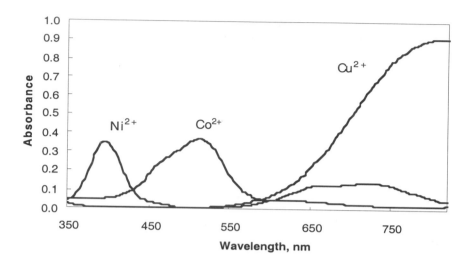

Figure 21-4. Spectra of Co^{2+}, Ni^{2+} and Cu^{2+} ions in aqueous solution (standards). (Spectrophotometric data provided by Dr. Lev Zompa.)

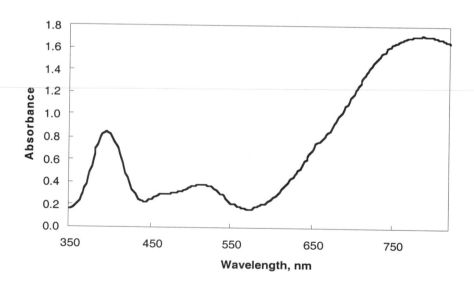

Figure 21-5. Spectrum of a mixture of Co^{2+}, Ni^{2+} and Cu^{2+} ions in aqueous solution.

	N	O	P	Q	R
2		STANDARDS			UNKNOWN
3		Molar Absorptivity, $M^{-1} cm^{-1}$			MIXTURE
4	λ/nm	Co^{2+}	Ni^{2+}	Cu^{2+}	(absorbance)
5	394	0.995	6.868	0.188	0.845
6	510	6.450	0.215	0.198	0.388
7	808	0.469	1.179	15.052	1.696

Figure 21-6. Data table for the determination of a mixture of Co^{2+}, Ni^{2+} and Cu^{2+} ions .

	O	P	Q
15	0.845	6.868	0.188
16	0.388	0.215	0.198
17	1.696	1.179	15.052

Figure 21-7. The determinant for calculating Co^{2+}.

Following the Cramer's rule procedures described in Chapter 9, we construct the determinant to determine Co^{2+} concentration shown in Figure 21-7.

Then the formula

=MDETERM(O15:Q17)/MDETERM(O5:Q7)

yields the value 0.05328 M for the Co^{2+} concentration. From similar formulas, $[Ni^{2+}] = 0.1125$ M and $[Cu^{2+}] = 0.1022$ M.

SOLUTION USING MATRIX INVERSION

A set of simultaneous linear equations can also be solved by using matrices, as shown in Chapter 9. The *solution matrix* is obtained by multiplying the matrix of constants by the inverse of the matrix of coefficients. Applying this simple solution to the spectrophotometric data used above, the inverted matrix is obtained by selecting a 3R × 3C array of cells, entering the array formula

=MINVERSE(O5:Q7)

The inverted matrix is shown in Figure 21-8.

The solution matrix is obtained by selecting a 3R × 1C array, then entering the array formula

=MMULT(O38:Q40,R5:R7)

The single array formula

=MMULT(MINVERSE(O5:Q7),R5:R7)

accomplishes the same result. The solution matrix is shown in Figure 21-9.

	O	P	Q
38	-0.0045315	0.15587674	-0.0019965
39	0.14657147	-0.0224973	-0.001539
40	-0.011342	-0.0030951	0.06661918

Figure 21-8. The inverted matrix.

	O	P
46	0.05328	(Co2+)
47	0.1125	(Ni2+)
48	0.1022	(Cu2+)

Figure 21-9. The solution matrix.

DECONVOLUTION OF SPECTRA

The resolution of a complex absorption spectrum into individual absorption bands may be necessary if information about the position, height or width of individual bands is required. There are a number of computer programs designed for the deconvolution of spectra, but you can do a reasonable job with Excel.

The procedures described next were developed for the deconvolution of electronic absorption spectra (UV–visible spectra) but are equally applicable to the deconvolution of infrared, Raman or NMR spectra. UV–visible spectra differ from vibrational spectra in that the number of bands is much smaller and the bandwidths are much wider. Band shape may also be different. UV–visible spectra are also usually recorded under conditions of high resolution and high signal-to-noise ratio. Spectra from older instruments usually require manual digitization from a spectrum on chart paper, at e.g., 10 nm intervals. With the widespread use of computer-controlled instruments, it is a simple matter to obtain a file of spectral data at, e.g., 1 nm intervals. In fact, it may be necessary to reduce the size of the data set to speed up calculations.

MATHEMATICAL FUNCTIONS FOR SPECTRAL BANDS[*]

A symmetrical spectral band is described by three parameters: position (wavelength or frequency corresponding to the absorption maximum), intensity (absorbance or molar absorptivity at the band maximum) and width (usually the bandwidth at half-height). The band shape functions most commonly used for deconvolution are the Gaussian function and the Lorentzian function. Both are symmetrical functions. UV–visible spectra generally have a Gaussian band shape. The Lorentzian function is useful for the simulation of NMR spectra. The

---

[*] P. Pelikán, M. Čeppan and M. Liška, *Applications of Numerical Methods in Molecular Spectroscopy*, CRC Press, Boca Raton FL, 1993.

log-normal band function has been applied to unsymmetrical spectral band shapes.

Many spectral bands can be closely approximated by a Gaussian line shape when the independent variable v is in energy units, e.g., cm^{-1}. The absorbance A at a wavenumber v is given by equation 21-1, where A_{max} is the band maximum, v_{max} is the wavenumber of the band maximum and Δv is the half-width.

$$A = A_{max} \exp\left[-(4 \ln 2)\frac{(v - v_{max})^2}{\Delta v^2}\right] \qquad (21\text{-}1)$$

The corresponding equation for a Lorentzian line shape is given by equation 21-2.

$$A = \frac{A_{max}}{1 + 4\frac{(v - v_{max})^2}{\Delta v^2}} \qquad (21\text{-}2)$$

For unsymmetrical bands, the equation for the log-normal line shape is

$$A = A_{max} \exp\left\{-\frac{\ln 2}{(\ln \rho)^2}\left[\ln\left(\frac{(v - v_{max})^2}{\Delta v^2}\frac{\rho^2 - 1}{\rho} + 1\right)\right]^2\right\} \qquad (21\text{-}3)$$

for the region $v \geq v_{max} - (\Delta x \rho / (\rho^2 - 1))$ and $A = 0$ elsewhere. The asymmetry parameter ρ is given by:

$$\rho = \frac{v_R - v_{max}}{v_{max} - v_L} \qquad (21\text{-}4)$$

where v_R and v_L are the right and left portions of the half-width.

A simpler form of the Gaussian band shape, where σ is simply treated as an adjustable parameter, is given in equation 21-5. This is the equation that will be used in the following treatment. By using an embedded chart to compare calculated and experimental data, you can fairly easily find a set of A_{max}, v_{max} and σ values that approximate the band shape, to use as initial guesses for the deconvolution procedure outlined in the box on the following page.

$$A = A_{max} \exp\left[-\frac{(v - v_{max})^2}{\sigma^2}\right] \qquad (21\text{-}5)$$

DECONVOLUTION OF A SPECTRUM: AN EXAMPLE

The spreadsheet shown in Figure 21-10 illustrates the deconvolution of the UV–visible spectrum of a mixed-ligand complex of nickel(II). Four bands are apparent in the spectrum, one a weak shoulder lying between relatively intense bands at approximately 350 and 550 nm. The fourth band appears only as the tail of a fairly intense band lying at longer wavelengths.

The formulas in cells C10 (converting wavelength λ to wavenumber ν) and D10 (calculating the Gaussian band profile of band 1) are:

=10000/A10

=band1 A_0*EXP(-(((C10-band1 max)/band1 s)^2)/2)

The Solver was used to vary the values in cells D4:F6 and G4:G5 to make cell I7 a minimum. Because the data did not permit a complete resolution of band 4, cell G6, the bandwidth parameter for band 4, was held constant at the reasonable value of 1.5. The results are shown on the spreadsheet. The resolved spectrum (solid line), with the four bands (broken lines), is shown in Figure 21-11.

The λ_{max} values for bands 2 and 4, from other experimental measurements, are 445 and 880 nm, respectively.

Deconvolution of a Spectrum

1. Start with a table of wavelength, absorbance data pairs.
2. Create a column of wavenumbers.
3. Determine the number of bands necessary to describe the spectrum. This can usually be arrived at by inspection: a strongly asymmetric band generally indicates one or more hidden bands; a band with a flat maximum indicates two strongly overlapped bands, etc. Alternatively, you can use the first derivative of the spectrum ($\Delta A/\Delta x$). Except for the most hidden shoulders, each $\Delta A/\Delta x = 0$ value indicates a band maximum.
4. Estimate the half-width of the bands by using one or more bands not overlapped by other bands. As first approximation, use this value for all bands in the spectrum
4. Set up a table of ν_{max}, A_{max} and σ for each band.
5. Calculate the band profile for each contributor.
6. Sum the individual band contributions.
7. Calculate the sum of squares of the residuals $(A_{obsd} - A_{calc})^2$.
8. Create an embedded chart, plotting A_{obsd} and A_{calc}.
9. Perform some manual adjustment of the parameters, attempting to make the calculated spectrum coincide with the observed. This is especially important if the spectrum is complicated (more than three or four bands, especially if they are overlapped strongly).
10. Use the Solver to minimize sum of squares of residuals by varying (ultimately) the $3N$ parameters for the N bands in the spectrum.

TACKLING A COMPLICATED SPECTRUM

For a complicated spectrum, it may be helpful to operate on a reduced-size data set. Many spectrometers record absorbance readings at 1-nm intervals; a complete UV–visible spectrum (200–700 nm) contains 500 data points. If the spectrum contains eight bands, you're performing calculations on more than 4000 cells. Start with a data set consisting of every 10th data point, for example. After you have achieved a reasonably good fit to this data set, use these values as initial parameters for the complete data set.

It may be necessary to first minimize portions of the spectrum separately.

	A	B	C	D	E	F	G	H	I
1				Ni(2,3,2-tet)(en)2+ spectrum					
2				(b = 5 cm, c = 0.0252 M, pH = 10.6)					
3				band1	band2	band3	band4		
4	(wavenumber)	n_{max}	29.25	22.72	18.56	11.69			
5	(absorbance)	A_{max}	1.12	0.15	0.87	0.77			
6	(bandwidth)	s	1.60	1.54	1.38	1.5		$\Sigma(\partial^2)$
7	(wavelength)	l_{max}	342	440	539	855		0.0140
8									
9	l,nm	Abs	n,cm^{-1}	band1	band2	band3	band4	Σ	d^2
10	300	0.173	33.33	0.043	0.000	0.000	0.000	0.043	
11	310	0.274	32.26	0.191	0.000	0.000	0.000	0.191	0.007
12	320	0.514	31.25	0.512	0.000	0.000	0.000	0.512	0.000
13	325	0.694	30.77	0.714	0.000	0.000	0.000	0.714	0.000
14	330	0.871	30.30	0.903	0.000	0.000	0.000	0.903	0.001
15	335	1.026	29.85	1.046	0.000	0.000	0.000	1.046	0.000
16	340	1.126	29.41	1.118	0.000	0.000	0.000	1.118	0.000
17	345	1.141	28.99	1.109	0.000	0.000	0.000	1.109	0.001
18	350	1.036	28.57	1.028	0.000	0.000	0.000	1.028	0.000
19	355	0.908	28.17	0.896	0.000	0.000	0.000	0.896	0.000
20	360	0.737	27.78	0.737	0.001	0.000	0.000	0.738	0.000
21	370	0.406	27.03	0.429	0.003	0.000	0.000	0.432	0.001
22	380	0.195	26.32	0.210	0.010	0.000	0.000	0.220	0.001
23	390	0.111	25.64	0.089	0.025	0.000	0.000	0.114	0.000

Figure 21-10. Deconvolution of the UV–visible spectrum.

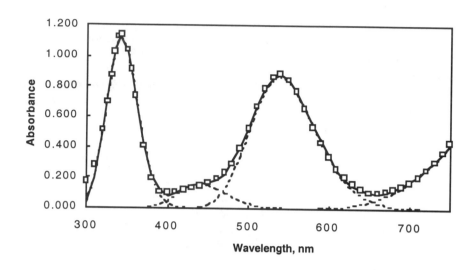

Figure 21-11. Deconvoluted spectrum.

22

CALCULATION
OF BINDING CONSTANTS

The measurement of binding constants (also called stability constants or formation constants) is of interest in many areas of chemistry. Quantitative information concerning the products of acid–base, metal–ligand or enzyme–substrate interactions is invaluable in analytical chemistry, industrial process chemistry, biochemistry, etc.

A wide range of experimental methods has been applied to the determination of binding constants. Three methods of most common use — potentiometry, spectrophotometry and NMR — are described here. The wide range of experimental methods has produced a wide range of types of experimental data and methods of calculation. Many "canned" computer programs are available, but for the occasional user, or for greater flexibility in tackling a non-standard situation, Microsoft Excel provides an ideal tool for the calculation of binding constants.

The binding constants (illustrated here as metal–ligand formation constants) may be either *stepwise formation constants* K_n (equation 22-1) or *cumulative formation constants* β_n (equation 22-2). Charges have been omitted for clarity.

$$ML_{n-1} + L \rightleftharpoons ML_n \qquad K_n = [ML_n]/\{[ML_{n-1}]\,[L]\} \qquad (22\text{-}1)$$

$$M + nL \rightleftharpoons ML_n \qquad \beta_n = [ML_n]/\{[M]\,[L]^n\} \qquad (22\text{-}2)$$

The relationship between stepwise and cumulative formation constants is

$$\beta_n = \prod_1^n K_n \qquad\qquad \text{e.g., } \beta_3 = K_1 K_2 K_3 \qquad (22\text{-}3)$$

Writing the equilibrium constant as a cumulative equilibrium constant does not indicate that two or more ligands are added simultaneously. All association reactions occur in a stepwise manner, although in a few cases the successive reactions overlap extensively.

DETERMINATION OF BINDING CONSTANTS BY pH MEASUREMENTS

The majority of ligands are weak bases. In aqueous solution the ligand base can be protonated, and thus metal-ligand complex formation involves competition between proton and metal ion for the ligand base. The progress of this competition reaction can be monitored by means of pH measurements. This method was pioneered by Bjerrum.[*] Advantages of the method include its precision and its applicability to a wide range of metal ions and ligands. Because potentiometry measures the activity of a species, three kinds of constant can be identified:

activity-product (thermodynamic) constants, e.g., $K_a = \{H^+\} \{A\} / \{HA\}$

concentration-product constants, e.g., $K_a = [H^+] [A] / [HA]$

mixed or Brønsted constants, e.g., $K_a = \{H^+\} [A] / [HA]$.

EXPERIMENTAL TECHNIQUES

The most common procedure is to carry out the measurements in the form of a titration. Most commonly a solution containing metal-ion, ligand and acid is titrated with base. Occasionally, when the rate of attaining equilibrium in the system is slow, a "batch" method is adopted: individual solutions of appropriate concentrations are prepared, sealed, placed in a constant-temperature bath and allowed to reach equilibrium, at which time the final pH measurements are made.

The following provides a brief description of a typical experimental set-up and methods for the determination of metal-ligand binding constants using the pH titration method.

PROCEDURE. Appropriate volumes of ligand solution, metal-ion solution, acid and distilled water are pipetted into a titration cell, usually a double-walled beaker through which thermostatted water is flowed. The solution is stirred magnetically and blanketed with nitrogen to prevent reaction with carbon dioxide (and occasionally oxygen). Increments of titrant (usually standard base) are added and the pH is recorded after the addition of each increment.

ACTIVITY COEFFICIENT CORRECTIONS. To eliminate uncertainties arising from activity constant variations, it is common practice to keep activity coefficients constant by use of a "background electrolyte" or "constant ionic atmosphere" (e.g., 0.10 M $NaClO_4$). Since the glass electrode measures (for practical purposes) hydrogen ion *activity*, i.e., $pH_{meas} = -\log\{H^+\} = -\log[H^+]\gamma_+$, it is necessary to convert activity to concentration in the calculations that follow. The relationship of equation 22-4 may be used, where the activity correction $C = \log \gamma_+$.

---

[*] J. Bjerrum, *Metal Ammine Formation in Aqueous Solution*, P. Haase and Son, Copenhagen, 1941.

$$-\log [H^+] = pH_{meas} + C \qquad (22\text{-}4)$$

The correction factor C may be determined from pH measurements on appropriate acid solutions, or calculated using the Debye-Hückel equation. For aqueous solutions at $25°C$ and ionic strength of 0.10, the correction factor is often taken to be -0.10 (calculated from the Debye-Hückel equation). Experimentally measured values are similar. For the calculation of [OH$^-$], the concentration-product value of K_w must be used; at $25°C$ and $I = 0.10$, $p_cK_w = 13.75$.

SEPARATION OF OVERLAPPING PROTONATION CONSTANTS FOR A POLYPROTIC ACID

In the case of a polyprotic acid for which the individual ionizations are well separated (ideally, by at least 3 log units), values for the individual constants can be calculated from data points in the appropriate regions of the titration curve. If the individual ionizations overlap, the Bjerrum $\bar{n}$ (n-bar) method may be used. This mathematical approach was introduced by Bjerrum for the calculation of stability constants of metal-ligand complexes, but it can also be applied to the determination of proton-ligand equilibrium constants.

The equilibrium constants to be determined in this example are *protonation constants*, introduced in Chapter 20. The protonation constant K^H of a base L is the reciprocal of the acid dissociation constant K_a for the corresponding conjugate acid HL. For a polyprotic acid, the general expression for the protonation constant is given by equation 22-5.

$$H_{n-1}L + H^+ \rightleftharpoons H_nL \qquad K_n^H = [H_nL]/\{[H_{n-1}L]\,[H^+]\} \qquad (22\text{-}5)$$

Protonation constants lend themselves much more readily to the systematic treatment of equilibria than do dissociation constants. Thus, for example, the definition of $\bar{n}_H$, the average number of protons bound per ligand, leads to the general expression (22-6) for a ligand base L derived from an acid of general formula H$_J$L. The β's are cumulative protonation constants (see Chapter 20).

$$\bar{n}_H = \frac{[HL] + 2\,[H_2L] + \cdots}{[L] + [HL] + [H_2L] + \cdots} = \frac{\beta_1^H[H^+] + 2\beta_2^H[H^+]^2 + \cdots + J\beta_J^H[H^+]^J}{1 + \beta_1^H[H^+] + \beta_2^H[H^+]^2 + \cdots + \beta_J^H[H^+]^J} \qquad (22\text{-}6)$$

To apply the $\bar{n}_H$ method to the determination of protonation constants, the quantity $\bar{n}_H$, the average number of protons bound, can be defined as (total available protons – free hydrogen-ion)/total ligand. At any point in a titration, the stoichiometric concentration of available protons is equal to the sum of the concentrations of dissociable protons from the ligand, from added strong acid and of hydrogen ions arising from the dissociation of water, less the concentration of added strong base. For the titration of an acid of general formula H_jL of concentration C_L plus added strong acid of concentration C_A, titrated with standard sodium hydroxide:

$$\bar{n}_H = \frac{j\,C_L + C_A + [OH^-] - [Na^+] - [H^+]}{C_L} \tag{22-7}$$

The quantity $\bar{n}_H$ is a normalized variable; for an acid of stoichiometry H_jL, it can have values $0 \leq \bar{n}_H \leq J$. A plot of $\bar{n}_H$ vs. pH is termed the formation curve or formation function. If protonation equilibria are well separated (by at least 3 log units), then the titration curve will exhibit a "break" between the two regions and the protonation constants can be calculated separately, using equation 22-8:

$$K_j^H = \frac{\bar{n}_H - j + 1}{(j - \bar{n}_H)\,[H^+]} \tag{22-8}$$

If two equilibria overlap, the constants can be obtained from the slope and intercept of a straight-line transformation of the $\bar{n}_H$ expression. If three or more equilibria overlap, either multiple linear regression or the Solver can be used to obtain the constants.

TWO OVERLAPPING PROTONATION CONSTANTS OF N-(2-AMINOETHYL)-1,4-DIAZACYCLOHEPTANE

The triamine N-(2-aminoethyl)-1,4-diazacycloheptane (aedach) was

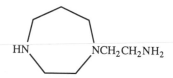

N-(2-aminoethyl)-1,4-diazacycloheptane

synthesized as a potential ligand for complexation of nickel(II) in a square-planar environment. The titration curve of N-(2-aminoethyl)-1,4-diazacycloheptane trihydrobromide with standard NaOH is shown in Figure 22-1. Two protonation equilibria overlap strongly in the pH 9–11 region, while the third protonation constant is much lower, occurring in the pH 3–4 region. The first two protonation constants are normal for aliphatic amines, while the third demonstrates the strong base-weakening effect of charge repulsion by the protonated primary and secondary amines.

It is clear that $\log K_1^H$ and $\log K_2^H$ cannot be calculated individually using equation 22-8. Instead, the general expression 22-6 for $\bar{n}_H$ can be used to obtain 22-9:

$$\bar{n}_H = \frac{\beta_1^H [H^+] + 2\beta_2^H [H^+]^2}{1 + \beta_1^H [H^+] + \beta_2^H [H^+]^2} \tag{22-9}$$

Then, rearranging in the form $y = mx + b$, results in equation 22-10. This equation can be employed, either graphically or by linear regression, to provide the constants β_1^H and β_2^H from the slope and intercept.

$$\frac{\bar{n}_H}{[H^+](1 - \bar{n}_H)} = \beta_2^H \frac{[H^+](2 - \bar{n}_H)}{(1 - \bar{n}_H)} + \beta_1^H \tag{22-10}$$

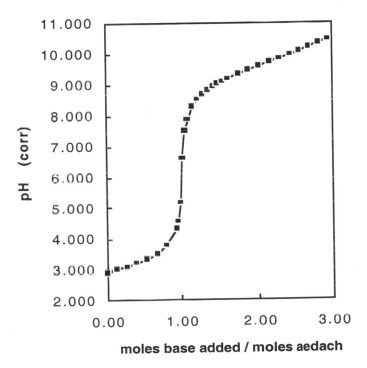

Figure 22-1. Titration curve of aedach·3HBr

THE SPREADSHEET

The constants table from the spreadsheet is shown in Figure 22-2. The names applied to the cell references are shown in column F of the data table; Create Names was used to assign names to the references.

A portion of the data table of the spreadsheet is shown in Figure 22-3. Columns A and B contain the experimental data. The expressions used in columns C, D and E are, respectively:

moles base added/moles aedach =v*CB/(V_0*CL)

pH$_{corr}$ =pH+C_

$\bar{n}_H$ =(3*CL*V_0/(V_0+v)+10^-(pcKw-

-10^-pH_corr)/(CL*V_0/(V_0+v))

The plot of $\bar{n}_H$ as a function of pH is shown in Figure 22-4. Since K_3^H does not overlap with the other protonation constants, the following expression, entered in column F, can be used to calculate it:

=LOG((n_bar-2)/(3-n_bar))+pH_corr

	B	C	D	E	F	G
4	Initial volume, mL				(V_0)	50.00
5	Initial Concentration of aedach.3HBr, M				(CL)	0.002056
6	pH correction factor C				(C_)	-0.11
7	Concentration of titrant NaOH, M				(CB)	0.1381
8	Buret calibration factor				(calib)	0.990
9	pcKw				(pcKw)	13.78

Figure 22-2. Constants table of the spreadsheet.

	A	B	C	D	E	F
12	v, mL	pH	mol b/a	pH(corr)	n-bar	log K
13	0.000	3.008	0.00	2.898	2.385	2.69
14	0.100	3.096	0.13	2.986	2.364	2.74
15	0.200	3.194	0.27	3.084	2.332	2.78
16	0.300	3.310	0.40	3.200	2.292	2.82
17	0.400	3.447	0.54	3.337	2.242	2.84
18	0.500	3.627	0.67	3.517	2.186	2.87
19	0.600	3.890	0.81	3.780	2.120	
20	0.700	4.438	0.94	4.328	2.046	
21	0.720	4.684	0.97	4.574	2.029	
22	0.740	5.270	0.99	5.160	2.012	
23	0.760	6.710	1.02	6.600	1.989	

Figure 22-3. Portion of the spreadsheet for calculating the protonation constants of aedach.

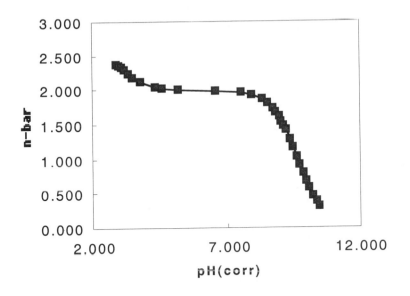

Figure 22-4. $\bar{n}_H$ as a function of pH

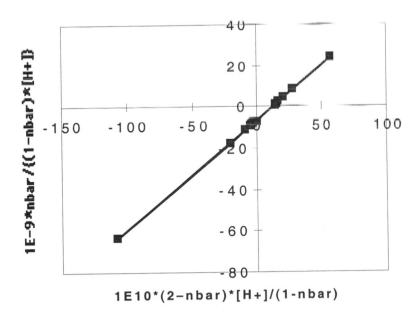

Figure 22-5. Linear transformation of the formation function equation used to obtain K_1^H and K_2^H from the slope and intercept.

The overlapping protonation constants K_1^H and K_2^H were resolved by using equation 22-10. The following scaled expressions were employed, to eliminate clutter on the y- and x axis labels:

=1E-9*n_bar/((1-n_bar)*10^-pH_corr)

=1E+10*(2-n_bar)*10^-pH_corr/(1-n_bar)

The chart (Figure 22-5) is used only to verify that the transformed data fit a linear relationship. LINEST was used to obtain the constants from the slope and intercept of the regression line.

	G	H
45	Slope & Intercept by LINEST	
46	*slope*	*intercept*
47	1.89E+19	1.15E+10
48	6.26E+16	1.13E+08
49	0.99984	438911125

Figure 22-6. Slope and Intercept using LINEST.

From the slope and intercept of Figure 22-6, the values log K_1^H = 10.06 ± 0.005 and log K_2^H = 9.22 ± 0.005 were obtained. The "literature" values, calculated from the same data using a FORTRAN program, are 10.10 ± 0.008 and 9.22 ± 0.01.[*]

THREE OVERLAPPING PROTONATION CONSTANTS OF A POLYAMINE USING LEAST-SQUARES CURVE FITTING AND THE SOLVER

There are a number of computer programs available for the determination of stability constants from pH titration data. The most general of these perform a least-squares fit of the data to a calculated titration curve. The programs are able to handle protonated complexes, polynuclear systems, etc. In this example least-squares curve fitting is applied to a somewhat simpler case, a polyprotic acid in which the equilibria overlap extensively. The method is that used in the computer program SCOGS[*] (Stability Constants of Generalized Species) and described in Chapter 20, namely using pH as the independent variable to calculate the volume of titrant. Then the Solver is used to minimize the sum of squares of residuals ($V_{calc} - V_{obsd}$) to find the best values of the set of equilibrium constants used to generate the curve.

The example used here is the determination of the protonation constants of 3,2,3-tet (1,5,8,12-tetraazadodecane, $H_2N(CH_2)_3NH(CH_2)_2NH(CH_2)_3NH_2$). The tetraprotonated amine was titrated with standard base. It can be seen from the titration curve shown in Figure 22-7 that while one proton is much more acidic

[*]   B. N. Patel and E. J. Billo, *Inorg. Nucl. Chem. Lett.* **1977**, *13*, 335.
[*]   I. G. Sayce, *Talanta* **1968**, *15*, 1397.

than the others and dissociates in the region from pH 4 to pH 6, the acidities of the other three protons are similar, so that three of the protonation regions overlap.

You can use equation 22-11 to calculate the volume of titrant as a function of the measured pH, then use the Solver to minimize the sum of squares of residuals $V_{exptl} - V_{calc}$.

$$V_{calc} = \frac{C_A + J\,C_L + [OH^-](V_0 + V) - [H^+](V_0 + V) - j\Sigma H_j L}{C_B} \qquad (22\text{-}11)$$

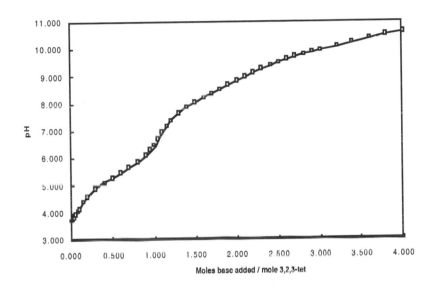

Figure 22-7. Titration curve of 3,2,3-tet. The line is calculated using the four protonation constants found by using the Solver.

THE SPREADSHEET

The spreadsheet header table is shown in Figure 22-8. Create Names was used to assign the names in cells D3:D9 to cells C3:C9. Figure 22-9 shows a small portion of the data (entered in columns A and C) and the corrected volume and pH values. V_{corr} is obtained by multiplying the nominal volume by the buret calibration factor, and pH_{corr} by adding the correction factor C to the measured pH. Columns of intermediate calculations are illustrated in Figure 22-10.

	A	B	C	D	E	F
1	Protonation Constants of 1,5,8,12-tetraazadodecane					
2	(N,N'-bis(3-aminopropyl)ethylenediamine or 3,2,3-tet)					
3		ligand, mmol		0.1037	(CL)	
4		strong acid, mmol		0.4232	(CA)	
5		titrant base, M		0.1050	(CB)	
6		total volume, mL		55	(V_0)	
7		buret calibration factor		0.990	(calib)	
8		pH correction factor C		-0.1	(C_)	
9		pcKw		13.78	(pcKw)	
10						

Figure 22-8. Data header table for the titration of protonated 3,2,3-tet with standard base

Expressions for the denominator of the α expressions, for α_4 and for V_{calc} are as follows:

=1+10^(logK1H-pH)+10^(logK1H+logK2H-2*pH)+10^(logK1H+logK2H +logK3H-3*pH) +10^(logK1H+logK2H+logK3H+logK4H-4*pH)

=10^(logK1H+logK2H+logK3H+logK4H-4*pH)/denom

=(CA+10^-(pKw-pH)*(V_0+v) -10^(-pH)*(V_0+v)- (G12+2*H12+3*I12+4*J12) *CL)/CB

	A	B	C	D
11	V, mL	V(corr)	pH	pH(corr)
12	0.000	0.000	3.798	3.698
13	0.020	0.020	3.862	3.762
14	0.040	0.040	3.951	3.851

Figure 22-9. Portion of the spreadsheet data table.

	Denom	alpha0	alpha1	alpha2	alpha3	alpha4	V(calc)	SUM =	d^2
10									0.0025
11									
12	1.1E+19	9.5E-20	5.0E-13	5.5E-07	1.8E-02	9.8E-01	-0.007		5.5E-05
13	5.9E+18	1.7E-19	7.7E-13	7.4E-07	2.1E-02	9.8E-01	0.010		1.0E-04
14	2.6E+18	3.8E-19	1.4E-12	1.1E-06	2.5E-02	9.7E-01	0.031		7.5E-05

Figure 22-10. Portion of the spreadsheet showing intermediate calculations and the sum of squares of residuals.

The standard deviations shown in Figure 22-11 were obtained by using the SolvStat.xls macro described in Chapter 12.

	G	H	I	J
3	Protonation constants:			± std.dev.
4	logK1H =		10.42	0.007
5	logK2H =		9.74	0.006
6	logK3H =		8.21	0.006
7	logK4H =		5.44	0.006

Figure 22-11. The final values of the protonation constants of 3,2,3-tet.

DETERMINATION OF BINDING CONSTANTS BY SPECTROPHOTOMETRY

The spectrophotometric method is the method of choice for the chemist with an occasional or one-time need to determine a binding constant. The basic concept is obvious, the apparatus is widely available and (usually) the chemist is experienced in the technique. By contrast, the potentiometric method requires significant preparation and familiarization before reliable results can be obtained. A further advantage of the spectrophotometric method is that it is applicable to non-basic ligands, such as halide ions.

In the spectrophotometric method, the molar absorptivity of the complex is an additional variable to be determined. As well, it is necessary to determine the stoichiometry of the complex before calculations can be performed. The *mole-ratio* and *continuous variations* methods are useful in determining the stoichiometry. The observation (or lack) of isosbestic points is also a useful guide to the complexity of the system.

In the mole-ratio method a series of solutions is prepared in which the concentration of one reactant (usually the metal ion) is held constant while the other reactant (usually the ligand) is varied. In the discussion that follows it is assumed that the ligand concentration is varied.

Absorbance measurements are made at a wavelength at which the complex absorbs strongly (it is convenient if neither the metal ion nor the ligand absorbs at that wavelength, but this is not a necessity). The absorbance is plotted versus concentration of the ligand (Figure 22-12). If only one complex of high stability is formed, the graph consists of two linear intersecting parts. The ratio of the concentration of ligand at the intersection point to the (fixed) concentration of metal ion gives the stoichiometry of the complex.

If the stability constant of the complex is high (curve A in Figure 22-12), there will be no appreciable dissociation of the complex at or near the stoichiometric point. If the complex is moderately stable (curve B in Figure 22-12), the plot will consist of two straight-line portions with a central curved portion. Extrapolation of the two straight-line portions yields the intersection point. If a complex of low stability is formed, a large excess of ligand will have to be used to drive the reaction to completion, and there will be no detectable break in the curve from which to obtain the stoichiometry (curves C and D in Figure 22-12).

Once the stoichiometry of the complex has been established, the stability constant(s) can be calculated, provided the data yields a curve showing some dissociation in the neighborhood of the stoichiometric point (curve B in Figure 22-12). Briefly, for any data point in the region of curvature, complex formation did not proceed to completion, as evidenced from the difference between the measured curve and the "theoretical" one. Here there is obviously an equilibrium between metal ion, ligand and complex, and from each data point a value of the stability constant can be calculated.

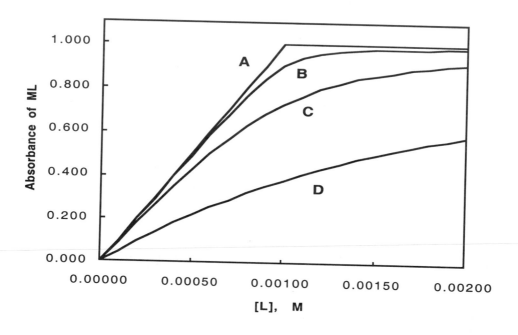

Figure 22-12. Absorbance curves for the formation of a complex ML in solutions containing a fixed stoichiometric concentration of metal ion M (0.00100 M) and varying concentrations of ligand L. Curves are shown for equilibrium constants $K = 1 \times 10^9$ (A), 1 $\times 10^5$ (B), 1×10^4 (C) and 1×10^3 (D).

The following example assumes the simplest case, in which a complex of stoichiometry ML is formed (equation 22-12), and M and L do not absorb at the

$$M + L \rightleftharpoons ML \qquad\qquad K = [M][L]/[ML] \qquad\qquad (22\text{-}12)$$

wavelength used. The concentrations of the three species can readily be calculated from the following relationships:

$$[M] = [M]_T - [ML] \tag{22-13}$$

$$[L] = [L]_T - [ML] \tag{22-14}$$

$$[ML] = A / (b\,\varepsilon) \tag{22-15}$$

where A is the measured absorbance reading, b is the path length (cm), ε is the molar absorptivity of the complex, $[M]_T$ is the analytical concentration of metal and $[L]_T$ is the analytical concentration of ligand.

Note that no valid calculation can be made for data points that are on or very close to the straight-line portions of the curve. Here the reaction has proceeded to completion (driven by excess metal ion or ligand), and the concentration of the other species (ligand or metal ion, respectively) is essentially zero.

EXPERIMENTAL TECHNIQUES

For maximum accuracy, the batch method (separate solutions prepared in volumetric flasks) is preferred, but a titration method (aliquots of ligand solution added to a single solution in a spectrophotometer cell) may be necessary in some situations. Spectrophotometry permits the use of much lower concentrations of metal ion and ligand than are feasible in the pH titration method, if the molar absorptivity of the complex is high.

Buffers can be used to control pH, provided it is determined that the buffer components do not interact, e.g., with the metal ion. However, buffers often absorb in the UV. In any event, a blank consisting of background electrolyte plus buffer should be used in the reference cell. Sodium or potassium nitrate usually can't be used as background electrolyte, because of the absorption band of nitrate centered at 300 nm.

CALCULATIONS

All equations derived for use in spectrophotometric methods are based on four fundamental relationships: Beer's law ($A = \varepsilon b C$), additivity of absorbances ($A = \Sigma A_i$), mass balance ($C_T = \Sigma C_i$), and equilibrium constant expressions; A is the absorbance, ε the molar absorptivity, b the path length and C the concentration.

Absorbance data from different experiments (where concentration and/or path length are different) may be combined by using the effective molar absorptivity $\varepsilon' = A_{obs}/bC_T$.

DETERMINATION OF TWO OVERLAPPING
PROTONATION CONSTANTS OF 4,5-DIHYDROXYACRIDINE

When spectrophotometric methods are used in cases involving multiple, overlapping equilibria, the situation becomes complicated.

The following example deals with the determination of two rather closely spaced protonation constants (equations 22-16 and 22-17) of 4,5-dihydroxyacridine[*].

4,5-dihydroxyacridine

$$H^+ + A^{2-} \rightleftharpoons HA^- \qquad\qquad K_1^H \qquad\qquad (22\text{-}16)$$

$$H^+ + HA^- \rightleftharpoons H_2A \qquad\qquad K_2^H \qquad\qquad (22\text{-}17)$$

Because of solubility considerations, the measurements were made in 50% (v/v) dioxane-water. 4,5-Dihydroxyacridine is essentially colorless in 50% (v/v) dioxane-water solution at pH 7 but becomes yellow as the pH is increased and the phenol groups ionize. UV-visible spectrophotometry revealed that the ionized compound absorbed at 450 nm. Solutions were prepared in volumetric flasks and the pH and A_{450} were measured, in 1.00-cm cells. A graph of absorbance vs. pH is shown in Figure 22-13.

The pH ranges for the gain or loss of the two protons overlap considerably, and as a result the graph does not exhibit a pH region where only the monoprotonated species HA^- absorbs, although the absorbances corresponding to the species A^{2-} and H_2A can readily be obtained. Estimates of the two protonation constants could be obtained by, for example, calculating log K_1^H from the data at high pH where, presumably, only the species A^{2-} and HA^- absorb. However, you can use the Solver to fit the complete range of absorbance-pH data with three parameters: K_1^H, K_2^H and A_1. The absorbance at any point is given by the equation

$$A_{obs} = \alpha_0 A_0 + \alpha_1 A_1 + \alpha_2 A_2 \qquad\qquad (22\text{-}18)$$

where the α's are the fractions in the forms containing 0, 1 or 2 protons and the A's are the corresponding absorbances. The equations given in Chapter 20 are used to calculate the α's .

[*]   A. Corsini and E. J. Billo, *J. Inorg. Nucl. Chem.* **1970**, *32*, 1241.

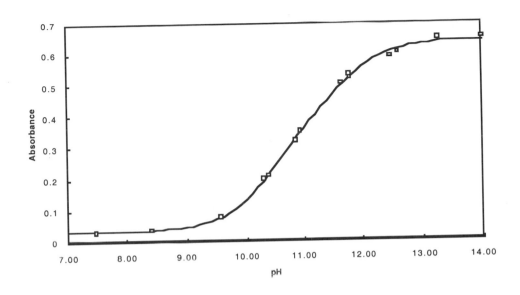

Figure 22-13. Spectrophotometric data for the protonation of 4,5-dihydroxyacridine. The line is calculated with the constants obtained by using the Solver.

THE SPREADSHEET.

Columns C, D and E of Figure 22-14 contain the expressions for α_0, α_1 and α_2, respectively:

=1/(1+10^(log_K1-A11)+10^(log_K1+log_K2-2*A11))

=10^(log_K1-A11)/(1+10^(log_K1-A11)+10^(log_K1+log_K2-2*A11))

=10^(log_K1+log_K2-2*A11)/(1+10^(log_K1-A11)+10^(log_K1+log_K2-2*A11))

and column F contains the expression for A_{calc}:

=A_0*C11+A_1*D11+A_2*E11

Column G contains the squares of the residuals, which are summed in $G$8. This is the target cell, to be minimized by variation of the changing cells $E$6 and $G$5:$G$6. Examination of the data suggested $A_1 = 0.4$, log $K_1^H = 12$, log $K_2^H = 10.5$ as initial values. $A_0 = 0.650$ and $A_2 = 0.032$ were held constant. Use of these values gives a satisfactory initial fit, as indicated by the residuals. The Solver obtained the solution shown in Figure 22-15.

	A	B	C	D	E	F	G	H
1		Protonation Constants of 4,5–Dihydroxyacridine						
2				Using the Solver				
3		(solvent 50%v/v dioxane–water, ionic strength 0.1, 25 deg C)						
4			Constants:	A_0 =	0.650	log K1 =	12	
5				A_1 =	0.4	log K2 =	10.5	
6				A_2 =	0.032			
7						SUM(d^2) =	7.7E-03	
8	pH	Abs	alpha0	alpha1	alpha2	Acalc	d^2	
9	7.48	0.032	0.000	0.001	0.999	0.032	1.2E-07	
10	8.42	0.037	0.000	0.008	0.992	0.035	3.9E-06	
11	9.56	0.077	0.000	0.103	0.897	0.070	4.7E-05	
12	10.30	0.200	0.008	0.384	0.608	0.178	4.8E-04	
13	10.39	0.209	0.011	0.432	0.557	0.198	1.3E-04	
14	10.87	0.320	0.049	0.666	0.284	0.308	1.5E-04	
15	10.95	0.349	0.062	0.693	0.246	0.325	5.8E-04	
16	11.65	0.502	0.294	0.659	0.047	0.456	2.1E-03	
17	11.78	0.530	0.364	0.604	0.032	0.479	2.6E-03	
18	11.79	0.520	0.370	0.600	0.031	0.481	1.5E-03	
19	12.48	0.590	0.749	0.248	0.003	0.586	1.3E-05	
20	12.61	0.602	0.802	0.197	0.002	0.600	4.6E-06	
21	13.27	0.648	0.949	0.051	0.000	0.637	1.2E-04	
22	14.00	0.650	0.990	0.010	0.000	0.648	6.1E-06	
23	.							Acalc
24	7.00		0.000	0.000	1.000			0.032
25	7.10		0.000	0.000	1.000			0.032
26	7.20		0.000	0.001	0.999			0.032

Figure 22-14. Spreadsheet for the determination of the two protonation constants of 4,5-dihydroxyacridine from spectrophotometric data.

	A	B	C	D	E	F	G
4		Constants:		A_0 =	0.650	log K1 =	11.86
5				A_1 =	0.432	log K2 =	10.51
6				A_2 =	0.032		
7						SUM(d^2) =	1.1E-03

Figure 22-15. Using the Solver for the determination of the two protonation constants of 4,5-dihydroxyacridine. Changing cells: $D$5, $F$4 and $F$5; target cell: $G$7.

THE BJERRUM pH-SPECTROPHOTOMETRIC METHOD

The heterocyclic ligand 1,10-phenanthroline forms an orange-red complex

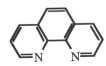

1,10-phenanthroline (phen)

with iron(II), $Fe(phen)_3^{2+}$, which absorbs at 510 nm (molar absorptivity 1.1×10^4 M^{-1} cm^{-1}). The stability of the complex is very high.[*] The mole-ratio plot (a plot of absorbance versus ligand concentration, with Fe(II) concentration constant) is a pair of intersecting straight lines, similar to curve A in Figure 22-12, intersecting at $n = 3$. This permits the determination of the stoichiometry with little uncertainty about the value of n but does not provide data from which the stability constant can be determined. However, the stability constant can be determined by making use of the fact that the ligand is a weak base. In acidic solution, protons compete with Fe^{2+} for the basic nitrogen donors, and at low pH the complex will be partially dissociated. The competition reaction (equation 22-19) can be used to determine the formation constant.

$$Fe^{2+} + 3\,HL^+ \rightleftharpoons FeL_3 + 3\,H^{+} \qquad (22\text{-}19)$$

The procedure used in this example begins with the preparation of a series of solutions, all containing the same concentration of metal ion and the same concentration of ligand, but having different low pH values. The absorbance of each solution is measured; a plot of absorbance versus pH provides a curve that shows the extent of formation of the complex as a function of pH, similar to Figure 22-16. From this data, the stability constant can be calculated in the following manner.

At a given pH the conditional constant β', so called because it is a constant valid for only a particular pH value, can be determined:

$$\beta' = \frac{[FeL_3]}{[Fe]\,[L']^3} \qquad (22\text{-}20)$$

where [L'], the total concentration of all forms of L, is $[L] + [HL] + [H_2L]$.

The pK_a values of the ligand and the measured pH, can be used to calculate the α_0 factor (see Chapter 20) for the ligand where α_0 is defined as the fraction of

---

[*]   Because of a change in spin state upon addition of the third ligand, the tris complex is much more stable than the mono or bis complex. As a result, the reaction essentially yields only the tris complex. The concentration of the mono or the bis complex is close to zero.

the total ligand in the unprotonated form, i.e., $\alpha_0 = $ [L]/[L']. For 1,10-phenanthroline, which effectively behaves as a mono-base,

$$\alpha_0 = \frac{K_a}{K_a + [\text{H}^+]} \tag{22-21}$$

The equivalent equation using protonation constants instead of pK_a values is given in equation 22-22.

$$\alpha_0 = \frac{1}{1 + K^{\text{H}} [\text{H}^+]} \tag{22-22}$$

Then [phen] = α_0 / [phen'] .

From this, equation 22-23 or 22-24 can be used to calculate the overall formation constant β_3

$$\beta_3 = \frac{[\text{Fe(phen)}_3]}{[\text{Fe}] \, [\text{phen}]^3} = \frac{\beta'}{\alpha_0{}^3} \tag{22-23}$$

$$\log \beta_3 = \log \beta' - 3 \log \alpha_0 \tag{22-24}$$

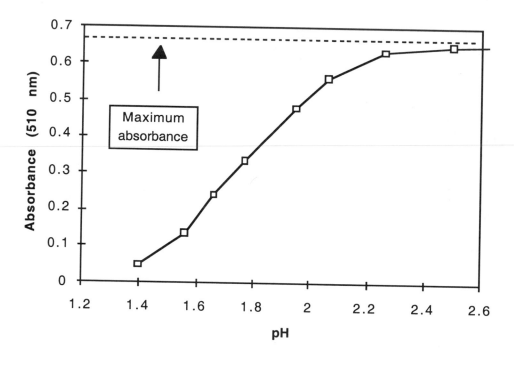

Figure 22-16. Absorbance data for Fe(II)-phen complex as a function of pH.

THE SPREADSHEET

The analytical concentrations of $[Fe^{2+}]_T$, $[phen]_T$, the molar absorptivity of the $Fe(phen)_3^{2+}$ complex (determined in a separate experiment), and the pK_a of 1,10-phenanthroline are entered as constants in cells B3, D3, F3 and H3, respectively. The experimental data (measured pH and absorbance at 525 nm) are in columns A and B of Figure 22-17. The expressions for $[Fe(phen)_3]$, free $[Fe]$, total $[phen']$, α_0, free $[phen]$ and $\log \beta_3$ are given in Figure 22-17.

The formation constant of the tris(1,10-phenanthroline)iron(II) complex has been determined by a number of methods, including potentiometry, a competitive spectrophotometric method, and a method involving partition between aqueous and organic solvents. The typical value for $\log \beta_3$ is about 21.3[*]

	A	B	C	D	E	F	G	H
1			Log beta for Fe(II)–1,10-phenanthroline					
2								
3	[Fe]T =	5.98E-05	[phen]T =	2.47E-04		E = 1.11E+04	pKa =	4.98
4								
5	pH	Abs.	[FeL3]	[Fe]	[L']	alpha	[L]	log b
6	1.56	0.134	1.21E-05	4.77E-05	2.10E-04	3.80E-04	7.99E-08	20.7
7	1.66	0.239	2.15E-05	3.83E-05	1.82E-04	4.78E-04	8.70E-08	20.9
8	1.76	0.337	3.04E-05	2.94E-05	1.55E-04	6.02E-04	9.36E-00	21.1
9	1.77	0.332	2.99E-05	2.99E-05	1.57E-04	6.16E-04	9.66E-08	21.0
10	1.93	0.480	4.32E-05	1.66E-05	1.17E-04	8.90E-04	1.04E-07	21.4
11	1.95	0.480	4.32E-05	1.66E-05	1.17E-04	9.32E-04	1.09E-07	21.3
12	2.06	0.561	5.05E-05	9.28E-06	9.49E-05	1.20E-03	1 14E-07	21.6
13	2.26	0.632	5.69E-05	2.88E-06	7.57E-05	1.90E-03	1 44E-07	21.8
14							Mean =	21.2
15			Equations used (in row 6):					
16	[FeL3] =	=B6/$F$3		alpha =	=10^-$H$3/(10^-$H$3+10^-A6)			
17	[Fe] =	=$B$3-C6		[L] =	=F6*E6			
18	[L'] =	=$D$3-3*C6		log beta =	=LOG(C6/(D6*G6^3))			

Figure 22-17. Spreadsheet for the determination of the overall formation constant of tris(1,10-phenanthroline)iron(II). (Student data, from E. J. Billo.)

[*]   T. S. Lee, I. M. Kolthoff and D. L. Leussing, *J. Am. Chem. Soc*, **1948**, *70*, 2348. H. Irving and D. H. Mellor, *J. Chem. Soc.* **1955**, 3457. H. Irving and D. H. Mellor, *J. Chem. Soc.* **1962**, 5222. G. Anderegg, *Helv. Chim. Acta* **1963**, *46*, 2397.

DETERMINATION OF BINDING CONSTANTS
BY NMR MEASUREMENTS

Nuclear magnetic resonance spectroscopy is a powerful tool for the determination of structural information of complexes in solution. It can also be used for the examination of solution equilibria. In addition to providing quantitative binding constant data, the NMR method can often yield information concerning the site of binding.

The majority of NMR studies of binding constants have employed proton NMR, and discussion will be restricted to that nucleus. Consider an organic ligand with one or more protons, forming a 1:1 complex with a metal ion. A ligand proton will experience different chemical environments when the ligand is free or bound, giving rise to two different chemical shifts, δ_{free} and δ_{bound}. The appearance of the NMR spectrum of the complex will depend on whether chemical exchange of the ligand between the free state and the bound state is slow or fast. If exchange is slow relative to the "NMR time scale" (i.e., lifetimes of minutes or longer), the spectrum of both species, the bound ligand and the free ligand, will be observable. If the system is truly at equilibrium, then the concentration of each of the species can be obtained from the NMR spectrum, and the equilibrium constant determined. This case is rarely encountered.

If the system is at the *fast exchange limit* (i.e., lifetimes of milliseconds or less), the magnetic environment experienced by the proton will be averaged over the environments of the free and bound states, and an NMR singlet will be observed, at a frequency that is the weighted average of the time spent in the two states. This leads to equation 22-25 for the position of the NMR singlet under conditions of fast exchange, where the α's are the fractions of the total ligand in the free and one or more bound states.

$$\delta_{obsd} = \alpha_{free} \delta_{free} + \sum \alpha_{bound} \delta_{bound} \qquad (22\text{-}25)$$

EXPERIMENTAL TECHNIQUES

For studies in aqueous solution, it is necessary to use D_2O as solvent. In this case, exchangeable protons (e.g., protons on O, N, etc.) are almost never observable; the chemical shift of protons on, e.g., carbon are monitored. If an organic solvent is used for the NMR study , then protons on N or O can be observed.

Often, a microliter pipet is used to add reagents to a single solution in the NMR tube. In this case, concentrations should be corrected for dilution. For the most precise work, individual solutions should be made up in e.g., 1-mL volumetric flasks.

Measurement of pH can be made in the NMR tube by using a micro combination pH electrode of approximately 3 mm diameter. For adjustment of pH in D_2O solutions, DCl, D_2SO_4, $DClO_4$ and/or NaOD solutions are used. The correction for pH measurements made in D_2O with a glass electrode calibrated against standard buffers solutions is $pD = pH_{meas} + 0.4.$[*]

CALCULATIONS

Only systems undergoing fast exchange, and only 1:1 binding, will be discussed here. Depending on the magnitude of the binding constant, the following situations may be observed:

Case I: Both δ_{free} and δ_{bound} can be measured independently. The calculations in this case are identical to those for the spectrophotometric determination of the pK_a of an indicator, and will not be discussed further.

Case II: Only δ_{free} can be measured independently. There are two unknowns to be determined, K and δ_{bound}, and these can be obtained as the slope and intercept of a linear transformation of the data.

Case III: Neither δ_{free} or δ_{bound} can be measured independently. A curve-fitting approach is necessary, using the Solver.

It will be recognized that these situations are the same as for the spectrophotometric method.

MONOMER-DIMER EQUILIBRIUM

As part of a study of host–guest complexation, the hydrogen-bonded dimerization of the substituted urea N-phenyl-N'-(2-pyridyl)urea (U) in 1:1 CH_2Cl_2/toluene was studied. The chemical shift of the high-field urea proton is especially sensitive to concentration (see Figure 22-20).

The variation in chemical shift was analyzed assuming dimer formation. The equations are as follows:

dimerization: $$2\,U \rightleftharpoons U_2 \qquad K = [U_2]\,/\,[U]^2 \qquad (22\text{-}26)$$

mass balance: $$[U]_T = [U] + 2\,[U_2] \qquad (22\text{-}27)$$

chemical shift: $$\delta_{obsd} = \alpha_1\,\delta_1 + \alpha_2\,\delta_2 \qquad (22\text{-}28)$$

where $\alpha_1 = [U]/[U]_T$ and $\alpha_2 = 2\,[U_2]/[U]_T$ are the fractions of U_T in the monomeric and dimeric form, respectively.

Combining equations 22-26 and 22-27 yields the following expressions for the concentration of free [U] and the chemical shift:

$$[U] = \frac{\sqrt{8\,K\,[U]_T + 1} - 1}{4\,K} \qquad (22\text{-}29)$$

$$\delta_{calc} = \frac{[U]\,\delta_1 + 2\,K\,[U]^2\,\delta_2}{[U]_T} \qquad (22\text{-}30)$$

---

[*]   P. K. Glasoe and F. A. Long, *J. Phys. Chem.* **1960**, *64*, 188.

The Solver can be used to minimize the sum of squares of residuals, $\Sigma(\delta_{obsd} - \delta_{calc})^2$, in order to find the best values of K, δ_1 and δ_2. The results are shown in the spreadsheets of Figures 22-18 and 22-19.

THE SPREADSHEET.

Figure 22-18 shows a portion of the spreadsheet and illustrates the layout used for generating the theoretical curve. The experimental data are in A11:A18 and B11:B18. Column C contains the expression for free [U]

=(SQRT(8*K*A11+1)-1)/(4*K)

and column D contains the expression for the calculated chemical shift:

=(C11*delta1+2*K*(C11^2)*delta2)/A11

Below the data section lies an extensive table (not shown), used to obtain the smooth calculated curve, in rows 20 to 118.

The Solver can be used to minimize the sum of squares of residuals (in cell E19), the changing cells being D3 (K), D4 (δ_1) and D5 (δ_2).

To plot the theoretical binding curve, generate δ_{calc} for a range of [U], as follows. In cell A20 type 0.000, in A21 type 0.001, then use AutoFill to generate values from 0.000 to 0.100 in steps of 0.001. **Copy** the expression for [U] and **Paste** in cell C20. **Fill Down** to end of table. Transfer the expression for δ_{calc} from any cell in row D to cell F20 (**Copy** to D20, **Cut** and **Paste** to F20 to transfer correctly). **Fill Down** to end of table.

	A	B	C	D	E
1	Dimerization of N-phenyl-N'-(2-pyridyl)urea (U) Studied by NMR				
2	(Chemical shift ∂ of high field urea proton)				
3			K=	50.0	
4			∂1 =	7.000	
5			∂2 =	10.000	
6	Equations used:				
7	free [U]	=(SQRT(8*K*A11+1)-1)/(4*K)			
8	∂(calc)	=(C11*delta1+2*K*(C11^2)*delta2)/A11			
9					
10	[U]total/M	∂(obsd)	[U] free	∂(calc)	resid^2
11	0.0010	7.074	0.0009	7.252	3.16E-02
12	0.0020	7.298	0.0017	7.438	1.95E-02
13	0.0050	7.718	0.0037	7.804	7.37E-03
14	0.0070	7.929	0.0047	7.966	1.34E-03
15	0.0100	8.128	0.0062	8.146	3.20E-04
16	0.0300	8.736	0.0130	8.697	1.50E-03
17	0.0800	9.358	0.0237	9.110	6.13E-02
18	0.1000	9.503	0.0270	9.190	9.83E-02
19				Σ =	2.21E-01

Figure 22-18. The spreadsheet before using the Solver, with initial estimates of K, δ_1 and δ_2. (Data provided by Dr. Steve Bell.)

	A	B	C	D	E
1	Dimerization of N-phenyl-N'-(2-pyridyl)urea (U) Studied by NMR				
2	(Chemical shift ∂ of high field urea proton)				
3		K=	39.1		
4		∂1 =	6.840		
5		∂2 =	10.588		
6	Equations used:				
7	free [U]	=(SQRT(8*K*A11+1)-1)/(4*K)			
8	∂(calc)	=(C11*delta1+2*K*(C11^2)*delta2)/A11			
9					
10	[U]total/M	∂(obsd)	[U] free	∂(calc)	resid^2
11	0.0010	7.074	0.0009	7.094	4.20E-04
12	0.0020	7.298	0.0018	7.293	2.59E-05
13	0.0050	7.718	0.0038	7.706	1.40E-04
14	0.0070	7.929	0.0050	7.897	1.01E-03
15	0.0100	8.128	0.0066	8.115	1.60E-04
16	0.0300	8.736	0.0142	8.813	5.86E-03
17	0.0800	9.358	0.0262	9.359	1.65E-06
18	0.1000	9.503	0.0299	9.466	1.37E-03
19				Σ =	8.99E-03

Figure 22-19. The spreadsheet after refinement of the constants.

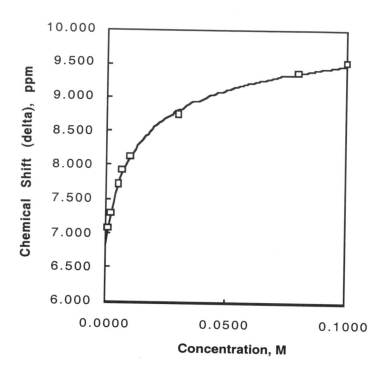

Figure 22-20. The fit of the chemical shift data. The curve is generated by using equations and constants found in the text and in Figure 22-21.

23

ANALYSIS OF KINETICS DATA

In this chapter you'll learn how to extract rate constant information from simple first-order processes, from biphasic processes and from complex rate processes.

EXPERIMENTAL TECHNIQUES

In principle, any measurable property of a reacting system that is proportional to the extent of reaction may be used to monitor the progress of the reaction. The most common techniques are spectrophotometric (UV–visible, fluorescence, IR, polarimetry and NMR) or electrochemical (pH, ion-selective electrodes, conductivity and polarography). Either a "batch" method can be used, in which samples are withdrawn from the reaction mixture and analyzed, or the reaction may be monitored *in situ*. By far the most widely used technique involves UV–visible spectrophotometry.

Since reaction rate is sensitive to temperature, the system must be thermostatted. For most reactions in aqueous solution, the ionic strength should be controlled at a fixed value (see "Experimental Techniques" in Chapter 22).

ANALYSIS OF MONOPHASIC KINETICS DATA

Most reactions are characterized by a change in reactant or product concentration that can be described by a single exponential. The differential form of the rate equation contains a single term; the integrated form yields a straight line from which the rate constant can be obtained. Some of the more common and useful cases are described.

FIRST-ORDER KINETICS

First-order reactions are by far the most common. They are also the simplest to study experimentally. For reactions of higher order, experimental conditions can usually be arranged so that they are first-order (see below). This simplifies the situation considerably.

For the reaction of species A to give product B, with rate constant k

$$A \xrightarrow{k} B$$

the rate of disappearance of A is proportional to the amount of A:

$$\frac{d[A]_t}{dt} = -k[A]_t \tag{23-1}$$

Of course the rate of appearance of product can also be used to monitor the reaction, since

$$\frac{-d[A]_t}{dt} = \frac{d[B]_t}{dt}$$

Integration of equation 23-1 leads to the relationship

$$\ln [A]_t = -kt + \ln [A]_0$$

or

$$\log [A]_t = -2.303 \, k \, t + \log [A]_0 \tag{23-2}$$

that is, a plot of the logarithm of the concentration of A, plotted vs. time, yields a straight line from which the rate constant k can be obtained. The intercept term is usually of no interest.

An alternative form of equation 23-1 that sometimes is useful is

$$[A]_t = [A]_0 \, e^{-kt} \tag{23-3}$$

Occasionally a first-order rate constant is obtained by experimental determination of the half-life $t_{1/2}$, the time required for the reactant concentration to decrease to one-half of its original value. From equation 23-2 it follows that $k = \ln(2)/t_{1/2} = 0.693/t_{1/2}$.

If a reaction is monitored by UV–visible spectrophotometry, for example, the concentration may be replaced by the absorbance (A) in equation 23-2. In the general case, both reactant and product may absorb at the monitoring wavelength, and thus the final absorbance is non-zero. Under these conditions the form of equation 23-4 that must be used is

$$\ln |A_t - A_\infty| = -k \, t + \ln |A_i - A_\infty| \tag{23-4}$$

where A_i is the initial absorbance reading and A_∞ is the absorbance value when the reaction is "complete". For first-order reactions the rule of thumb is that 10 half-lives must elapse before the reaction can be considered to be complete. After 10 half-lives a first-order reaction is $(1 - 0.5^{10})$ or 99.9% complete.

Figure 23-1 illustrates the application of equation 23-4 in the determination of the hydrolysis of a substrate by the enzyme thermolysin. The parameters returned by the SLOPE and INTERCEPT functions were used to calculate the theoretical line in column D of Figure 23-1. The formula in cell D8 is

=$B$26+A8*$B$25.

The first-order behavior is verified by the straight-line fit of the data, shown in Figure 23-2.

	A	B	C	D
1	Enzymatic hydrolysis of N-α-furyl acryloylglycyl-L-leucinamide substrate (FAGLA) by Thermolysin			
2	Run No:	XIII-28A	Wavelength:	340 nm
3	Date:	4/12/83	Pathlength:	1 cm
4	Operator:	M. McCarthy	Concentration:	1.00 mM
5			pH:	7.5
7	time, min	A(340)	$-\ln(A - A\infty)$	$-\ln A(\text{calc})$
8	0	1.731	0.4	0.357
9	2	1.581	0.644	0.627
10	4	1.457	0.914	0.897
11	6	1.367	1.168	1.167
12	8	1.292	1.444	1.437
13	10	1.254	1.619	1.707
14	12	1.201	1.931	1.977
15	14	1.158	2.283	2.247
16	16	1.138	2.501	2.517
17	18	1.12	2.749	2.787
18	20	1.103	3.058	3.057
19	22	1.091	3.352	3.327
20	24	1.082	3.650	3.597
21	26	1.077	3.863	3.867
22	28	1.072	4.135	4.137
23	∞	1.056		
24				
25		SLOPE =	0.13499118	
26		INTERCEPT =	0.35708031	

Figure 23-1. Data table for the enzymatic hydrolysis of FAGLA by thermolysin.

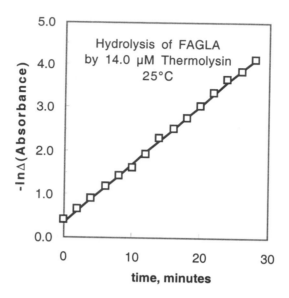

Figure 23-2. First-order plot for the hydrolysis of FAGLA.

REVERSIBLE FIRST-ORDER REACTIONS

If the reaction is reversible, e.g.,

$$A \underset{k_r}{\overset{k_f}{\rightleftharpoons}} B$$

then the rate of approach to equilibrium is a first-order process. If the A_∞ value is denoted by A_{eq}, then the first-order rate expression is simply

$$\ln|A_t - A_{eq}| = -k_{obsd}\, t + \ln|A_i - A_\infty| \tag{23-5}$$

and $k_{obsd} = k_f + k_r$. Only the experimental constant k_{obsd} can be obtained from the first-order plot. If the equilibrium constant is known, the values of k_f and k_r can be calculated, since $k_f / k_r = K_{eq}$.

WHEN THE FINAL READING IS UNKNOWN

Occasionally it is not possible to obtain A_∞ — for example, if the reaction is very slow, if secondary reactions occur toward the end of the primary reaction or if the experiment was terminated before the final reading was obtained. Obviously if a reaction has a half-life of one year it may not be practical to wait for the reaction to be complete.

Several ways have been developed to deal with a reaction for which the A_∞ value is not available. The Guggenheim method, for example, uses paired readings at t and $t + \Delta t$ to calculate the rate constant. By now you probably realize that a much simpler and direct method will be to use the Solver to find both the rate constant k and the A_∞ value by non-linear least-squares.

The worksheet in Figure 23-3 illustrates a case of a reaction so slow that it was necessary to use the Solver to find the final absorbance reading. The unstable *cis*-octahedral isomer of the nickel(II) complex of the macrocyclic ligand cyclam (1,4,8,11-tetraazacyclotetradecane) isomerizes to the planar complex [Ni(cyclam)]$^{2+}$, which absorbs at 450 nm.* In acidic solution the reaction is slow.

Note the use of date and time arithmetic to calculate the elapsed time between readings. The formula in cell B7 is:

=1440*(A7-$A$7)

Because the absorbance of the product is being monitored, the formula in cell D7 is:

=Af-Af*EXP(-k_obsd*t)

The Solver was used to minimize the value in the target cell (E18, sum of squares of residuals) by varying the values of the changing cells (C19 and C20, A_∞ and k_{obsd}).

	A	B	C	D	E
1	Isomerization of cis–[Ni(cyclam)(H$_2$O)$_2$]$^{2+}$ in 0.50 M HClO$_4$				
2	Run No:	XII–21E	Wavelength:	450 nm	
3	Date:	1/15/81	Pathlength:	10 cm	
4	Operator:	J. Billo	Concentration: 2.41 mM		
5			pH:	0.50 M HClO$_4$	
6	Date & Time	t, min	Absorbance	A(calc)*	∂^2
7	01/15/81 10:05 AM	0	0.040	0.000	1.60E-03
8	01/15/01 12:20 PM	135	0.057	0.027	9.24E-04
9	01/16/81 09:10 AM	1385	0.270	0.240	9.06E-04
10	01/19/81 09:15 AM	5710	0.660	0.661	1.58E-06
11	01/20/81 09:20 AM	7155	0.724	0.735	1.25E-04
12	01/23/81 09:45 AM	11500	0.840	0.859	3.66E-04
13	01/26/81 09:20 AM	15795	0.906	0.908	4.07E-06
14	01/28/81 09:30 AM	18685	0.933	0.923	9.88E-05
15	02/02/81 09:10 AM	25865	0.950	0.937	1.71E-04
16	02/05/81 09:30 AM	30205	0.937	0.939	5.11E-06
17				* =Af-Af*EXP(-k_obsd*t)	
18				$\Sigma(\partial^2)$ =	4.20E-03
19			A∞ =	0.941	(Af)
20			k_{obsd}/min^{-1} =	2.13E-04	(k_obsd)

Figure 23-3. Using the Solver to obtain the rate constant when the A_∞ value is unknown.

---

* E. J. Billo, *Inorg. Chem.* **1984**, *23*, 236.

SECOND-ORDER KINETICS

For the bimolecular reaction of species A and B to give product or products, with rate constant k

$$A + B \xrightarrow{k} C$$

the reaction is second-order and the rate depends on the concentration of both A and B:

$$\frac{d[A]_t}{dt} = -k\,[A]_t\,[B]_t \qquad (23\text{-}6)$$

Integration of equation 23-6 yields equation 23-7, which can be used to demonstrate that a reaction is second-order and to obtain the rate constant:

$$\frac{1}{[A]_0 - [B]_0} \ln \frac{[B]_0\,[A]_t}{[A]_0\,[B]_t} = kt$$

or

$$\frac{2.303}{[A]_0 - [B]_0} \log \frac{[B]_0\,[A]_t}{[A]_0\,[B]_t} = kt \qquad (23\text{-}7)$$

For the special case $[A] = [B]$, equation 23-7 fails (since the denominator term becomes zero) and the alternate second-order expression 23-8 must be used:

$$\frac{1}{[A]_t} - \frac{1}{[A]_0} = kt \qquad (23\text{-}8)$$

The same equation applies if the reaction is second-order in a single reactant, e.g.,

$$\frac{d[A]_t}{dt} = -k\,[A]_t^2 \qquad (23\text{-}9)$$

PSEUDO-FIRST-ORDER KINETICS

If the concentration of species B (for example) is large relative to A, it will remain essentially unchanged during the course of the reaction, and the rate expression 23-6 is simplified to 23-10, a form of equation 23-1. The reaction is said to be run under *pseudo-first-order* conditions:

$$\frac{d[A]_t}{dt} = -k\,[B]_i\,[A]_t = -k_{obsd}\,[A]_t \qquad (23\text{-}10)$$

and thus

$$k_{obsd} = k\,[B]_i \qquad (23\text{-}11)$$

Once the first-order behavior with respect to [A] has been verified, the reaction can be run with varying concentrations of B (B still in large excess over A). A graph of k_{obsd} as a function of [B] should be linear; the slope is the rate constant k. For large variations in [B], resulting in large variations in k_{obsd}, it is often useful to plot log k_{obsd} vs. log [B]. The slope of the plot gives the order of the reaction with respect to [B}, in this case 1.0.

ANALYSIS OF BIPHASIC KINETICS DATA

Often a plot of concentration vs. time, or the monitored parameter vs. time, or the rate plot, will not be monophasic. This can arise from a number of different situations, the more common of which are described below.

CONCURRENT FIRST-ORDER REACTIONS

If, in a mixture of A and B, these components react by parallel first-order processes to give a common product C, and A and B do not interconvert, then a first-order plot of the rate of appearance of P will be curved, having a fast and a slow component.

$$A \xrightarrow{k_1} C$$
$$B \xrightarrow{k_2} C$$

This situation is commonly encountered in the measurement of radioactive decay of a mixture of radioisotopes.

CONSECUTIVE FIRST-ORDER REACTIONS

For consecutive first-order processes,

$$A \xrightarrow{k_1} B \xrightarrow{k_2} C$$

the rate expressions are

$$\frac{d[A]_t}{dt} = -k_1[A]_t$$

$$\frac{d[B]_t}{dt} = k_1 A]_t - k_2 [B]_t$$

$$\frac{d[C]_t}{dt} = k_2 [B]_t$$

which lead to the following expressions for the concentrations:

$$[A]_t = [A]_0 e^{-k_1 t} \tag{23-12}$$

$$[B]_t = [A]_0 \frac{k_1}{k_2 - k_1} \left[e^{-k_1 t} - e^{-k_2 t} \right] \tag{23-13}$$

$$[C]_t = [A]_0\left[1 - \frac{k_2}{k_2 - k_1}e^{-k_1 t} + \frac{k_1}{k_2 - k_1}e^{-k_2 t}\right] \qquad (23\text{-}14)$$

The concentrations of A, B and C for a typical series first-order process are shown in Figure 23-4.

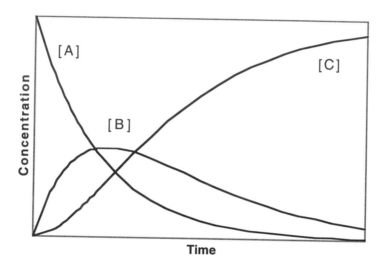

Figure 23-4. Concentration vs. time for consecutive first-order reactions.

The disappearance of A is purely first-order and can be used to determine the rate constant k_1. The species B is formed and then decays in an unmistakable series-first-order manner (Figure12-5 is an example of this). The appearance of C may seem to be pure first-order if the slight deviation from first-order behavior at the beginning of the reaction is missed. In addition, more than one species may absorb at a particular wavelength, complicating and confusing the situation. In the example that follows, both B and C absorb at the same wavelength. This results in behavior that is similar to, and difficult to distinguish from, concurrent first-order reactions.

AN EXAMPLE

The unstable *cis*-octahedral isomer of the nickel(II) complex of the macrocyclic ligand 13aneN$_4$ (1,4,7,10-tetraazacyclotridecane) isomerizes to an intermediate planar isomer, which then converts to the stable planar isomer

[Ni(13aneN$_4$)$^{2+}$; both planar isomers absorb at 425 nm.[*] The reaction exhibits a fast and a slow component, as illustrated in Figure 23-5.

The rate constants for the fast and slow reactions can be obtained in the following manner: the rate constant for the slow reaction is obtained from the data in the latter part of the reaction, by the usual first-order plot. The intercept of this plot at $t = 0$ is used to obtain A_∞ for the fast reaction; the early-time data is then used to construct a second first-order plot. The first-order plots of $\ln(A_\infty - A_t)$ vs. t for the data are shown in Figure 5-23.

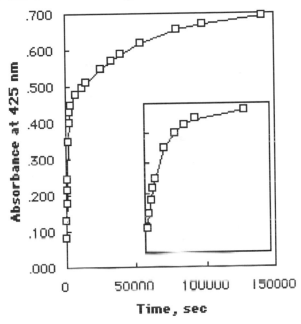

Figure 23-5. Fast (inset) and slow reactions in the isomerization of *cis*-[Ni(13aneN$_4$)(H$_2$O)$_2$]$^{2+}$.

---

[*]   Anne M. Martin, Kenneth J. Grant and E. Joseph Billo, *Inorg. Chem.* **1986**, *25*, 4904.

	A	B	C	D
1	Rate of Isomerization of cis-[Ni(13aneN$_4$)(H$_2$O)$_2$]$^{2+}$			
2	Run No:	XIII-28A	Wavelength:	425 nm
3	Date:	7/19/82	Pathlength:	5 cm
4	Operator:	E. J. Billo	Concentration:	1.43 mM
5			pH:	2.34
7	t, sec	A(obs)	ln(A∞−A)	ln(A'∞−A)
8	160	.080	−0.465	−1.023
9	320	.131	−0.550	−1.176
10	480	.177	−0.633	−1.338
11	640	.216	−0.709	−1.499
12	800	.246	−0.772	−1.643
13	1600	.350	−1.027	−2.414
14	2400	.400	−1.178	
15	3200	.429	−1.277	
16	4000	.449	−1.351	
17	8000	.477	−1.465	
18	12000	.496	−1.551	
19	15200	.510	−1.619	
20	26000	.548	−1.833	
21	33200	.570	−1.981	
22	40400	.590	−2.137	
23	54800	.618	−2.408	
24	80000	.655	−2.937	
25	98000	.670	−3.270	
26	141200	.692	−4.135	
27	∞	0.708		

Figure 23-6. Data table for the isomerization of cis-[Ni(13aneN$_4$)(H$_2$O)$_2$]$^{2+}$.

	A	B	C
28	Slow Reaction		
29	(LINEST on Rows 17-26)		
30		−kobsd	intercept
31		−2.00E−05	−1.315
32		8E−08	0.005
33		0.99988	0.01024
34			
35	A'∞ (from intercept):		0.439

Figure 23-7. Results of first-order plot of the slow part of the isomerization of cis-[Ni(13aneN$_4$)(H$_2$O)$_2$]$^{2+}$.

Applying **LINEST** to the data in the straight-line portion of the slow process (rows 17–26 of Figure 23-6) yields the rate constant for the slow process and permits the calculation of A'_∞ from the intercept value (A_0 for the slow process is A_∞ for the fast process). From $\ln (A_\infty - A_0) = -1.315$, $A_\infty - A_0 = 0.269$, from which $A'_\infty = A_0 = 0.439$, as shown in Figure 23-7.

	A	B	C
37		**Fast Reaction**	
38		**(LINEST on Rows 17-13)**	
39		*-kobsd*	*intercept*
40		-9.66E-04	-0.872
41		5E-06	0.004
42		0.99990	0.00554

Figure 23-8. Results of first-order plot of the fast part of the isomerization of cis-[Ni(13aneN4)(H$_2$O)$_2$]$^{2+}$.

Having established the value for A'_∞, the data from the early stages of the reaction can be analyzed as a first-order process by plotting $\ln |A_t - A'_\infty|$ (see the inset in Figure 5-23). The results are shown in Figure 23-8.

CONSECUTIVE REVERSIBLE FIRST-ORDER REACTIONS

The sequence of coupled reversible first-order processes

$$A \underset{k_2}{\overset{k_1}{\rightleftharpoons}} B \underset{k_4}{\overset{k_3}{\rightleftharpoons}} C$$

yields the following set of differential equations:

$$\frac{d[A]_t}{dt} = -k_1 [A]_t + k_2 [B]_t$$

$$\frac{d[B]_t}{dt} = k_1 [A]_t - k_2 [B]_t - k_3 [B]_t + k_4 [C]_t$$

$$\frac{d[C]_t}{dt} = k_3 [B]_t - k_4 [C]_t$$

The differential equations can be solved to yield analytical expressions for the concentrations of A, B and C. For the case where $[B]_0 = [C]_0 = 0$ at $t = 0$, the expressions are[*] :

$$[A]_t = [A]_0 \left[\frac{k_2 k_4}{\lambda_2 \lambda_3} + \frac{k_1 (\lambda_2 - k_3 - k_4)}{\lambda_2 (\lambda_2 - \lambda_3)} e^{-\lambda_2 t} + \frac{k_1 (k_3 + k_4 - \lambda_3)}{\lambda_3 (\lambda_2 - \lambda_3)} e^{-\lambda_3 t} \right] \quad (21\text{-}15)$$

[*]   John W. Moore and Ralph G. Pearson, *Kinetics and Mechanism*, 3rd edition, Wiley-Interscience, New York, 1982, p. 298.

$$[B]_t = [A]_0 \left[\frac{k_1 \, k_4}{\lambda_2 \, \lambda_3} + \frac{k_1 \, (k_4 - \lambda_2)}{\lambda_2 \, (\lambda_2 - \lambda_3)} \, e^{-\lambda_2 \, t} + \frac{k_1 \, (\lambda_3 - k_4)}{\lambda_3 \, (\lambda_2 - \lambda_3)} \, e^{-\lambda_3 \, t} \right] \qquad (23\text{-}16)$$

$$[C]_t = [A]_0 \left[\frac{k_1 \, k_3}{\lambda_2 \, \lambda_3} + \frac{k_1 \, k_3}{\lambda_2 \, (\lambda_2 - \lambda_3)} \, e^{-\lambda_2 \, t} - \frac{k_1 \, k_3}{\lambda_3 \, (\lambda_2 - \lambda_3)} \, e^{-\lambda_3 \, t} \right] \qquad (23\text{-}17)$$

where
$$\lambda_2 = (p + q)/2 \qquad (23\text{-}18)$$
$$\lambda_3 = (p - q)/2 \qquad (23\text{-}19)$$
$$p = (k_1 + k_2 + k_3 + k_4) \qquad (23\text{-}20)$$
$$q = [p^2 + 4(k_1 k_3 + k_2 k_4 + k_1 k_4)]^{1/2} \qquad (23\text{-}21)$$

A system such as this can readily be solved by using the Solver. The previous example of the isomerization of cis-[Ni(13aneN4)(H2O)2]$^{2+}$ (Figure 23-5),is actually a case of coupled reversible first-order processes. There are two observable rate processes, k_{fast} and k_{slow}, but each is reversible. Treatment of the data according to the consecutive reversible first-order scheme is shown in Figure 23-6.

The expressions for $[A]_t$, $[B]_t$, $[C]_t$ and Abs$_{calc}$ in cells C14, D14, E14 and F14 are:

```
=C_T*(k_2*k_4/(L_2*L_3)+k_1*(L_2-k_3-k_4)*EXP(-L_2*t)/
    L_2*(L_2-L_3))+k_1*(k_3+k_4-L_3)*EXP(-L_3*t)/
    (L_3*(L_2-L_3)))

=C_T*(k_1*k_4/(L_2*L_3)+k_1*(k_4-L_2)*EXP(-L_2*t)/
    (L_2*(L_2-L_3))+k_1*(L_3-k_4)*EXP(-L_3*t)/(L_3*(L_2-L_3)))

=C_T*(k_1*k_3/(L_2*L_3)+k_1*k_3*EXP(-L_2*t)/(L_2*(L_2-L_3))-
    k_1*k_3*EXP(-L_3*t)/(L_3*(L_2-L_3)))

=b*(E_a*C14+E_b*D14+E_c*E14)
```

The quantities p, q, λ_2 and λ_3 were calculated by using the formulas:

```
=k_1+k_2+k_3+k_4

=SQRT(P*P-4*(k_1*k_3+k_2*k_4+k_1*k_4))

=(P+Q)/2

=(P-Q)/2
```

The Solver was used to minimize the sum of squares of residuals (cell G12, Figure 23-9) by changing the values of k_1, k_2, k_3 and k_4, and the molar absorptivity of the intermediate species B (ε_A and ε_C could be determined separately). The Solver provided a reasonable fit to the data. All five changing cells could be varied at once, although approximate estimates of the parameters had to be provided before the Solver could proceed to a solution. The five parameters of the Solver solution are shown in bold in Figure 23-9.

	A	B	C	D	E	F	G
1	\multicolumn{7}{c|}{Rate of Isomerization of cis-[Ni(13aneN₄)(H₂O)₂]²⁺}						
2	Run No:		XIII-28A		Wavelength:		425 nm
3	Date:		7/19/82		Pathlength:		5 cm
4	Operator:		E. J. Billo		Concentration:		1.43 mM
5					pH:		2.34
6			(Values in bold are the changing cells, $G$12 is the target cell.)				
7			Rate constants:		Molar Absorptivities:		
8			k_1 =	5.27E-04	E_a =		0
9			k_2 =	5.14E-04	E_b =		122
10			k_3 =	3.07E-05	E_c =		115
11			k_4 =	7.19E-06			
12						$\Sigma\partial^2$ =	4.52E-04
13	t, sec	A(obs)	[A]*	[B]*	[C]*	A(calc)	∂^2
14	160	.080	1.32E-03	1.11E-04	2.80E-07	0.067	1.6E-04
15	320	.131	1.23E-03	2.04E-04	1.06E-06	0.125	4.2E-05
16	480	.177	1.15E-03	2.82E-04	2.26E-06	0.173	1.7E-05
17	640	.216	1.08E-03	3.48E-04	3.81E-06	0.214	4.3E-06
18	800	.246	1.02E-03	4.04E-04	5.65E-06	0.249	7.3E-06
19	1600	.350	8.40E-04	5.73E-04	1.79E-05	0.358	6.8E-05
20	2400	.400	7.57E-04	6.40E-04	3.28E-05	0.408	6.4E-05
21	3200	.429	7.17E-04	6.65E-04	4.86E-05	0.432	8.5E-06
22	4000	.449	6.95E-04	6.71E-04	6.47E-05	0.445	1.8E-05
23	8000	.477	6.45E-04	6.42E-04	1.43E-04	0.472	2.1E-05
24	12000	.496	6.09E-04	6.07E-04	2.14E-04	0.492	1.5E-05
25	15200	.510	5.82E-04	5.81E-04	2.67E-04	0.507	1.2E-05
26	26000	.548	5.05E-04	5.05E-04	4.20E-04	0.548	1.4E-07
27	33200	.570	4.63E-04	4.64E-04	5.03E-04	0.571	1.3E-06
28	40400	.590	4.27E-04	4.29E-04	5.73E-04	0.591	2.7E-07
29	54800	.618	3.71E-04	3.74E-04	6.85E-04	0.621	9.3E-06
30	80000	.655	3.07E-04	3.11E-04	8.11E-04	0.656	4.1E-07
31	98000	.670	2.80E-04	2.84E-04	8.66E-04	0.671	6.0E-07
32	141200	.692	2.45E-04	2.50E-04	9.35E-04	0.690	5.4E-06
33	\multicolumn{7}{c|}{* (Equations for A-B-C reversible system are from Moore & Pearson, p. 298)}						

Figure 23-9. Data table for the isomerization of *cis*-[Ni(13aneN4)(H2O)2]²⁺.

SIMULATION OF KINETICS
BY NUMERICAL INTEGRATION

Complex systems like that of Figure 23-5 can also be analyzed by Runge-Kutta numerical integration discussed in Chapter 9. The big advantage of direct numerical integration is its generality. The method can be applied to reaction sequences of (in theory, at least) any complexity. The following example applies the Runge-Kutta method to a more complex case.

In the experiment described in the preceding section, the reactant species A was prepared from the final product C. The absorbance of the experiment at $t = 0$ showed that the solution contained a small amount of unconverted C. Instead of $[A]_0 = 1.43$ mM and $[B]_0 = [C]_0 = 0$, the initial conditions were actually $[A]_0 = 1.37$ mM, $[B]_0 = 0$, $[C]_0 = 0.06$ mM. This set of initial condition cannot be accurately treated by using equations 23-15, 23-16 and 23-17, but it can be treated by using RK integration.

The worksheet of Figure 23-9 was modified to use the RK method, as described in Chapter 9.

The Runge-Kutta formulas entered in row 18, for TA1, TA2, TA3, TA4 and $[A]_{t+\Delta t}$, and for TB1, etc, are:

=(-k_1*C_A+k_2*C_B)*(A19-A18)

=(-k_1*(C_A+TA1/2)+k_2*(C_B+TB1/2))*(A19-A18)

=(-k_1*(C_A+TA2/2)+k_2*(C_B+TB2/2))*(A19-A18)

=(-k_1*(C_A+TA3)+k_2*(C_B+TB3))*(A19-A18)

=C_A+(TA1+2*TA2+2*TA3+TA4)/6

=(k_1*C_A-(k_2+k_3)*C_B+k_4*C_C)*(A19-A18)

=(k_1*(C_A+TA1/2)-(k_2+k_3)*(C_B+TB1/2)
  +k_4*(C_C+TC1/2))*(A19-A18)

=(k_1*(C_A+TA2/2)-(k_2+k_3)*(C_B+TB2/2)
  +k_4*(C_C+TC2/2))*(A19-A18)

=(k_1*(C_A+TA3)-(k_2+k_3)*(C_B+TB3) +k_4*(C_C+TC3))*(A19-A18)

=C_B+(TB1+2*TB2+2*TB3+TB4)/6

=(k_3*C_B-k_4*C_C)*(A19-A18)

=(k_3*(C_B+TB1/2)-k_4*(C_C+TC1/2))*(A19-A18)

=(k_3*(C_B+TB2/2)-k_4*(C_C+TC2/2))*(A19-A18)

=(k_3*(C_B+TB3)-k_4*(C_C+TC3))*(A19-A18)

=C_C+(TC1+2*TC2+2*TC3+TC4)/6

Following the layout suggested in Chapter 9, the initial concentration of A was entered in cell C18; cell C19 contains the formula =H18. Similar formulas were used for $[B]_t$ and $[C]_t$.

When the RK method (or any method involving numerical integration) is used, it is important to chose time increments small enough to assure accuracy in calculations. In the RK formulas above, a calculated time increment, e.g. (A19-A18) in row 18, was used rather than a constant value, so that the interval between successive data points could be varied. In this way larger time increments can be used at the end of the reaction, when concentrations are changing slowly. Only a few cells in column B contain A_{obsd} measurements; only one is shown in the spreadsheet fragment of Figure 23-10. The data table of time and A_{obsd} values was located elsewhere in the worksheet; VLOOKUP was used in column B to enter the A_{obsd} values for the appropriate time values. The formula is as follows:

=IF(ISERROR(VLOOKUP(t,DataTable,1,0)),"",VLOOKUP(t,DataTable,2,0))

The ISERROR function was used with *range_lookup* = 0; otherwise VLOOKUP returns #N/A! for all values of *t* for which there is no corresponding entry in DataTable.

An IF statement was used to calculate the squares of residuals only for rows that contained an A_{obsd} value.

The Solver was used in the same way as in the preceding example, although it was necessary to use subsets of the complete set of parameters to perform the initial stages of the refinement. Note the difference in the results when the small amount of product species that was present at the beginning of the reaction is taken into acount: the values of the regression parameters are significantly different, and the sum of squares of residuals (1.38×10^{-4}) is significantly smaller than in the preceding treatment (4.52×10^{-4}).

Calculation time will be significantly longer for a solution by direct numerical integration than for a solution using analytical expressions.

	A	B	C	D	E	F	G	H
1			Rate of Isomerization of cis-[Ni(13aneN$_4$)(H$_2$O)$_2$]$^{2+}$					
2			(Simulation by RUNGE-KUTTA method)					
3	Run No:		XIII-28A		Wavelength:		425 nm	
4	Date:		7/19/82		Pathlength:	(b)	5 cm	
5	Operator:		E. J. Billo		Concentration:		1.43E-03 M	
6					pH:		2.34	
7		(Values in bold are the changing cells, $H$15 is the target cell.)						
8			Rate constants:		Molar Absorptivities:			
9		k_1 =	4.34E-04		E_a =	0		
10		k_2 =	4.39E-04		E_b =	122		
11		k_3 =	2.98E-05		E_c =	115		
12		k_4 =	6.43E-06					
13								
14		Concentrations corrected for		[A]i =	1.37E-03			
15		small amount of C present:		[C]i =	6.00E-05		$\Sigma\delta^2$ =	1.38E-04
16								
17	t, sec	Abs	[A](t)	TA1	TA2	TA3	TA4	[A](t+Δt)
18	0		1.37E-03	-5.95E-06	-5.92E-06	-5.92E-06	-5.90E-06	1.36E-03
19	10		1.36E-03	-5.90E-06	-5.87E-06	-5.87E-06	-5.85E-06	1.36E-03
20	20		1.36E-03	-5.85E-06	-5.82E-06	-5.82E-06	-5.79E-06	1.35E-03
21	30		1.35E-03	-5.79E-06	-5.77E-06	-5.77E-06	-5.74E-06	1.35E-03
22	40		1.35E-03	-5.74E-06	-5.72E-06	-5.72E-06	-5.69E-06	1.34E-03
23	50		1.34E-03	-5.69E-06	-5.67E-06	-5.67E-06	-5.64E-06	1.34E-03
24	60		1.34E-03	-5.64E-06	-5.62E-06	-5.62E-06	-5.60E-06	1.33E-03
25	70		1.33E-03	-5.60E-06	-5.57E-06	-5.57E-06	-5.55E-06	1.32E-03
26	80		1.32E-03	-1.11E-05	-1.10E-05	-1.10E-05	-1.09E-05	1.31E-03
27	100		1.31E-03	-1.09E-05	-1.08E-05	-1.08E-05	-1.07E-05	1.30E-03
28	120		1.30E-03	-1.07E-05	-1.06E-05	-1.06E-05	-1.05E-05	1.29E-03
29	140		1.29E-03	-1.05E-05	-1.04E-05	-1.04E-05	-1.03E-05	1.28E-03
30	160	0.080	1.28E-03	-2.07E-05	-2.03E-05	-2.03E-05	-2.00E-05	1.26E-03
31	200		1.26E-03	-2.00E-05	-1.96E-05	-1.96E-05	-1.93E-05	1.24E-03

Figure 23-10. Data table for the isomerization of cis-[Ni(13aneN$_4$)(H$_2$O)$_2$]$^{2+}$ analyzed by the RK method. Columns I through Y (not shown) contain formulas for calculation of [B] and [C].

APPENDICES

SELECTED WORKSHEET FUNCTIONS BY CATEGORY

Excel 2000 provides more than 300 worksheet functions, in 11 categories: Engineering, Financial, Date & Time, Information, Math & Trig, Statistical, Lookup & Reference, Database, Text, Logical, External. This book uses functions from seven of those categories. This appendix lists selected worksheet functions by category. In Appendix B, the functions are listed alphabetically, with comments, examples and related functions.

DATABASE & LIST MANAGEMENT FUNCTIONS

DAVERAGE	Returns the average of selected values.
DCOUNT	Returns the number of cells that contain numbers in a specified field, according to specified criteria.
DCOUNTA	Returns the number of cells that contain values in a specified field, according to specified criteria.
DGET	Returns a single record that matches the specified criteria.
DMAX	Returns the maximum value in the range *Database*.
DMIN	Returns the minimum value in the range *Database*.
DPRODUCT	Returns the product of values in a specified field that match specified criteria in a database.
DSTDEV	Returns the standard deviation of values in a specified field that match specified criteria in a database.
DSTDEVP	Returns the standard deviation of values in a specified field that match specified criteria in a database.
DSUM	Returns the sum of numbers in the field column of records in the database that match the criteria.
DVAR	Returns the variance of values in a specified field that match specified criteria in a database.
DVARP	Returns the variance of values in a specified field that match specified criteria in a database.
SUBTOTAL	Returns a subtotal in a database.

DATE & TIME FUNCTIONS

DATE	Returns the serial number of a date.
DATEVALUE	Converts a date in the form of text to a serial number.
DAY	Converts a serial number to a day of the month.
MONTH	Converts a serial number to a month.
NOW	Returns the serial number of the current date and time.
TODAY	Returns the serial number of today's date.
WEEKDAY	Converts a serial number to a day of the week.
YEAR	Converts a serial number to a year.

INFORMATION FUNCTIONS

CELL	Returns information about the formatting, location or contents of a cell.
COUNTBLANK	Counts the number of blank cells within a range.
ERROR.TYPE	Returns a number corresponding to an error value.
INFO	Returns information about the current operating environment.
ISBLANK	Returns TRUE if the cell is blank.
ISERR	Returns TRUE if the argument is any error value except #N/A.
ISERROR	Returns TRUE if the argument is an error value.
ISEVEN	Returns TRUE if the argument is even.
ISLOGICAL	Returns TRUE if the argument is a logical value.
ISNA	Returns TRUE if the argument is #N/A.
ISNONTEXT	Returns TRUE if the argument is not text.
ISNUMBER	Returns TRUE if the argument is a number.
ISODD	Returns TRUE if the argument is odd.
ISREF	Returns TRUE if the argument is a reference.
ISTEXT	Returns TRUE if the argument is text.
N	Returns a value converted to a number. Excel usually does this automatically when necessary.
NA	Returns the error value #N/A
TYPE	Returns a number indicating the data type of a value.

LOGICAL FUNCTIONS

AND	Returns TRUE if all arguments are true, otherwise FALSE.
FALSE	Returns the logical value FALSE.
IF	Returns one value if logical_test is TRUE, another value if logical_test is FALSE.
NOT	Reverses the logic of its argument.
OR	Returns TRUE if any argument is TRUE.
TRUE	Returns the logical value TRUE.

LOOKUP & REFERENCE FUNCTIONS

ADDRESS	Returns a reference in the form of text.
AREAS	Returns the number of areas in a multiple selection.
CHOOSE	Chooses a value from a list of values, based on index number.
COLUMN	Returns the column number of a reference.
COLUMNS	Returns the number of columns in a reference.
HLOOKUP	Finds the value in the first row of an array that is equal to or less than *lookup_value*. Returns the associated value in the *n*th row, as determined by *offset_num*.
INDEX	Returns a value from a reference or array, using a specified index.
INDIRECT	Returns a reference specified by a text value.
LOOKUP	Looks up values in an array.
MATCH	Looks up a value in an array and returns its relative position.
OFFSET	Returns a reference offset from a base reference by specified number of rows and columns.
ROW	Returns the row number of a reference.
ROWS	Returns the number of rows in a reference.
TRANSPOSE	Returns the transpose of an array.
VLOOKUP	Finds the value in the first column of an array that is equal to or less than *lookup_value*. Returns the associated value in the *n*th column, as determined by *offset_num*.

MATH & TRIG FUNCTIONS

ABS	Returns the absolute value of a number.
ACOS	Returns the angle corresponding to a cosine value.
ACOSH	Returns the inverse hyperbolic cosine of a number.
ASIN	Returns the angle corresponding to a sine value.
ASINH	Returns the inverse hyperbolic sine of a number.
ATAN	Returns the angle corresponding to a tangent value.
ATAN2	Returns the angle defined by a pair of x- and y coordinates.
ATANH	Returns the inverse hyperbolic tangent of a number.
COS	Returns the cosine of a given angle.
COSH	Returns the hyperbolic cosine of a number.
COUNTIF	Returns the number of non-blank cells within a range that meet the given criteria.
DEGREES	Converts a value in radians to degrees.
EXP	Returns the value of e raised to the power *number*.

FACT	Returns the factorial of a number, i. e., 1*2*3*... *number*.
INT	Rounds a number down to the nearest integer.
LN	Returns the base-e logarithm of a number.
LOG	Returns the logarithm of a number to the specified base.
LOG10	Returns the base-10 logarithm of a number.
MDETERM	Returns the determinant of an array.
MINVERSE	Returns the inverse of an matrix .
MMULT	Returns the product of two matrices.
MOD	Returns the remainder of the division of number by divisor.
MROUND	Returns a number down to the nearest specified level.
PI	Returns π.
PRODUCT	Returns the product of the specified arguments.
RADIANS	Converts an angle in degrees to radians.
RAND	Returns a random number between 0 and 1.
RANDBETWEEN	Returns a random number within a specified range.
ROMAN	Converts an Arabic number to Roman numerals.
ROUND	Rounds a number to a specified number of digits.
ROUNDDOWN	Rounds a number down.
ROUNDUP	Rounds a number up.
SIGN	Returns the sign of a number.
SIN	Returns the sine of a given angle.
SINH	Returns the hyperbolic sine of a number.
SQRT	Returns the square root of a number.
SUM	Returns the sum of all the numbers in the reference.
SUMIF	Returns the sum of all the numbers in the reference that satisfy the specified criteria.
SUMPRODUCT	Returns the sum of the products of corresponding array components.
SUMSQ	Returns the sum of the squares of arguments.
TAN	Returns the tangent of a given angle.
TANH	Returns the hyperbolic tangent of a number.
TRUNC	Truncates a number.

STATISTICAL FUNCTIONS

AVEDEV	Returns the average absolute deviation of data points from the mean.
AVERAGE	Returns the average of all the numbers in the reference.
CORREL	Returns the correlation coefficient between two data sets.
COUNT	Returns the number of numbers in the reference.
COUNTA	Returns the number of non-blank values in the reference.
INTERCEPT	Returns the intercept of the linear regression line $y = mx + b$.
LARGE	Returns the kth largest value in a list of values.
LINEST	Returns the parameters of multiple linear regression.
MAX	Returns the maximum value in a list of arguments.
MEDIAN	Returns the median value in a list of arguments.
MIN	Returns the minimum value in a list of arguments.
PEARSON	Returns the Pearson correlation coefficient.
SLOPE	Returns the slope of the linear regression line $y = mx + b$..
SMALL	Returns the kth smallest value in a list of values.
STDEV	Returns the standard deviation of a sample.
VAR	Returns the variance of a sample.

TEXT FUNCTIONS

CHAR	Returns the character corresponding to the character code number.
CLEAN	Removes all non-printable characters from a text string.
CODE	Returns the numeric code corresponding to a character.
CONCATENATE	Concatenates text values into a single text string.
DOLLAR	Converts a number to a text value in currency format.
EXACT	Compares two text strings; returns TRUE if they are the same.
FIND	Returns the position at which one text string occurs within another text string.
FIXED	Rounds a number to the specified number of decimal places and returns the result as text.
LEFT	Returns the specified number of characters from a text string, beginning at the left.
LEN	Returns the number of characters in a text string.
LOWER	Converts a text string to lowercase.
MID	Returns the specified number of characters from a text string, beginning at the specified position.
PROPER	Capitalizes the first letter in each word of a text string.

REPLACE	Replaces characters at a specified position within a text string.
REPT	Repeats a text value a specified number of times.
RIGHT	Returns the specified number of characters from a text string, beginning at the right.
SEARCH	Finds the position of a text value within a text string.
SUBSTITUTE	Finds and substitutes characters within a text string.
T	Converts a value to text. Excel usually does this automatically when necessary.
TEXT	Formats a number and returns it as text.
TRIM	Removes spaces from a text string, except for single spaces between words.
UPPER	Converts a text string to uppercase.
VALUE	Converts a text argument to a number. Excel usually does this automatically when necessary.

B

ALPHABETICAL LIST OF SELECTED WORKSHEET FUNCTIONS

This appendix lists selected worksheet functions that will be useful in creating advanced worksheet formulas for chemical applications. For each function, a description of the function and its syntax are given; most cases include an example of the use of the function, and any related functions are listed. Function arguments are in italics; arguments are in boldface type if the argument is required and plain text if the argument is optional. The required data type for an argument is indicated by the suffixes *_num*, *_ref*, *_logical*, *_text*. If a function can accept an argument of more than one type, then no suffix is attached.

ABS
Returns the absolute value of a number
Syntax: ABS(***number***)
Example: =ABS(-7.3) returns 7.3
Related Function: SIGN

ACOS
Returns the angle corresponding to a cosine value.
Syntax: ACOS(***number***)
Number must be from −1 to +1. The returned angle is in radians, in the range 0 to π. To convert the result to degrees, multiply by $180/\pi$.
Example: =ACOS(0) returns 1.570796327, or 90 degrees.
Related Functions: COS, other trigonometric functions

ADDRESS
Returns a reference in the form of text.
Syntax: ADDRESS(***row_num***,***column_num***,*abs_num*,*a1*,*sheet_text*)
Abs_num specifies the type of reference to return. If *abs_num* is 1 or omitted, returns an absolute reference; if 2, absolute row and relative column; if 3, relative row and absolute column; if 4, relative. A1 is a logical value that specifies the A1 or R1C1 reference style. If A1 is TRUE or omitted, ADDRESS returns an A1-style

reference; if FALSE, ADDRESS returns an R1C1-style reference. Sheet_text is text specifying the name of the worksheet to be used as the external reference. If sheet_text is omitted, no sheet name is used.

Examples:

=ADDRESS(2,3) returns "$C$2"

=ADDRESS(2,3,2) returns "C$2"

=ADDRESS(2,3,2,FALSE) returns "R2C[3]"

=ADDRESS(2,3,1,FALSE,"[Book1]Sheet1") returns "[Book1]Sheet1!R2C3"

Related Functions: COLUMN, INDIRECT, ROW

AND

Returns TRUE if all arguments are TRUE, otherwise returns FALSE.

Syntax: AND(*logical1,logical2,...*)

Up to 30 logical conditions can be tested.

Example: If A1 contains 0 and A2 contains 110, then =AND(A1=0,A2>100) returns TRUE.

Related Functions: NOT, OR

AREAS

Returns the number of areas in a multiple selection.

Syntax: AREAS(*reference*)

Related Functions: COLUMN, COLUMNS, ROW, ROWS

ASIN

Returns the angle corresponding to a sine value.

Syntax: ASIN(*number*)

Number must be from -1 to $+1$. The returned angle is in radians, in the range $-\pi/2$ to $+\pi/2$. To convert the result to degrees, multiply by $180/\pi$.

Example: If A1 contains 0.7071, then =ASIN(A1) returns 0.785388573, or 45 degrees.

Related Functions: SIN, other trigonometric functions

ATAN

Returns the angle corresponding to a tangent value.

Syntax: ATAN(*number*)

The returned angle is in radians, in the range 0 to π. To convert the result to degrees, multiply by $180/\pi$.

Example: =ATAN(0) returns 0.785388573 or 45 degrees.

Related Functions: ATAN2, TAN, other trigonometric functions

ATAN2

Returns the angle defined by a pair of x- and y coordinates. The angle is between the x axis and the line connecting the origin (0,0) and the point (x,y).

Syntax: ATAN2(*x_num,y_num*)

The returned angle is in radians, in the range $-\pi$ to π. To convert the result to degrees, multiply by $180/\pi$. A negative result represents a clockwise angle from the x-axis.

Example: =ATAN2(3,4) returns 0.927295218, or 53.13 degrees.

Related Functions: ATAN, TAN, other trigonometric functions

AVERAGE

Returns the average of all the numbers in the reference.

Syntax: AVERAGE(*number1*,*number2*,...)

The arguments may be numbers, or names, arrays or references that contain numbers. Up to 30 separate arguments can be listed. Only numbers in the array or range are counted.

Related Function: MEDIAN

CELL

Returns information about location, formatting and contents of a cell.

Syntax: CELL(*info_type_text*,*reference*)

Info_type_text specifies the cell information to be returned. If *reference* is omitted, information about the active cell is returned. For further information, see *Microsoft Excel 97 Worksheet Function Reference.*

Example: If the cell format is 0.00E+00, =CELL("format") returns S2.

Related Function: INFO

CHAR

Returns the character corresponding to the character code number, in either the Macintosh character set or the ANSI character set (Windows).

Syntax: CHAR(*number*)

Number must be between 1 and 255.

Example: =CHAR(65) returns A.

Related Function: CODE

CHOOSE

Chooses a value from a list of values, based on index number.

Syntax: CHOOSE(*index_number*, *value1*,*value2*,...)

Index_number must be a number between 1 and 29, or a formula or reference that evaluates to same. The arguments may be numbers, names, formulas or references.

Example: If cell H8 contains 1, then

=CHOOSE(H8+1,$F$2,$G$2,$H$2,$I$2)/SUM($F$2:$I$2) calculates the fraction $G$2/SUM($F$2:$I$2).

Related Functions: INDEX, MATCH

CLEAN

Removes all non-printable characters from a text string.

Syntax: CLEAN(*text*)

Use on text imported from other applications that may contain non-printing

characters.
Related Function: TRIM

CODE
Returns the numeric code for the first character of text,
Syntax: CODE(*text*)
Example: =CODE ("A") returns 65.
Related Function: CHAR

COLUMN
Returns the column number of reference.
Syntax: COLUMN(*reference*)
If *reference* is a range of cells, returns the column number of the upper left cell
of the range. *Reference* cannot be a multiple selection. If *reference* is omitted, it
is assumed to be the reference in which the COLUMN function appears.
Example: If C3:N10 is selected, then =COLUMN(SELECTION()) returns 3.
Related Functions: COLUMNS, ROW, ROWS

COLUMNS
Returns the number of columns in reference.
Syntax: COLUMNS(*reference*)
If *reference* is a multiple selection, use INDEX to select a specified area within the
selection.
Example: If C3:N10 is selected, then =COLUMNS(SELECTION()) returns 12.
Related Functions: COLUMN, ROW, ROWS

CORREL
Returns the correlation coefficient between two data sets.
Syntax: CORREL(*array1,array2*)
Array1 and *array2* must have the same number of data points.
Related Functions: SLOPE, INTERCEPT, COVAR, PEARSON, RSQ

COS
Returns the cosine of a given angle.
Syntax: COS(*number*)
Number is the angle in radians. To convert an angle in degrees to one in radians,
multiply by $\pi/180$.
Related Functions: ACOS, SIN, TAN, other trigonometric functions

COUNT
Returns the number of numbers in the list or range.
Syntax: COUNT(*value1,value2,...*)
Up to 30 arguments are allowed. Any value, other than an empty cell, is
counted.
Related Functions: COUNTA, COUNTBLANK, COUNTIF

COUNTA
Returns the number of non-blank values in the list or range.

Syntax: COUNTA(**value1**,value2,...)
Up to 30 arguments are allowed. Nulls (""), logical values, numbers formatted as dates, or text values that can be converted to numbers, are counted.
Related Functions: COUNT, COUNTBLANK, COUNTIF

COUNTBLANK
Returns the number of blank cells within a range.

Syntax: COUNTBLANK(**reference**)
Cells containing null values are counted, but cells containing zero values are not counted.
Related Functions: COUNT, COUNTA, COUNTIF

COUNTIF
Returns the number of cells within a range that meet the specified criteria.

Syntax: COUNTIF(**reference,criteria**)
Example: =COUNTIF(year_of_graduation,<1990)
Related Functions: COUNT, COUNTA, COUNTBLANK, SUMIF

DATE
Returns the serial number of a particular date.

Syntax: DATE(**year, month, day**)
Year is a number from 1900 (Windows) or 1904 (Macintosh) to 9999. *Year* can be specified using two digits if in the range 1920 to 2019; otherwise use four digits. *Month* or *day* can be greater than 12 or 31 (see example).
Example: DATE(94,7,4) corresponds to July 4, 1994. DATE(94,7,90) corresponds to Sept. 28, 1994.
Related Functions: DATEVALUE, DAY, MONTH, YEAR, NOW, TODAY

DATEVALUE
Converts a date in the form of text to a serial number.

Syntax: DATEVALUE(**text**)
Example: =DATEVALUE("3 Aug 1938") returns 12633 if the 1904 Date System is in effect.
Related Functions: NOW, TODAY

DAY
Converts a serial number to a day of the month.

Syntax: DAY(**serial_number**)
Serial_number can also be given as text, such as "Jul-4-1994". Returns a value in the range 1 to 31.
Related Functions: NOW, TODAY, WEEKDAY, YEAR, MONTH

DEGREES

Converts an angle in radians to degrees

Syntax: DEGREES(*number*)

Available in Excel 4.0 only as an add-in function.

Related Functions: RADIANS, trigonometric functions

EXACT

Compares two text strings; returns TRUE if they are the same.

Syntax: EXACT(*text1,text2*)

EXACT is case-sensitive.

Related Functions: FIND, SEARCH

EXP

Returns the value of e raised to a power.

Syntax: EXP(*number*)

Returns the value of e raised to the power *number*.

Related Functions: LN, LOG

FACT

Returns the factorial of a number.

Syntax: FACT(*number*)

Returns the factorial of a number, i.e., 1*2*3*...*number*. *Number* must be positive. If *number* is not an integer, it is truncated.

FIND

Returns the position at which one text string occurs within another text string.

Syntax: FIND(*find_text, within_text, start_number*)

Find_text is the text string you want to find. *Within_text* is the string you are searching. *Start_number* is the character at which to start searching. If *start_number* is omitted, it is assumed to be 1. FIND is case-sensitive.

Example: If cell A3 contains BILLO, E JOSEPH, the formula =FIND(",",A3) returns 6.

Related Functions: EXACT, SEARCH

FIXED

Rounds a number to the specified number of decimal places and returns the result as text.

Syntax: FIXED(*number*, decimals, no_commas_flag)

Number is the number to be converted. *Decimals* is the number of decimal places desired; if *decimals* is omitted, it is assumed to be 2. If *decimals* is negative, *number* is rounded to the left of the decimal point. If *no_commas_flag* is TRUE, commas are not included in the formatted number.

Example: =FIXED(12345,3,1) returns 12345.000. =FIXED(12345) returns 12,345.00. =FIXED(PI(),4) returns 3.1416.

Related Functions: ROUND, TEXT

HLOOKUP

Finds the value in the first row of an array that is equal to or less than *lookup_value*. Returns the associated value in the *n*th row, as determined by *offset_num*.

Syntax: HLOOKUP(***lookup_value,   array,   offset_num***, *match_logical*)

The values in the first row must be in ascending order. If *match_logical* is TRUE or omitted, returns the largest value that is less than or equal to *lookup_value*. If *match_logical* is FALSE, returns #N/A if an exact match is not found.

Example: See example under VLOOKUP.

Related Functions: INDEX, MATCH, VLOOKUP

I F

Returns one value if *logical_test* is TRUE, another value if *logical_test* is FALSE.

Syntax: IF(***logical_test,   value_if_TRUE***, *value_if_FALSE*)

Value_if_TRUE and/or *value_if_FALSE* can be IF functions; up to seven IF functions can be nested.

Example: =IF(result=0, 0.08*estimate, 0.08*result)

Related Functions: AND, NOT, OR, IS functions

INDEX

Chooses a value from an array, based on row and column number pointers.

Syntax: INDEX(***array**, row_number, column_number*)

If the array is one-dimensional, only one pointer argument need be specified.

Example: If the reference $C$2:$C$102 contains the atomic weights of the elements, arranged in order of atomic number, then =INDEX($C$2:$C$102,51) returns the atomic weight of element number 51.

Related Functions: CHOOSE, HLOOKUP, MATCH, VLOOKUP

INDIRECT

Returns a reference specified by a text string.

Syntax: INDIRECT(***reference_text**,A1_logical*)

Reference_text can be an A1- or R1C1-style reference, or a name. *A1_logical* specifies the form of *reference_text*: if *reference_text* is TRUE or omitted, *reference_text* is interpreted as an A1-style reference; if *reference_text* is FALSE, *reference_text* is interpreted as an R1C1-style reference.

Example: If cell B3 contains "Sheet2!A5" then the formula =INDIRECT(B3) displays the contents of Sheet2!A5.

Related Functions: OFFSET

INFO

Returns information about the operating environment.

Syntax: INFO(***type_text***)

Type_text specifies what information is to be returned. Some useful values of *type_text*: "directory" returns path of the current directory or folder, "memavail" returns the amount of memory available, "release" returns Microsoft Excel version, "system" returns mac or pcdos. See *Microsoft Excel 97 Worksheet Function Reference* for details.

Example: The formula =INFO("release") returns 9.0 if Excel 2000 is being used.

Related Function: CELL

INT

Rounds a number down to the nearest integer.

Syntax: INT(*number*)

INT rounds downward. Use TRUNC to return the integer part of a number, irrespective of sign.

Example: =INT(3.897) returns 3. =INT(-0.05) returns −1.

Related Functions: ROUND, TRUNC

INTERCEPT

Returns the intercept of the linear regression line $y = mx + b$.

Syntax: INTERCEPT(*known_ys, known_xs*)

See LINEST for details.

Related Functions: LINEST, SLOPE

ISBLANK

Returns TRUE if the value is blank.

Syntax: ISBLANK(*value*)

Related Functions: other IS functions

ISERR

Returns TRUE if the value is an error value other than #N/A.

Syntax: ISERR(*value*)

Related Functions: other IS functions

ISERROR

Returns TRUE if the value is any error value.

Syntax: ISERROR(*value*)

Related Functions: other IS functions

ISNA

Returns TRUE if the value is #N/A.

Syntax: ISNA(*value*)

Related Functions: other IS functions

ISNUMBER

Returns TRUE if the value is a number.

Syntax: ISNUMBER(*value*)

Does not convert text representation of a number.

Example: =ISNUMBER(5) returns TRUE. =ISNUMBER("9") returns FALSE.

Related Functions: other IS functions

ISTEXT
Returns TRUE if the value is text.
Syntax: ISTEXT(*value*)
Related Functions: other IS functions

LARGE
Returns the *k*th largest value in a list of values.
Syntax: LARGE(*array,k*)
Related Functions: MAX, MIN, SMALL

LEFT
Returns the specified number of characters from a text string, beginning at the left.
Syntax: LEFT(*text, num_chars*)
If *num_chars* is omitted, it is assumed to be 1.
Example: =LEFT("CHEMISTRY",4) returns CHEM.
Related Functions: LEN, MID, RIGHT

LEN
Returns the number of characters in a text string.
Syntax: LEN(*text*)
Example: =LEN("CHEMISTRY") returns 9.
Related Functions: LEFT, MID, RIGHT

LINEST
Returns an array of linear regression parameters.
Syntax: LINEST(*known_ys, known_xs, const_logical, stats_logical*)
See Chapter 11 for details.
Related Functions: CORREL, INTERCEPT, SLOPE

LN
Returns the natural (base-e) logarithm of a number.
Syntax: LN(*number*)
Related Functions: EXP, LOG

LOG
Returns the logarithm of a number to the specified base.
Syntax: LOG(*number, base*)
If *base* is omitted, returns the base-10 logarithm.
Related Functions: LN, LOG10

LOG10
Returns the base-10 logarithm of a number.
Syntax: LOG10(*number*)

Related Functions: LN, LOG

LOOKUP

Looks up a value in an array.

The LOOKUP function has two syntax forms: vector and array. The vector form of LOOKUP looks in a one-row or one-column range for a value and returns a value from the same position in a second one-row or one-column range. See the On-line Help for information about the array form of LOOKUP.

Syntax: LOOKUP(***lookup_value,lookup_vector,result_vector***)

The values in *lookup_vector* must be placed in ascending order; if LOOKUP can't find the *lookup_value,* it matches the largest value in *lookup_vector* that is less than or equal to *lookup_value*.

Related Functions: HLOOKUP, MATCH, VLOOKUP

LOWER

Converts a text string to lowercase.

Syntax: LOWER(***text***)

Related Functions: PROPER, UPPER

MATCH

Looks up a value in an array and returns its relative position.

Syntax: MATCH(***lookup_value, array***, *match_type*)

If *match_type* = 0, the function finds the first value that is exactly equal to *lookup_value; array* can be in any order. If *match_type* = 1, finds the largest value that is less than or equal to *lookup_value; array* must be in ascending order. If *match_type* = −1, finds the smallest value that is greater than or equal to *lookup_value; array* must be in descending order.

Example: =MATCH(MAX(Spectrum),Spectrum,0) returns the relative position in the array *Spectrum* of the maximum value in the array.

Related Functions: HLOOKUP, INDEX, VLOOKUP

MAX

Returns the maximum value in a list of arguments.

Syntax: MAX(***number1***, *number2,...*)

There may be up to 30 arguments. If an argument is a reference, only numbers in the reference are examined.

Related Function: MIN

MDETERM

Returns the determinant of an array.

Syntax: MDETERM(***array***)

Array must have an equal number of rows and columns. If any cells in *array* do not contain numbers, returns #VALUE!.

Example: See Chapter 9 for details.

Related Functions: MINVERSE, MMULT, SUMPRODUCT, TRANSPOSE

MEDIAN
Returns the median value in a list of arguments.
Syntax: MEDIAN(**number1**, number2,...)
If there are an even number of numbers in the set, returns the average of the two median values.
Related Function: AVERAGE

MID
Returns the specified number of characters from a text string, beginning at the specified position.
Syntax: MID(**text, start_num, num_chars**)
If *num_chars* extends beyond the end of text, returns characters to the end of text.
Example: If cell A4 contains H2SO4, =MID(A4,2,1) returns 2.
Related Functions: LEFT, LEN, RIGHT

MIN
Returns the minimum value in a list of arguments.
Syntax: MIN(**number1**, number2,...)
There may be up to 30 arguments. If an argument is a reference, only numbers in the reference are examined.
Related Function: MAX

MINVERSE
Returns the inverse of a matrix.
Syntax: MINVERSE(**array**)
Array must have an equal number of rows and columns. If any cells in *array* do not contain numbers, MINVERSE returns #VALUE!. If MDETERM for the array returns 0, the array cannot be inverted; MINVERSE will return #NUM! error.
Example: See Chapter 9 for details.
Related Functions: MDETERM, MMULT, SUMPRODUCT, TRANSPOSE

MMULT
Returns the product of two matrices.
Syntax: MMULT(**array1, array2**)
COLUMNS for *array1* must equal ROWS for *array2*.
Example: See Chapter 9 for details.
Related Functions: MDETERM, MINVERSE, SUMPRODUCT, TRANSPOSE

MOD
Returns the remainder of the division of *number* by *divisor*.
Syntax: MOD(**number, divisor**)
If *divisor* is greater than *number*, returns *number*.
Use MOD(number,1) to return the decimal part of a floating-point number.
Example: =MOD(2.3333,2) returns 0.3333. =MOD(2.3333,3) returns

2.3333. =MOD(10.37,1) returns 0.37.
Related Functions: INT, ROUND, TRUNC

MONTH
Converts a serial number to a month.
Syntax: MONTH(*serial_number*)
Serial_number can also be given as text, such as "Jul-4-1994". Returns a number between 1 and 12.
Related Functions: YEAR, DAY, WEEKDAY

NA
Returns the error value #N/A
Syntax: NA()

NOT
Reverses the logical value of its argument.
Syntax: NOT(*logical*)
Example: =IF(NOT(ISERROR(F2/G2)),F2/G2,"")
Related Functions: AND, OR

NOW
Returns the serial number of the current date and time.
Syntax: NOW()
The integer part of the serial number represents the day, month and year; the decimal part represents the time. NOW is recalculated whenever the sheet is recalculated.
Related Functions: DATE, TODAY, YEAR, MONTH, DAY

OFFSET
Returns a reference offset from a base reference by specified number of rows and columns.
Syntax: OFFSET(*reference, rows, cols*, height, width)
Rows or *columns* can be negative. *Height* and *width* must be positive. If *height* and *width* are omitted, they are assumed to be the same as in *reference*.
Example: =OFFSET(table, rows, cols, 1,1) returns a reference to a single cell within the reference *table*.
Related Function: INDEX

OR
Returns TRUE if any argument is TRUE.
Syntax: OR(*logical1*, logical2,...)
Up to 30 logical conditions can be tested.
Example: If cell G2 contains 0.75, the formula
=IF(OR(G2<=0,G2>=1),"",F2/G2) returns 1.33333333; if cell G2 contains 1.75, the formula returns a null.

Related Functions: AND, NOT

PEARSON
Returns the Pearson product moment correlation coefficient between two data sets.

Syntax: PEARSON(*array1, array2*)

Array1 is the array of independent (x) values; *array2* is the array of dependent (y) values. Returns a value of R between -1 and 1.

Example: See *Microsoft Excel 97 Worksheet Function Reference* for details.

Related Functions: CORREL, RSQ

P I
Returns π.

Syntax: PI()

PRODUCT
Returns the product of the specified arguments.

Syntax: PRODUCT(*number1,number2,...*)

Up to 30 numbers can be multiplied.

Related Function: SUM

PROPER
Capitalizes the first letter in each word of a text string.

Syntax: PROPER(*text*)

Example: =PROPER("JOHN Q. PUBLIC") returns John Q. Public.

Related Functions: LOWER, UPPER

RADIANS
Converts an angle in degrees to radians.

Syntax: RADIANS(*number*)

Available in Excel 4.0 only as an add-in function.

Related Functions: DEGREES, trigonometric functions

RAND
Returns a random number between 0 and 1.

Syntax: RAND()

A new random number is generated each time the worksheet is recalculated. You may need to **Copy** and **Paste Special** (Values) to prevent a random-number series from changing each time you change anything on a worksheet.

Example: =0.1*(0.5-RAND()) returns a number between -0.05 and $+0.05$.

REPLACE
Replaces characters at a specified position within a text string.

Syntax: REPLACE(*old_text, start_num, num_chars, new_text*)

Old_text is the text in which characters are to be replaced. *Start_num* is the position of the first character in *old_text* to be replaced. *Num_chars* is the number of characters to be replaced. *New_text* is the text to replace characters in *old_text*.

Example: If cell A1 contains the text General Chemistry I Lab, the formula =REPLACE(A1,19,1,"II") returns General Chemistry II Lab.

Related Functions: SEARCH, SUBSTITUTE

REPT

Repeats a text string a specified number of times.

Syntax: REPT(***text, number***)

Example: If *decimals* = 3, the formula ="0."&REPT("0",decimals) returns 0.000 as text.

RIGHT

Returns the specified number of characters from a text string, beginning at the right.

Syntax: RIGHT(***text***, *num_chars*)

If *num_chars* is omitted, it is assumed to be 1.

Example: =RIGHT(303585842,4) returns 5842.

Related Functions: MID, LEFT, LEN

ROMAN

Converts an Arabic number to Roman numerals.

Syntax: ROMAN(***number***, *form*)

If *form* = 0, TRUE or omitted, the "classic" form is returned. *Form* can also be 1, 2, or 3, in which case successively more "concise" forms are returned. If *form* = 4 or FALSE, a "simplified" form is returned.

Example: =ROMAN(1997,0) returns MCMXCVII.

ROUND

Rounds a number to a specified number of digits.

Syntax: ROUND(***number, digits***)

If the result ends in a zero, the zero is not displayed.

Example: If cell A1 contains 0.001736, the formula =ROUND(A1,5) returns 0.00174. If cell A1 contains 0.007702, the same formula returns 0.0077, i.e., the number is not padded out with trailing zeros.

Related Functions: INT, TRUNC, TEXT

ROW

Returns the row number of reference.

Syntax: ROW(*reference*)

If *reference* is a range of cells, returns the row number of the upper left cell of the range. *Reference* cannot be a multiple selection.

Example: =ROW(B5:C7) returns 5.

Related Functions: COLUMN, ROWS

ROWS

Returns the number of rows in reference.

Syntax: ROWS(*reference*)

If *reference* is a multiple selection, can use INDEX to select a specified area within the selection.

Example: =ROWS(B5:C7) returns 3.

Related Functions: ROW, COLUMNS

RSQ

Returns the correlation coefficient R^2.

Syntax: RSQ(*known_ys, known_xs*)

This is the correlation coefficient returned by LINEST. For more details, see LINEST.

Related Functions: LINEST, CORREL

SEARCH

Finds the position of a text value within a text string.

Syntax: SEARCH(*find_text, within_text, start_num*)

Find_text is the text to be found. Include the ? wildcard character in *find_text* to match any single character or * to match any sequence of characters within the source string *within_text*. *Start_num* is the position in *within_text* to begin searching. SEARCH is not case-sensitive, FIND is case-sensitive.

Example: See example under FIND.

Related Functions: FIND, REPLACE, SUBSTITUTE

SIGN

Returns the sign of a number.

Syntax: SIGN(*number*)

Returns 1, 0, –1 if the number is positive, zero or negative, respectively

Related Function: ABS

SIN

Returns the sine of a given angle.

Syntax: SIN(*number*)

Number is the angle in radians. To convert an angle in degrees to one in radians, multiply by $\pi/180$.

Related Functions: ASIN, COS, TAN, other trigonometric functions

SLOPE

Returns the slope of the linear regression line $y = mx + b$.

Syntax: SLOPE(*known_ys, known_xs*)

See LINEST for details.

Related Functions: INTERCEPT, LINEST

SMALL
Returns the *k*th smallest value in a list of values.
Syntax: SMALL(***array,k***)
Related Functions: LARGE, MAX, MIN

SQRT
Returns the square root of a number.
Syntax: SQRT(***number***)

STDEV
Returns the standard deviation of a sample.
Syntax: STDEV(***value1***, *value2,...*)
The sample can consist of up to 30 separate arguments or arrays.
Related Functions: AVERAGE, AVDEV

SUBSTITUTE
Finds and substitutes characters within a text string.
Syntax: SUBSTITUTE(***text, old_text, new_text***, *instance_num*)
Text is the string in which characters are to be substituted. *Old_text* is the text to be replaced. *New_text* is the text to replace *old_text*. *Instance_num* specifies which occurrence of *old_text* is to be replaced; if omitted, all are replaced.
Example: =SUBSTITUTE(A3," ","",2) replaces the second occurrence of a space in the string in cell A3 with a null .
Related Function: REPLACE

SUM
Returns the sum of all the numbers in the reference.
Syntax: SUM(***number1***, *number2,...*)
The arguments may be numbers, names, arrays or references that contain numbers. Up to 30 separate arguments. Only numbers in the array or range are counted.
Related Functions: AVERAGE, COUNT, COUNTA

SUMIF
Returns the sum of all the numbers in the reference that satisfy the specified criteria.
Syntax: SUMIF(***range, criteria***, *sum_range*)
Range is the range of cells in which criteria will be evaluated. *Sum_range* contains the cells that will be summed. If *sum_range* is omitted, cells in *range* are summed.
Example: =SUMIF(B3:B553,>100,D3:D553)
Related Functions: AVERAGE, COUNT, COUNTA

SUMPRODUCT
Returns the sum of the products of corresponding array components.
Syntax: SUMPRODUCT(***array1, array2***,...)

Up to 30 arrays may be included. All arrays must have the same dimensions.
Related Functions: MMULT, TRANSPOSE

TAN
Returns the tangent of a given angle.
Syntax: TAN(*number*)
Number is the angle in radians. To convert an angle in degrees to one in radians, multiply by $\pi/180$.
Related Functions: ATAN, ATAN2, other trigonometric functions

TEXT
Formats a number and returns it as text.
Syntax: TEXT(*value, format_text*)
Format_text is a number format, similar to those in the Number format dialog box. TEXT converts a number to text; the result will sometimes fail to be calculated as a number.
Example: The formula =TEXT(PI(),"0.000") in cell B1 returns 3.142. The formula =2*B1 returns 6.284, but the formula =SUM(B1) returns 0.
Related Function: FIXED

TIME
Returns the serial number of a time.
Syntax: TIME(*hour, minute, second*)
Hour is a number from 0 to 23; *minute* is a number from 0 to 59; *second* is a number from 0 to 59.
Related Function: NOW

TODAY
Returns the serial number of today's date.
Syntax: TODAY()
TODAY returns the integer part of the serial number returned by NOW.
Related Functions: DATE, DAY, NOW

TRANSPOSE
Returns the transpose of an array
Syntax: TRANSPOSE(*array*)
Must be entered as an array formula in a range with number of columns equal to number of rows in *array*, and number of rows equal to number of columns in *array*.
Example: Does the same thing as **Paste Special** (Transpose) from the **Edit** menu.
Related Functions: MDETERM, MINVERSE, MMULT, SUMPRODUCT

TRIM
Removes spaces from a text string, except for single spaces between words.
Syntax: TRIM(*text*)
Related Function: CLEAN

TRUNC
Truncates a number.
Syntax: TRUNC(*number*, *num_digits*)
Num_digits is the number of digits after the decimal point.   If omitted, *num_digits* is taken to be zero.   TRUNC does not round up.
Example: The formula =TRUNC(-8.913) returns -8.   The formula =TRUNC(PI(),3) returns 3.141.
Related Functions: INT, ROUND

TYPE
Returns a number indicating the data type of a value.
Syntax: TYPE(*value*)
Returns 1 (if *value* is a number), 2 (text), 4 (logical value), 16 (error value) or 64 (array).
Related Functions: CELL, IS functions

UPPER
Converts a text string to uppercase.
Syntax: UPPER(*text*)
Related Functions: LOWER, PROPER

VALUE
Converts a text argument to a number.
Syntax: VALUE(*text*)
Not usually necessary, since Excel automatically converts text to number.
Example: =ISNUMBER(MID("H2SO4",2,1)) returns FALSE but =ISNUMBER(VALUE(MID("H2SO4",2,1))) returns TRUE.
Related Functions: FIXED, TEXT

VAR
Returns the variance of a sample.
Syntax: VAR(*number1*,*number2*,...)
The sample can consist of up to 30 separate arguments or arrays.
Example: See *Microsoft Excel 97 Worksheet Function Reference* for details.
Related Function: STDEV

VLOOKUP
Finds the value in the first column of *array* that is equal to or less than *lookup_value*. Returns the associated value in the *n*th column, as determined by *offset_num*.

Syntax: VLOOKUP(*lookup_value, array, offset_num, match_logical*)
The values in the first column must be in ascending order. If *match_logical* is TRUE or omitted, returns the largest value that is less than or equal to *lookup_value*. If *match_logical* is FALSE, returns an exact match, #N/A if an exact match is not found.
Example: See example under HLOOKUP.
Related Functions: HLOOKUP, INDEX, MATCH

WEEKDAY
Converts a serial number to a day of the week.
Syntax: WEEKDAY(*serial_number, return_type_num*)
Returns a number indicating the day of the week. If *return_type_num* = 1 or omitted, returns 1 (Sunday) to 7 (Saturday). If *return_type_num* = 2, returns 1 (Monday) to 7 (Sunday). If *return_type_num* = 3, returns 0 (Monday) to 6 (Sunday).
Related Functions: DAY, NOW, TODAY

YEAR
Converts a serial number to a year.
Syntax: YEAR(*serial_number*)
Related Function: NOW

C

SELECTED
VISUAL BASIC KEYWORDS
BY CATEGORY

This appendix lists selected VBA keywords (reserved words) for functions, statements, methods and properties. See Excel's On-line Help for a complete list of keywords.

FUNCTIONS

Abs	Returns the absolute value of a number.
Array	Returns a **Variant** containing an array.
Asc	Returns the numeric code for the first character of text.
Atn	Returns the angle corresponding to a tangent value.
Chr	Returns the character corresponding to a code.
Cos	Returns the cosine of an angle.
Exp	Returns *e* raised to a power.
Fix	Truncates a number to an integer.
Format	Formats a value according to a formatting code expression.
InputBox	Displays an input dialog box and waits for user input.
Int	Rounds a number to an integer.
IsArray	Returns **True** if the variable is an array.
IsDate	Returns **True** if the expression can be converted to a date.
IsEmpty	Returns **True** if the variable has been initialized.
IsMissing	Returns **True** if an optional argument has not been passed to a procedure.
IsNull	Returns **True** if the expression is null (i.e., contains no valid data).
IsNumeric	Returns **True** if the expression can be evaluated to a number.
LBound	Returns the lower limit of an array dimension.
LTrim	Returns a string without leading spaces.
LCase	Converts a string into lowercase letters.
Left	Returns the leftmost characters of a string.

Len	Returns the length (number of characters) in a string.
Log	Returns the natural (base-*e*) logarithm of a number.
Mid	Returns a specified number of characters from a text string, beginning at a specified position.
MsgBox	Displays a message box.
Now	Returns the current date and time.
Right	Returns the rightmost characters of a string.
Rnd	Returns a random number between 0 and 1.
RTrim	Returns a string without trailing spaces.
Sgn	Returns the sign of a number.
Sin	Returns the sine of an angle.
Sqr	Returns the square root of a number.
Str	Converts a number to a string.
Tan	Returns the tangent of an angle.
Trim	Returns a string without leading or trailing spaces.
UBound	Returns the upper limit of an array dimension.
UCase	Converts a string into uppercase letters.
Val	Converts a string to a number.

STATEMENTS (COMMANDSS)

Beep	Makes a "beep" sound.
Call	Transfers control to a **Function** or **Sub** procedure.
Dim	Declares an array and allocates storage for it.
Do...Loop	Delineates a block of statements to be repeated.
Else	Optional part of **If...Then** structure.
ElseIf	Optional part of **If...Then** structure.
End	Terminates a procedure or block.
Exit	Exits a **Do**..., **For**..., **Function**... or **Sub**... structure.
For Each...Next	Delineates a block of statements to be repeated.
For...Next	Delineates a block of statements to be repeated.
Function	Marks the beginning of a **Function** procedure.
GoSub	Branches to a subroutine within a procedure.
GoTo	Unconditional branch within a procedure.
If...Then...End If	Delineates a block of conditional statements.
On...GoSub	Branches to one of several specified subroutines, depending on the value of an expression.
On...GoTo	Branches to one of several specified lines, depending on the value of an expression.
Option Base	Used at module level to declare lower bound for an array.
Preserve	Preserves data in an existing array when using **ReDim**.
Private	Indicates that the procedure is available to all other procedures.

Public	Indicates that the procedure is available only to procedures in the same module.
ReDim	Allocates or re-allocates dynamic array storage.
Return	Delineates the end of a subroutine within a procedure.
Select Case	Executes one of several blocks of statements, depending on the value of an expression.
Set	Assigns an object reference to a variable.
Static	Preserves a procedures local variables between calls.
Stop	Stops execution, but does not close files or clear variables.
Sub	Marks the beginning of a **Sub** procedure.
Until	Optional part of **Do...Loop** structure
While	Optional part of **Do...Loop** structure
With...End With	Delineates a block of statements to be executed on a single object.

METHODS

Activate	Activates an object.
Address	Returns a reference, as text
Cells	Returns a single cell by specifying the row and column.
Clear	Clears formulas and formatting from a range of cells.
Close	Closes a window, workbook or workbooks.
Columns	Returns a **Range** object that represents a single column or multiple columns.
Copy	Copies the selected object to the Clipboard or to another location.
Cut	Cuts the selected object to the Clipboard or to another location.
Delete	Deletes the selected object.
FillDown	Copies the contents and format(s) of the top cell(s) of a specified range into the remaining rows.
FillRight	Copies the contents and format(s) of the leftmost cell(s) of a specified range into the remaining columns.
InputBox	Displays an input dialog box and waits for user input.
Insert	Inserts a range of cells in a worksheet.
MacroOptions	Sets options in the Macro Options dialog box.
Paste	Pastes the contents of the Clipboard onto a worksheet.
Quit	Quits Microsoft Excel.
Range	Returns a **Range** object that represents a cell or range of cells.
Rows	Returns a **Range** object that represents a single row or multiple rows.
Save	Saves changes to active workbook.

SaveAs	Saves changes to active workbook or other document with a different filename.
Select	Selects an object.
Sort	Sorts a range of cells.

PROPERTIES

ActiveCell	Returns the active cell of the active window.
ActiveSheet	Returns the active sheet of the active workbook.
Bold	Returns **True** if the font is Bold. Sets the Bold font.
Column	Returns a number corresponding to the first column in the range.
Count	Returns the number of items in the collection.
Font	Returns the font of the object.
Formula	Returns or sets the formula associated with an object.
FontStyle	Returns or sets the font of the object.
Italic	Returns **True** if the font is Italic. Sets the Italic font.
Name	Returns or sets the name of an object.
NumberFormat	Returns or sets the number format code of a cell.
Row	Returns a number corresponding to the first row in the range.
Selection	Returns the selected object.
Text	Returns or sets the text associated with an object.
Value	Returns the value of an object.

OTHER KEYWORDS AND OPERATORS

False	Boolean keyword.
True	Boolean keyword.
And	Logical operator.
Or	Logical operator.

ALPHABETICAL LIST
OF SELECTED
VISUAL BASIC KEYWORDS

This listing of VBA functions, statements, methods and properties will be useful when creating your own macros. For each VBA keyword, the required syntax is given, along with some comments on the required and optional arguments, one or more examples and a list of related keywords. See Excel's On-line Help for further information.

Abs Function
Returns the absolute value of a number.
Syntax: **Abs**(*number*)
Example: **Abs**(-7.3) returns 7.3
See also: **Sgn**

Activate Method
Activates an object.
Syntax: *object*.**Activate**
Object can be **Chart, Worksheet** or **Window.**
Example: **Workbooks**("BOOK1.XLS").**Worksheets**("Sheet1").**Activate**
See also: **Select**

ActiveCell Property
Returns the active cell of the active window. Read-only.
Syntax: **ActiveCell** and **Application.ActiveCell** are equivalent.
See also: **Activate, Select**

ActiveSheet Property
Returns the active sheet of the active workbook. Read-only.
Syntax: *object*.**ActiveSheet**
Object can be **Application, Window** or **Workbook.**
Example: **Application.ActiveSheet.Name** returns the name of the active

sheet of the active workbook. Returns **None** if no sheet is active.
See also: **Activate, Select**

Address Method
Returns a reference, as text
*Syntax: object.***Address** *(rowAbsolute,columnAbsolute, referenceStyle, external, relativeTo)*
All arguments are optional. If *rowAbsolute* or *columnAbsolute* are True or omitted, returns that part of the address as an absolute reference. *ReferenceStyle* can be xlA1 or xlR1C1. If *external* is True, returns an external reference. See On-line Help for information about the *relativeTo* argument.
See also: **Offset**

Array Function
Returns a **Variant** containing an array.
Syntax: **Array** *(arglist)*
Example: **Array** *(31,28,31,30,31,30,31,31,30,31,30,31)*
See also: **Dim**

Asc Function
Returns the numeric code for the first character of text.
Syntax: **Asc**(*character*)
Example: **Asc** ("A") returns 65.
See also: **Chr**

Atn Function
Returns the angle corresponding to a tangent value.
Syntax: **Atn**(*number*)
Number can be in the range $-\infty$ to $+\infty$. The returned angle is in radians, in the range $-\pi/2$ to $+\pi/2$ ($-90°$ to $90°$). To convert the result to degrees, multiply by $180/\pi$.
Example: **Atn**(1) returns 0.785388573 or 45 degrees.
See also: **Cos, Sin, Tan**

Beep Command
Makes a "beep" sound.
Syntax: **Beep**

Bold Property
Returns **True** if the font is Bold. Sets the Bold font. Read-write.
*Syntax: object.***Bold**
Object must be **Font**.
Example: **Range**("A1:E1").**Font.Bold = True** makes the cells bold.
See also: **Italic**

Call Command
Transfers control to a **Function** or **Sub** procedure.
Syntax: **Call** *name (argument1, ...)*
Name is the name of the procedure. *Argument1*, etc., are the names assigned to the arguments passed to the procedure. **Call** is optional; if omitted, the parentheses around the argument list must also be omitted.
Example: **Call** Task1(argument1,argument2)
See also: **Sub, Function**

Cells Method
Returns a single cell by specifying the row and column.
*Syntax: object.***Cells**(*row, column*)
Object is optional; if not specified, **Cells** refers to the active sheet.
Example: **Cells**(2,1).**Value** = 5 enters the value 5 in cell A2.
See also: **Range**

Chr Function
Returns the character corresponding to a code.
Syntax: **Chr**(*number*)
Number must be between 1 and 255.
Example: **Chr**(65) returns A.
See also: **Asc**

Clear Method
Clears formulas and formatting from a range of cells.
*Syntax: object.***Clear**
Object can be **Range** (or **ChartArea**).
Example: **Range**("A1:C10").**Clear**
See also: **ClearContents, ClearFormats** in Excel's On-line Help.

Close Method
Closes a window, workbook or workbooks.
Syntax: For workbooks, use *object.***Close**. For a workbook or window, use *object.***Close**(*SaveChangesLogical, FileName*).
Object can be **Window, Workbook** or **Workbooks**. If *SaveChangesLogical* is **False**, does not save changes; if omitted, displays a "Save Changes?" dialog box.
Example: **Workbooks**("BOOK1.XLS").**Close**
See also: **Open, Save, SaveAs**

Column Property
Returns a number corresponding to the first column in the range. Read-only.
*Syntax: object.***Column**
Object must be **Range**.
See also: **Columns, Row, Rows**

Columns Method

Returns a **Range** object that represents a single column or multiple columns

*Syntax: object.***Columns**(*index*)

Object can be **Worksheet** or **Range**. *Index* is the name or number (column A = 1, etc.) of the column.

Example: **Selection.Columns.Count** returns the number of columns in the selection.

See also: **Range, Rows**

ColumnWidth Property

Returns or sets the width of all columns in the range. If columns in the range have different widths, returns **Null**.

Example: **Worksheets**("Sheet1").**Columns**("C").**ColumnWidth** = 30

See also: **RowHeight**

Copy Method

Copies the selected object to the Clipboard or to another location.

*Syntax: object.***Copy**(*destination*)

Object can be **Range, Worksheet, Chart** and many other objects. *Destination* specifies the range where the copy will be pasted. If omitted, copy goes to the Clipboard.

Example: **Worksheets**("Sheet1").**Range**("A1:C50").**Copy**

See also: **Cut, Paste**

Cos Function

Returns the cosine of an angle.

Syntax: **Cos**(*number*)

Number is the angle in radians; it can be in the range $-\infty$ to $+\infty$. To convert an angle in degrees to one in radians, multiply by $\pi/180$. Returns a value between -1 and 1.

See also: **Atn, Sin, Tan**

Count Property

Returns the number of items in the collection. Read-only.

*Syntax: object.***Count**

Object can be any collection.

Example: The statement N = array.**Count** counts the number of values in the range array.

Cut Method

Cuts the selected object to the Clipboard or to another location.

*Syntax: object.***Cut**(*destination*)

Object can be **Range, Worksheet, Chart** or one of many other objects. *Destination* specifies the range where the copy will be pasted. If omitted, copy goes to the Clipboard.

Example: **Worksheets**("Sheet1").**Range**("A1:C50").**Cut**

See also: **Copy, Paste**

CVErr Function

Returns a Variant containing an error value specified by the user.

Syntax: **CVErr**(*number*)

CVErr can return either Excel's built-in worksheet error values, or a user-defined error value. The values of *number* for built-in worksheet error values are xlErrDiv0, xlErrNA, xlErrName, xlErrNull, xlErrNum, xlErrRef, xlErrValue.

See also: **IsError**

Delete Method

Deletes the selected object.

Syntax: *object*.**Delete***(shift)*

Object can be **Range**, **Worksheet**, **Chart** and many other objects.   *Shift* specifies how to shift cells when a range is deleted from a worksheet (xlToLeft or xlUp). Can also use *shift* = 1 or 2, respectively. If *shift* is omitted, Excel moves the cells without displaying the Shift Cells? dialog box.

Example: **Worksheets**("Sheet12").**Range**("A1:A10").**Delete** (xlToLeft) deletes the indicated range and shifts cells to left.

Dim Command

Declares an array and allocates storage for it.

Syntax: **Dim** *variable* *(subscripts)*

Variable is the name assigned to the array. *Subscripts* are the dimensions of the array; an array can have up to 60 dimensions. Each dimension has a default lower value of zero; a single number for a dimension is taken as the upper limit. Use *lower* **To** *upper* to specify a range that does not begin at zero. Use **Dim** with empty parentheses to specify an array whose dimensions are defined within a procedure by means of the **ReDim** statement.

Example: **Dim** Matrix (5,5) creates a 6 × 6 array.

See also: **ReDim**

Do...Loop Command

Delineates a block of statements to be repeated.

Syntax: The beginning of the loop is delineated by **Do** or **Do Until** *condition* or **Do While** *condition*. The end of the loop is delineated by **Loop** or **Loop Until** *condition* or **Loop While** *condition*.   *Condition* must evaluate to **True** or **False**.

Example: See examples of **Do...Loop** structures in Chapter 14.

See also: **Exit, For, Next, Wend, While**

Else Command
Optional part of **If...Then** structure.

ElseIf Command
Optional part of **If...Then** structure.

End Command
Terminates a procedure or block.
Syntax: **End** terminates a procedure. **End Function** is required to terminate a
Function procedure. **End If** is required to terminate a block **If** structure. **End
Select** is required to terminate a **Select Case** structure. **End Sub** is required
to terminate a **Sub** procedure. **End With** is required to terminate a **With**
structure.
Example: See examples under **Select Case**.
See also: **Exit, Function, If, Then, Else, Select Case, Sub, With**

Exit Command
Exits a **Do...**, **For...**, **Function**... or **Sub**... structure.
Syntax: **Exit Do, Exit For, Exit Function, Exit Sub**
From a **Do** or **For** loop, control is transferred to the statement following the
Loop or **Next** statement, or, in the case of nested loops, to the loop that is one
level above the loop containing the **Exit** statement. From a **Function** or **Sub**
procedure, control is transferred to the statement following the one that called
the procedure.
Example: See examples of **Exit** procedures in Chapter 14.
See also: **Do, For...Next, Function, Stop, Sub**

Exp Function
Returns *e* raised to a power.
Syntax: **Exp**(*number*)
Returns the value of *e* raised to the power *number*.
See also: **Log**

False Keyword
Use the keywords **True** or **False** to assign the value **True** or **False** to Boolean
(logical) variables.
When other numeric data types are converted to Boolean values, 0 becomes
False while all other values become **True**. When Boolean values are converted
to other data types, **False** becomes 0 while **True** becomes -1.
Example: **If** SubFlag = **False Then**...
See also: **True**

FillDown Method
Copies the contents and format(s) of the top cell(s) of a specified range into the remaining rows.
*Syntax: object.***FillDown**
Object must be **Range**.
Example: **Worksheets**("Sheet12").**Range**("A1:A10").**FillDown**
See also: **FillLeft, FillRight, FillUp** in Excel's On-line Help.

FillRight Method
Copies the contents and format(s) of the leftmost cell(s) of a specified range into the remaining columns.
*Syntax: object.***FillDown**
Object must be **Range**.
Example: **Worksheets**("Sheet12").**Range**("A1:A10").**FillRight**
See also: **FillDown, FillLeft, FillUp** in Excel's On-line Help.

Fix Function
Truncates a number to an integer.
Syntax: **Fix**(*number*)
If *number* is negative, **Fix** returns the first negative integer greater than or equal to *number*.
Example: **Fix**(-2.5) returns −2.
See also: **Int**

Font Property
Returns the font of the object. Read-only.
*Syntax: object.***Font**
Example: ActiveCell.Font.Bold − **True** makes the characters in the active cell bold.
See also: **FontStyle**

FontStyle Property
Returns or sets the font of the object. Read-write.
*Syntax: object.***FontStyle**
Example: **Range**("A1:E1").**Font.FontStyle** = "Bold"
See also: **Font**

For...Next Command
Delineates a block of statements to be repeated.
Syntax: **For** *counter = start* **To** *end* **Step** *increment*
                 (statements)
                 Next *counter*
Step *increment* is optional; if not included, the default value 1 is used. *Increment* can be negative, in which case *start* should be greater than *end*.

Example: See examples of **For...Next** procedures in Chapter 14.
See also: **Do...Loop, Exit, For Each...Next, While...Wend**

For Each...Next Command
Delineates a block of statements to be repeated.
Syntax: **For Each** *element* **In** *group*
                   (statements)
                   Next *element*
Group must be a collection or array. *Element* is the name assigned to the variable used to step through the collection or array. *Group* must be a collection or array.
Example: See examples of **For Each...Next** procedures in Chapter 14.
See also: **Do...Loop, Exit, For...Next, While...Wend**

Format Function
Formats a value according to a formatting code expression.
Syntax: **Format(*expression,formattext*)**
Expression is usually a number, although strings can also be formatted. *Formattext* is a built-in or custom format. Additional information can be found in *Microsoft Excel/Visual Basic Reference,* or VBA On-line Help.
Example: **Format**(TelNumber,"(###) ###-####") formats the value TelNumber in the form of a telephone number.

Formula Property
Returns or sets the formula in a cell.
If a cell contains a value, returns the value; if the cell contains the formula, returns the formula as a string.
See also: **Text, Value**

Function Command
Marks the beginning of a **Function** procedure.
Syntax: **Function** *name argument1, ...*
Name is the name of the variable whose value is passed back to the caller. *Argument1,* etc., are the names assigned to the arguments passed from the caller to the procedure.
Example: See examples of **Function** procedures in Chapter 14.
See also: **Call, Sub**

GoSub Command
Branches to a subroutine within a procedure.
Syntax: **GoSub** *label*
The beginning of the subroutine is delineated by *label,* which can be a name or a line number. The end of the subroutine is delineated by one or more **Return** statements. The subroutine must be within the calling procedure.
Example: See examples of subroutines in Chapter 14.

See also: GoTo, On...GoSub, On...GoTo, Return, Sub

GoTo Command
Unconditional branch within a procedure.
Syntax: GoTo *label*
Label can be a name or a line number.
See also: GoSub

If...Then...Else...End If Command
Delineates a block of conditional statements.
Syntax: If *condition* Then ... Else ... End If
The statement can be all on one line (e.g., If *condition* Then *statement*).
Alternatively, a block If structure can be used, in which case the first line
consists of If *condition* Then; the end of the structure is delineated by End If.
Condition must evaluate to True or False. The ellipsis following Then and
Else can represent a single statement or several statements separated by colons;
these are executed if *condition* is True or False, respectively.
Examples: If Char = "." Then GoTo 2000
               If (Char >= "0" And Char <= "9") Then
               (statements)
               End If
See also: ElseIf, End

InputBox Function
Displays an input dialog box and waits for user input.
Syntax: InputBox(*prompt,title,default,xpos,ypos,helpfile,context*)
See *Microsoft Excel/Visual Basic Reference* or On-line Help for details.
See also: InputBox Method, MsgBox

InputBox Method
Displays an input dialog box and waits for user input.
*Syntax: object.*InputBox(*prompt,title,default,left,top,helpfile,context,
type*)
Object must be Application. See *Microsoft Excel/Visual Basic Reference* or On-
line Help for details.
See also: InputBox Function, MsgBox

Insert Method
Inserts a range of cells in a worksheet.
*Syntax: object.*Insert*(shift)*
Object is a Range object. *Shift* specifies how to shift cells when a range is
inserted in a worksheet (xlToRight or xlDown). Can also use *shift* = 1 or 2,
respectively. If *shift* is omitted, the Shift Cells? dialog box is not displayed.
Examples: Worksheets("Sheet12").Range("A1:A10").Insert (1) inserts

the indicated range and shifts cells to right.
Worksheets("Sheet1").**Columns**(4).**Insert** inserts a new column to the left of column D.
See also: **Delete**

Int Function
Rounds a number to an integer.
Syntax: **Int**(*number*)
If *number* is negative, **Int** returns the first negative integer less than or equal to *number*.
Example: **Int**(-2.5) returns –3.
See also: **Fix**

IsArray Function
Returns **True** if the variable is an array.
Syntax: **IsArray**(*name*)
See also: other **Is** functions

IsDate Function
Returns **True** if the expression can be converted to a date.
Syntax: **IsDate**(*expression*)
See also: other **Is** functions

IsEmpty Function
Returns **True** if the variable has been initialized.
Syntax: **IsEmpty**(*expression*)
See also: other **Is** functions

IsMissing Function
Returns **True** if an optional argument has not been passed to a procedure.
Syntax: **IsMissing**(*name*)
See also: other **Is** functions

IsNull Function
Returns **True** if the expression is null (i.e., contains no valid data).
Syntax: **IsNull**(*expression*)
See also: other **Is** functions

IsNumeric Function
Returns **True** if the expression can be evaluated to a number.
Syntax: **IsNumeric**(*expression*)
See also: other **Is** functions

Italic Property
Returns **True** if the font is Italic. Sets the Italic font. Read-write.
*Syntax: object.***Italic**
Object must be **Font**.
Example: **Range("A1:E1").Font.Italic = True** makes the cells italic.
See also: **Bold**

LBound Function
Returns the lower limit of an array dimension.
Syntax: **LBound**(*array,dimension*)
Array is the name of the array. *Dimension* is an integer (1, 2, 3, etc.) specifying the dimension to be returned; if omitted, the value 1 is used.
Example: If the array table was dimensioned using the statement **Dim** table (1 **To** 3, 1000), **LBound**(table,1) returns 1, **LBound**(table,2) returns 0.
See also: **Dim, UBound**

LCase Function
Converts a string into lowercase letters.
Syntax: **LCase**(*string*)
See also: **UCase**

LTrim Function
Returns a string without leading spaces.
Syntax: **LTrim**(*string*)
See also: **RTrim**

Left Function
Returns the leftmost characters of a string.
Syntax: **Left**(*string,number*)
If *number* is zero, a null string is returned. If *number* is greater than the number of characters in *string*, the entire string is returned.
Example: **Left("CHEMISTRY",4)** returns CHEM
See also: **Len, Mid, Right**

Len Function
Returns the length (number of characters) in a string.
Syntax: **Len**(*string*)
Example: **Len("CHEMISTRY")** returns 9.
See also: **Left, Mid, Right**

Log Function
Returns the natural (base-e) logarithm of a number.
Syntax: **Log**(*number*)
Number must be a value or expression greater than zero. VBA does not provide base-10 logarithms; use **Log**(value)/**Log**(10).

See also: **Exp**

MacroOptions Method
Sets options in the Macro Options dialog box.
Syntax: **Application.MacroOptions**(*macro, description, hasMenu, menuText, hasShortcutKey, shortcutKey, category, statusbar, helpContext, helpFile*)

macro is the name of the macro. *description* is the description that appears in the dialog box. *category* is the function category that the macro appears in: Financial, 1; Date & Time, 2; Math & Trig, 3; Statistical, 4; Lookup & Reference, 5; Database, 6; Text, 7; Logical, 8; Information, 9; User Defined, 14; Engineering, 15.
Example: **Application.MacroOptions** macro:="FtoC", Description:= "Converts Fahrenheit temperature to Celsius", Category:=3
provides a description for the macro FtoC and assigns it to the Math & Trig category.

Mid Function
Returns the specified number of characters from a text string, beginning at the specified position.
Syntax: **Mid**(*string,start,number*)
If *start* is greater than the number of characters in *string*, returns a null string. If *number* is omitted, all characters from *start* to the end of the string are returned.
Example: **Mid**("H2SO4",2,1) returns 2.
See also: **Left, Len, Right**

MsgBox Function
Displays a message box.
Syntax: **MsgBox**(*prompt,buttons,title,helpfile,context*)
See *Microsoft Excel/Visual Basic Reference* or On-line Help for details.
See also: **InputBox**

Name Property
Returns or sets the name of an object.
Example: SeriesName = **Selection.Name** assigns the name of the selected chart series to the variable SeriesName.
See also: **NameLocal, Names**

Now Function
Returns the current date and time.
Syntax: **N o w**
See also: other date and time functions.

NumberFormat Property
Returns or sets the number format code of a cell.
Example: **Range**("A1:A10").**NumberFormat=** "0.00" sets the number

format of the specified range of cells.
See also: **GoSub, GoTo, Return, Select Case**

On...GoSub Command
Branches to one of several specified subroutines, depending on the value of an expression.
Syntax: **On** *expression* **GoSub** *label1, ...*
Control is transferred to the label (line number or label) determined by the value of *expression*. *Expression* is any expression that evaluates to a number; if not an integer, it is rounded to the nearest whole number. The number must be between 0 and 255.
Example: See examples of **On...GoSub** procedures in Chapter 14.
See also: **GoSub, GoTo, Return, Select Case**

On...GoTo Command
Branches to one of several specified lines, depending on the value of an expression.
Syntax: **On** *expression* **GoTo** *label1, ...*
See explanation under **On...GoSub** Command.
Example: See examples of **On...GoTo** procedures in Chapter 14.
See also: **GoSub, GoTo, Return, Select Case**

Open Method
Opens a workbook.
Syntax: *object.***Open**(*filename, ...*)
Object must be **Workbooks**. *Filename* is required. See On-line Help for the remaining arguments.
Example: **Workbooks.Open**("SOLVSTAI.XLS")
See also: **Close, Save, SaveAs**

Option Base Statement
Used at module level to declare lower bound for an array.
Can be **Option Base** 0 or 1. The statement can appear only once in a module and must precede all **Dim** or equivalent declaration.
See also: **Dim, LBound, ReDim**

Paste Method
Pastes the contents of the Clipboard onto a worksheet.
Syntax: *object.***Paste**(*destination*)
Object must be **Worksheet.** There are other **Paste** methods, with different syntax, for **Chart** and many other objects. *Destination* specifies the range where the copy will be pasted. If omitted, copy is pasted to the current selection.
Example: **Worksheets**("Sheet1").**Range**("A1:C50").**Copy**
        ActiveSheet.Paste
See also: **Copy, Cut**

ParamArray Command
Allows the use of an indefinite number of arguments for a function. The specified argument is an **Optional** array of **Variant** elements.
See also: **Dim, Function**

Preserve Command
Preserves data in an existing array when using **ReDim**.

Private Command
Indicates that the procedure is available to all other procedures.

Public Command
Indicates that the procedure is available only to procedures in the same module.

Quit Method
Quits Microsoft Excel.
*Syntax: object.***Quit**
Object must be **Application**.
Example: **Application.Quit**
See also: **Close, Save**

Range Method
Returns a **Range** object that represents a cell or range of cells.
*Syntax: object.***Range**(*reference*)
Object is required if it is **Worksheet**. *Reference* must be an A1-style reference, in quotes, or the name of the reference.
Example: **Worksheets**("Sheet12").**Range**("A1").**Value** = 5
See also: **Cells**

ReDim Command
Allocates or re-allocates dynamic array storage.
Syntax: **ReDim** *variable* *(subscripts)*
For discussion of *variable* and *subscripts*, see comments under the entry for **Dim**. You can use **ReDim** repeatedly to change the number of elements in an array, or the number or dimensions.
Example: **Dim** Matrix()
          (statements)
          ReDim Matrix (5,5)
          (statements)
          ReDim Matrix (15,25)
See also: **Dim**

Return Command
Delineates the end of a subroutine within a procedure.
See also: **GoSub**

Right Function
Returns the rightmost characters of a string.
Syntax: **Right**(*string,number*)
If *number* is zero, a null string is returned. If *number* is greater than the number
of characters in *string,* the entire string is returned.
Example: **Right**(303585842,4) returns 5842.
See also: **Left, Len, Mid**

Rnd Function
Returns a random number between 0 and 1.
Syntax: **Rnd**

Row Property
Returns a number corresponding to the first row in the range. Read-only.
*Syntax: object.***Row**
Object must be **Range**.
Example: **If ActiveCell.Row = 10 Then**
ActiveCell.Interior.ColorIndex = 27
changes the interior color of the active cell to yellow if it is in row 10.
See also: **Column, Columns, Rows**

RowHeight Property
Returns or sets the height of all rows in the range.
Example: **Worksheets**("Sheet1").**Rows**(1).**RowHeight** = 1 5
See also: **ColumnWidth**

Rows Method
Returns a **Range** object that represents a single row or multiple rows.
*Syntax: object.***Rows**(*index*)
Object can be **Worksheet** or **Range**. *Index* is the name or number of the row.
Example: **Selection.Rows.Count** returns the number of rows in the
selection.
See also: **Columns, Range**

RTrim Function
Returns a string without trailing spaces.
Syntax: **RTrim**(*string*)
See also: **LTrim, Trim**

Save Method
Saves changes to active workbook.
*Syntax: object.***Save**(*filename*)
Object must be **Workbook**. If *filename* is omitted, uses a default name.
Example: **ActiveWorkbook.Save**
See also: **Close, Open, SaveAs**

SaveAs Method
Saves changes to active workbook or other document with a different filename.
Syntax: *object*.**SaveAs**(*filename, ...*)
Object can be **Worksheet, Workbook, Chart** or other document types. See
Microsoft Excel/Visual Basic Reference or On-line Help for details.
Example: NewChart.**SaveAs**("New Chart")
See also: **Close, Open, Save**

Select Method
Selects an object.
Syntax: *object*.**Select**
Object can be **Chart, Worksheet** or one of many other objects.
Example: **Range**("A1:C50").**Select**
See also: **Activate**

Select Case Command
Executes one of several blocks of statements, depending on the value of an
expression.
Syntax: **Select Case** *expression*
            Case *expression1*
            (statements)
            Case *expression2*
            (statements)
            End Select
You can also use the **To** keyword in *expression* , e.g., **Case** "A" **To** "M".
Expression can also be a logical expression. Use **Case Else** (not required) to
handle all cases not covered by the preceding **Case** statements.
Example: See examples of **Select Case** procedures in Chapter 14.
See also: **If...Then...Else, On...GoSub, On...GoTo**

Selection Property
Returns the selected object. The object returned depends on the type of selection.
See also: **Activate, ActiveCell, Select**

Set Command
Assigns an object reference to a variable.
See also: **Dim, ReDim**

Sgn Function
Returns the sign of a number.
Syntax: **Sgn**(*number*)
Returns 1, 0 or −1 if *number* is positive, zero or negative, respectively.
Example: **Sgn**(-7.3) returns −1.
See also: **Abs**

Sin Function
Returns the sine of an angle.
Syntax: **Sin**(*number*)
Number is the angle in radians; it can be in the range −∞ to +∞. To convert an angle in degrees to one in radians, multiply by π/180. Returns a value between −1 and 1.
See also: **Atn, Cos, Tan**

Sort Method
Sorts a range of cells.
Syntax: *object*.**Sort**(*sortkey1,order1,sortkey2,order2, ...*)
Object must be **Range.** See *Microsoft Excel/Visual Basic Reference* or On-line Help for details.

Sqr Function
Returns the square root of a number.
Syntax: **Sqr**(*number*)
Number must be greater than or equal to zero.

Stop Command
Stops execution, but does not close files or clear variables.
See also: **End**

Str Function
Converts a number to a string.
Syntax: **Str**(*number*)
A leading space is reserved for the sign of the number; if the number is positive, the string will contain a leading space.
See also: **Format**

Sub Command
Marks the beginning of a **Sub** procedure.
Syntax: **Sub** *name argument1, ...*
Name is the name of the procedure. *Argument1*, etc., are the names assigned to the arguments passed from the caller to the procedure. The end of the procedure is delineated by **End Sub**
Example: See examples of **Sub** procedures in Chapter 14.
See also: **Call, Function**

Tan Function
Returns the tangent of an angle.
Syntax: **Tan**(*number*)
Number is the angle in radians; it can be in the range −∞ to +∞. To convert an angle in degrees to one in radians, multiply by π/180. Returns a value between −∞ and +∞.

See also: **Atn, Cos, Sin**

Text Property
Returns or sets the text associated with an object.
The text can be associated with a chart, button, textbox, control or range. For all except range, this property is read-write, but for a range, it is read-only.
Example: **Worksheets**("Sheet1").**Buttons**(1).**Text** = "Undo"
See also: **Formula, Value**

Trim Function
Returns a string without leading or trailing spaces.
Syntax: **Trim**(*string*)
See also: **LTrim, RTrim**

True Keyword
Use the keywords **True** or **False** to assign the value **True** or **False** to Boolean (logical) variables.
When other numeric data types are converted to Boolean values, 0 becomes **False** while all other values become **True**. When Boolean values are converted to other data types, **False** becomes 0 while **True** becomes –1.
Example: **If** FirstFlag = **True Then GoTo** 2000
See also: **False**

UBound Function
Returns the upper limit of an array dimension.
Syntax: **UBound**(*array*, *dimension*)
Array is the name of the array. *Dimension* is an integer (1, 2, 3, etc.) specifying the dimension to be returned; if omitted, the value 1 is used.
Example: If the array table was dimensioned using the statement **Dim** table (1 **To** 3, 1000), **UBound**(table,3) returns 1, **UBound**(table,2) returns 1000.
See also: **Dim, LBound**

UCase Function
Converts a string into upper case letters.
Syntax: **UCase**(*string*)
See also: **LCase**

Until Command
Optional part of **Do...Loop** structure.
Syntax: See explanation under **Do...Loop**.

Val Function
Converts a string to a number.
Syntax: **Val**(*string*)
Val stops at the first non-numeric character other than the period.

Example: **Val**("21 Lawrence Avenue") returns 21.
See also: **Str**

Value Property
Returns the value of an object.
*Syntax: object.***Value**
If *object* is **Range**, returns or sets the value(s) of the cell(s). Read-write.
If **Range** contains more than one cell, returns an array of values.
Example: **Worksheets**("Sheet12").**Range**("A1").**Value** = "Volume, mL"

Wend Command
Delineates the end of a **While...Wend** procedure.
Syntax: See explanation under **Do...Loop.**
See also: **Do...Loop, While...Wend**

While Command
Executes a series of statements as long as a specified condition is true.
Syntax: See explanation under **Do...Loop.**
See also: **Do...Loop, Wend**
With...End With Command
Delineates a block of statements to be executed on a single object.
Syntax: **With** *object*
            (statements)
            End With
Example: **With** ActiveSheet.PageSetup
            .CenterHeader = "DRAFT VERSION"
            .CenterFooter = ""
            End With
See also: **Do...Loop, While...Wend**

E

SHORTCUT KEYS
FOR PC AND MACINTOSH

Key	Applies to*	PC (Excel 97, Excel 2000)	Macintosh (Excel 98, Excel 2001)
arrow keys	W	Move one cell in direction of arrow	Move one cell in direction of arrow
arrow keys	D	Move between options in a group of options	Move between options in a group of options
arrow keys	W	Move selected object(s)	Move selected object(s)
right/left arrow keys	C	Select the next/previous element within a group of chart elements	Select the next/previous element within a group of chart elements
right/left arrow keys	D	Move one character to the right or left in the entry in an edit box	Move one character to the right or left in the entry in an edit box
up/down arrow keys	C	Select the next/previous chart element	Select the next/previous chart element
up/down arrow keys	D	Move between options in a list box	Move between options in a list box
BACKSPACE	F	Delete character to the left of the insertion point, or selected character(s)	N/A
COMMAND	W	N/A	Make a non-adjacent selection
CONTROL		Make a non-adjacent selection	Display a shortcut menu

(W = Worksheet, F = Formula bar, C = Chart, D = Dialog box, V = Visual Basic editor)

CONTROL	W	When using **AutoFill,** copy a cell or range rather than create a series	When using **AutoFill,** display the shortcut menu
DELETE	F	Delete character to the right of the insertion point	Clear selected character(s) to the left of the insertion point
DELETE	W	Delete contents of selected cells	
END	D	Move to the end of the entry in an edit box	Move to the end of the entry in an edit box
END	F	Extend selection to last-used cell (lower-right corner)	N/A
END	W	Toggle END mode on/off	N/A
ENTER	F, W	Enter data and move down one cell	Enter data and move down one cell
ENTER/ RETURN	D	Perform the action assigned to the default button (usually the OK button)	Perform the action assigned to the default button (usually the OK button)
ESC	D	Cancel and close the dialog box or shortcut menu	Cancel and close the dialog box or shortcut menu
ESC	F	Undo editing	Undo editing
ESC	W,V	Halt execution of a macro	Halt execution of a macro
HOME	D	Move to the beginning of the entry in an edit box	Move to the beginning of the entry in an edit box
HOME	F	Move insertion point to beginning	Move insertion point to beginning
HOME	W	Select cell in column A in same row as active cell	Select cell in column A in same row as active cell
INS	W	Toggle between Insert and Typeover modes	N/A
OPTION	W	N/A	When using **AutoFill,** copy a cell or range rather than create a series
SHIFT	W	Select multiple objects on worksheet	Select multiple objects on worksheet
TAB	D	Move to the next option	Move to the next option

(W = Worksheet, F = Formula bar, C = Chart, D = Dialog box, V = Visual Basic editor)

TAB	D	Move to the next input box	Move to the next input box
TAB	W	Move right one cell, or move to next unlocked cell in a protected worksheet	Move right one cell, or move to next unlocked cell in a protected worksheet
TAB	W	Move to the next toolbutton (when toolbar is active)	Move to the next toolbutton (when toolbar is active)
(space)	D	Perform the action assigned to the active button, or select or clear the active check box	Perform the action assigned to the active button, or select or clear the active check box

CONTROL+(key)

arrow keys	W	Jump to end of block of data in direction of arrow	Jump to end of block of data in direction of arrow
right/ left arrow keys	F	Move to beginning of next element in formula, in direction of arrow	Move to beginning of next element in formula, in direction of arrow
right/ left arrow keys	D	Move one word to the right or left in the entry in an edit box	Move one word to the right or left in the entry in an edit box
BACKSPACE	W	Display active cell	N/A
BREAK	V	Halt macro execution	N/A
DELETE	F	Clear from insertion point to end	Clear from insertion point to end
END	W	Select bottom right cell	N/A
ENTER/ RETURN	W	Enter the contents of the formula bar in the selected range (references are adjusted)	Enter the contents of the formula bar in the selected range (references are adjusted)
PAGE DOWN	W	Select next sheet in workbook	Select next sheet in workbook
PAGE DOWN	D	Switch to next tab	Switch to next tab
PAGE UP	W	Select previous sheet in workbook	Select previous sheet in workbook
PAGE UP	D	Switch to previous tab	Switch to previous tab

(W = Worksheet, F = Formula bar, C = Chart, D = Dialog box, V = Visual Basic editor)

TAB	W	Move to the previous toolbutton (when toolbar is active)	Move to the previous toolbutton (when toolbar is active)
'(apostrophe)	W	Copy value or formula from cell above (references are not adjusted)	Copy value or formula from cell above (references are not adjusted)
(space)	V	Display global list of Methods and Properties (with cursor on blank line)	
(space)	V	Jump to end of next element in line of code	
(space)	W	Select entire column	Select entire column
* (numeric keypad)	W	Select block of data containing the active cell	
-(minus)	W	Display the **Delete...** dialog box	Display the **Delete...** dialog box
. (period)	W	Move clockwise to the next corner of the selection	Move clockwise to the next corner of the selection
/(slash)	W	Select all cells of the array containing the active cell	Select all cells of the array containing the active cell
:(colon)	W	Insert current time	Insert current time
;(semicolon)	W	Insert current date	Insert current date
[	W	Select cells that are direct precedents of cells in selection	Select cells that are direct precedents of cells in selection
]	W	Select cells that are direct dependents of cells in selection	Select cells that are direct dependents of cells in selection
~ (tilde)	W	Toggle between values and formulas	Toggle between values and formulas
0	W	**Hide** column(s) containing selection	**Hide** column(s) containing selection
1	W	Display the Format Cells dialog box	
2	W	Apply or remove Bold formatting	

(W = Worksheet, F = Formula bar, C = Chart, D = Dialog box, V = Visual Basic editor)

3	W	Apply or remove Italic formatting	
4	W	Apply or remove Underline formatting	
5	W	Apply or remove Strikethrough formatting	
6	W	Toggle between displaying objects, placeholders for objects or hiding objects	Toggle between displaying objects, placeholders for objects or hiding objects
7	W	Toggle display of Standard toolbar	Toggle display of Standard toolbar
8	W	Toggle display of outline symbols	Toggle display of outline symbols
9	W	**Hide** row(s) containing selection	**Hide** row(s) containing selection
A	F	Display the Formula Palette (the Insert Function Step 2 dialog box) after typing a function name in a formula	Display the Formula Palette (the Insert Function Step 2 dialog box) after typing a function name in a formula
A	W	Select entire worksheet	
B	W,F	Apply or remove Bold formatting	Apply or remove Bold formatting
C	W,F	**Copy**	**Copy**
D	W	**Fill Down**	**Fill Down**
F	W	Display the **Find...** dialog box	Display the **Find...** dialog box
G	V	Display Immediate window	
G	W	Display the **Go To...** dialog box	Display the **Go To...** dialog box
H	W	Display the **Replace...** dialog box	Display the **Replace...** dialog box
I	W,F	Apply or remove Italic formatting	Apply or remove Italic formatting

(W = Worksheet, F = Formula bar, C = Chart, D = Dialog box, V = Visual Basic editor)

J	V	Display global list of Methods and Properties (with cursor on blank line)	
J	V	Display list of Methods and Properties for selected object	
K	W	Display the **Insert Hyperlink** dialog box	Display the **Insert Hyperlink** dialog box
N	W	**New** (Open a new workbook)	**New** (Open a new workbook)
O	W	Display the **Open...** dialog box	Display the **Open...** dialog box
P	W	Display the **Print...** dialog box	Display the **Print...** dialog box
R	V	Display the Project Explorer window	
R	W	**Fill Right**	**Fill Right**
S	W	**Save** the active workbook	**Save** the active workbook
U	W	Apply or remove Underline formatting	Apply or remove Underline formatting
V	W,F	**Paste**	**Paste**
X	W,F	**Cut**	**Cut**
Y	V	Delete line in a macro	
Y	W	**Repeat**	**Repeat**
Z	W	**Undo**	**Undo**

SHIFT+(key)

arrow keys	W	Extend selection by one row/column	Extend selection by one row/column
arrow keys	W	With END mode on, extend selection to end of block in direction of arrow	N/A
right/left arrow keys	D	Select or unselect one character to the right or left in the entry in an edit box	

(W = Worksheet, F = Formula bar, C = Chart, D = Dialog box, V = Visual Basic editor)

(space)	W	Select entire row	Select entire row
END	D	Select from the insertion point to the end of the entry in an edit box	N/A
END	F	Extend selection from insertion point to end of current line in formula	N/A
END	W	Extend selection to end of row	N/A
ENTER	W	Enter data and move up one cell	Enter data and move up one cell
HOME	D	Select from the insertion point to the beginning of the entry in an edit box	
HOME	F	Extend selection from insertion point to beginning of current line in formula	Extend selection from insertion point to beginning of current line in formula
HOME	W	Extend selection to beginning of row	Extend selection to beginning of row
HOME	W	With END mode on, extend selection to last-used cell (lower-right corner)	N/A
RETURN	W	With END mode on, extend selection to last cell in row	N/A
TAB	D	Move to the previous option	Move to the previous option
TAB	W	Enter data and move left one cell	Enter data and move left one cell

CONTROL+SHIFT+(key)

arrow keys	F	Select to end of block of data in direction of arrow (inserts reference in formula)	Select to end of block of data in direction of arrow (inserts reference in formula)
arrow keys	W	Select to end of block of data in direction of arrow	Select to end of block of data in direction of arrow
right/ left arrow keys	F	Extend selection, one term at a time	Extend selection, one term at a time

(W = Worksheet, F = Formula bar, C = Chart, D = Dialog box, V = Visual Basic editor)

right/ left arrow keys	D	Select or unselect one word to the right or left in the entry in an edit box	
END	F	Extend selection from insertion point to end of formula	N/A
END	W	Extend selection to end of worksheet (bottom right corner)	N/A
ENTER/ RETURN	W	Enter an array formula in the selected range	Enter an array formula in the selected range
HOME	F	Extend selection from insertion point to beginning of formula	Extend selection from insertion point to beginning of formula
HOME	W	Extend selection to beginning of worksheet (upper left corner)	Extend selection to beginning of worksheet (upper left corner)
PAGE DOWN	W	Select current and next sheet in workbook	Select current and next sheet in workbook
PAGE UP	W	Select current and previous sheet in workbook	Select current and previous sheet in workbook
TAB	D	Switch to previous tab	
!	W	Apply Decimal number format (two decimal places, thousands separator, minus sign for negative numbers)	Apply Decimal number format (two decimal places, thousands separator, minus sign for negative numbers)
#	W	Apply Date number format (d-mmm-yy)	Apply Date number format (d-mmm-yy)
$	W	Apply Currency number format (two decimal places, negative numbers in parentheses)	Apply Currency number format (two decimal places, negative numbers in parentheses)
%	W	Apply Percent number format (no decimal places)	Apply Percent number format (no decimal places)
&	W	Apply Date number format (d-mmm-yy)	Apply Date number format (d-mmm-yy)

(W = Worksheet, F = Formula bar, C = Chart, D = Dialog box, V = Visual Basic editor)

(	W	**Unhide** rows	**Unhide** rows
)	W	**Unhide** columns	**Unhide** columns
* (asterisk)	W	Select block of data containing the active cell	Select block of data containing the active cell
+	W	Display the **Insert** dialog box	Display the **Insert** dialog box
@	W	Apply Time number format (hours and minutes, AM and PM)	Apply Time number format (hours and minutes, AM and PM)
^	W	Apply Scientific number format (two decimal places)	Apply Scientific number format (two decimal places)
{	W	Select cells that are direct or indirect precedents of cells in selection	Select cells that are direct or indirect precedents of cells in selection
}	W	Select cells that are direct or indirect dependents of cells in selection	Select cells that are direct or indirect dependents of cells in selection
~ (tilde)	W	Apply General number format	Apply General number format
(space)	W	Select entire worksheet	Select entire worksheet
A	F	Insert placeholder arguments and closing parenthesis after typing a function name in a formula	Insert placeholder arguments and closing parenthesis after typing a function name in a formula
F	W	Activate the **Font** list box	Display the **Format Cells** dialog box
J	V	Display list of constants	
O	W	Select all cells with comments	
P	W	Activate the **Font Size** list box	Display the **Format Cells** dialog box

(W = Worksheet, F = Formula bar, C = Chart, D = Dialog box, V = Visual Basic editor)

COMMAND+(key) [Macintosh only]

arrow keys	W	N/A	Jump to end of block of data in direction of arrow
ENTER/ RETURN	W	N/A	Enter the contents of the formula bar in the selected range
-(hyphen)	W	N/A	Insert current date
.(period)	W	N/A	Halt execution of a macro
/		N/A	**Help**
;(semicolon)	W	N/A	Insert current time
=	F	N/A	Calculate and display result of selection
=	W	N/A	Recalculate all open workbooks
?	W	N/A	Display **Help**
~(tilde)	W	N/A	Toggle between values and formulas
1	W	N/A	Display **Format Cells** dialog box
6		N/A	**Find Next**
7	W	N/A	Toggle display of Standard toolbar
8	W	N/A	Toggle display of outline symbols
A	W	N/A	Select entire worksheet
B	W	N/A	**Clear Contents**
C	W,F	N/A	**Copy**
D	W	N/A	**Fill Down**
E, G, J, M	W	N/A	Not used
F	W	N/A	Display **Find...** dialog box
G	V	N/A	Display Immediate window
H	W	N/A	Display **Replace...** dialog box

(W = Worksheet, F = Formula bar, C = Chart, D = Dialog box, V = Visual Basic editor)

I	W	N/A	Display **Insert Cells...** dialog box
K	W	N/A	Display **Delete...** Rows/Columns dialog box
L	W	N/A	Display **Define Name** dialog box
N	W	N/A	Insert **New** workbook
O	W	N/A	Display **Open...** dialog box
P	W	N/A	Display **Print...** dialog box
Q	W	N/A	**Quit**
R	V	N/A	Display the Project Explorer window
R	W	N/A	**Fill Right**
S	W	N/A	**Save**
T	F	N/A	Toggle between relative and absolute reference
U	W	N/A	Begin Edit mode
V	W,F	N/A	**Paste**
W	W	N/A	**Close**
X	W,F	N/A	**Cut**
Y	V	N/A	Delete line in a macro
Y	W	N/A	**Repeat**
Z	W	N/A	**Undo**

COMMAND+SHIFT+(key) [Macintosh only]

3	W	N/A	Create a picture file of the entire screen and save to hard drive as PictureN
4	W	N/A	Create a picture file of a selected region of the screen and save to hard drive as PictureN

(W = Worksheet, F = Formula bar, C = Chart, D = Dialog box, V = Visual Basic editor)

A	W	N/A	Move clockwise to the next corner of the selection
B	W,F	N/A	Apply or remove Bold formatting
C	W	N/A	Display **Copy Picture** dialog box.
D	W,F	N/A	Apply or remove **Outline** formatting
G	W	N/A	**Find Next**
I	W,F	N/A	Apply or remove Italic formatting
J	W	N/A	Ungroup
K	W	N/A	Group
L	W	N/A	Display the **Style** dialog box.
N	W	N/A	Insert a **Cell Comment**
O	W	N/A	Select all cells with comments
P	W,F	N/A	Remove all formatting
S	W	N/A	**Save**
T	W	N/A	Insert the **SUM** function in a cell
U	W,F	N/A	Apply or remove Underline formatting
W	W,F	N/A	Apply or remove Shadow formatting
Z	W	N/A	Select only visible cells in current selection

(W = Worksheet, F = Formula bar, C = Chart, D = Dialog box, V = Visual Basic editor)

ALT+(key) [PC only]

down arrow	D	Open the selected drop-down list box	N/A
ENTER	F	Insert a line break	N/A
BACKSPACE	W	**Undo**	N/A
'(apostrophe)	W	Display the **Style** dialog box.	N/A
-(minus)	W	Display the Workbook Icon menu to Maximize/ Minimize/Restore a window	N/A
;(semicolon)	W	Select only visible cells in the current selection	N/A
(letter)	D	Select the menu or dialog box option indicated by the underlined letter	N/A

COMMAND+OPTION+(key) [Macintosh only]

arrow keys	W	N/A	Apply border

CONTROL+ALT+(key) [PC only]

right/ left arrow keys	W	Move between non-adjacent selections	N/A

Function Keys

F1	W	**Help**	**Undo**
F2	W	Begin Edit. Position the insertion point at the end of the line.	**Cut**
F3	W	Display the **Paste Name** dialog box.	**Copy**
F4	W	**Repeat**	**Paste**

(W = Worksheet, F = Formula bar, C = Chart, D = Dialog box, V = Visual Basic editor)

F4	F	Toggle Absolute/Relative references	
F5	W	Display the **Go To** dialog box	Display the **Go To** dialog box
F6	W	Move to the next pane in a split workbook	Move to the next pane in a split workbook
F7	W	Display the **Spell Check** dialog box	Display the **Spell Check** dialog box
F8	W	Activate/deactivate Extend mode (then use arrow keys)	Activate/deactivate Extend mode (then use arrow keys)
F9	W	Recalculate all sheets in all open workbooks	
F9	F	Calculate and display result of selection	Calculate and display result of selection
F10	W	Activate menu bar	Activate menu bar
F11	W	Create a new chart that uses the current selection	Create a new chart that uses the current selection
F12	W	**Save** the active workbook	**Save** the active workbook

SHIFT+(function key)

F1	W	**Help**	**Help**
F1	V	Get Help on a selected VBA keyword	**Undo**
F2	W	Insert a **Cell Comment**	Insert a **Cell Comment**
F2	V	Display the Object Browser window	
F3	W	Display the **Paste Function** dialog box	Display the **Paste Function** dialog box
F4	W	Display the **Find Next** dialog box	Display the **Find Next** dialog box
F4	V	Display the Properties window	
F5	W	Display the **Find** dialog box	Display the **Find** dialog box

(W = Worksheet, F = Formula bar, C = Chart, D = Dialog box, V = Visual Basic editor)

F5	V	Run a macro	Run a macro
F6	W	Move to the previous pane in a workbook that has been split	Move to the previous pane ir a workbook that has been split
F8	W	Toggle Modify Selection mode	Toggle Modify Selection mode
F8	V	Step through macro	Step through macro
F9	W	Recalculate the active worksheet	Recalculate the active worksheet
F9	V	Toggle breakpoint	Toggle breakpoint
F10	W	Display a shortcut menu	Display a shortcut menu
F11	W	Insert a new worksheet	Insert a new worksheet
F12	W	Display the **Save** As dialog box	Display the **Save** As dialog box

CONTROL+(function key)

F3	W	Display the **Define Name** dialog box	Display the **Define Name** dialog box
F4	W	Close the active window	Close the active window
F5	W	Restore window size	Restore window size
F6	W	Move to the next workbook active window	Move to the next workbook active window
F7	W	Move the window (use arrow keys)	Move the window (use arrow keys)
F8	V	Run to cursor	
F8	W	Re-size the window (use arrow keys)	Re-size the window (use arrow keys)
F9	V	Select statement to execute next	
F9	W	Minimize the window	
F10	W	Maximize the window	Toggle Window Maximize
F11	W	Insert Excel 4 macro Sheet	Insert Excel 4 macro Sheet
F12	W	Display **Open** dialog box	Display **Open** dialog box

(W = Worksheet, F = Formula bar, C = Chart, D = Dialog box, V = Visual Basic editor)

ALT+(function key) [PC only]

F1	W	Create a new chart that uses the current selection	N/A
F2	W	**Save** the active workbook	N/A
F4	W	Close	N/A
F6	W	Switch to next window	N/A
F8	W	Display **Macro** dialog box	N/A
F11	W	Toggle between **Visual Basic Editor** and **Excel**	N/A

CONTROL+ALT+(function key) [PC only]

F9	W	Recalculate all sheets in the active workbook	N/A

OPTION+(function key) [Macintosh only]

F8	W	N/A	Display **Macro** dialog box.
F10	W	N/A	Make toolbar active (then use arrow keys)
F11	W	N/A	Toggle between Visual Basic Editor and Excel

CONTROL+SHIFT+(function key)

F3	W	Display the **Create Names** dialog box	Display the **Create Names** dialog box
F4	W	Display the **Find Next** dialog box	
F6	W		Move to the previous workbook active window
F12	W	Display **Print...** dialog box	Display **Print...** dialog box

ALT+SHIFT+(function key) [PC only]

F1	W	Insert a **New** worksheet	N/A
F2	W	Display the **Save As** dialog box	N/A

(W = Worksheet, F = Formula bar, C = Chart, D = Dialog box, V = Visual Basic editor)

SELECTED SHORTCUT KEYS
FOR PC AND MACINTOSH,
BY CATEGORY

	PC (Excel 97, Excel 2000)	**Macintosh** (Excel 98, Excel 2001)
Cut, Copy, Paste, Undo		
Undo	CONTROL+Z or ALT+BACKSPACE	CONTROL+Z or COMMAND+Z or F1
Cut	CONTROL+X	CONTROL+X or COMMAND+X or F2
Copy	CONTROL+C	CONTROL+C or COMMAND+C or F3
Paste	CONTROL+V	CONTROL+V or COMMAND+V or F4
Clear contents of selection (formula bar or worksheet)	DELETE	COMMAND+B
Repeat	CONTROL+Y or F4	CONTROL+Y or COMMAND+Y
Moving and Selecting		
Jump to end of block of data in direction of arrow	CONTROL+arrow key	CONTROL+arrow key or COMMAND+arrow key
Select to end of block of data in direction of arrow (inserts reference in formula)	CONTROL+SHIFT+ arrow key	CONTROL+SHIFT+ arrow key
Select entire row	SHIFT+(space)	SHIFT+(space)

Select entire column	CONTROL+(space)	CONTROL+(space)
Select all cells of the array containing the active cell	CONTROL+/(slash)	CONTROL+/(slash)
Extend selection by one row/column	SHIFT+arrow key	SHIFT+arrow key
Extend selection in a formula, one term at a time	CONTROL+SHIFT+right or left arrow key	CONTROL+SHIFT+right or left arrow key
Hide column(s) containing selection	CONTROL+0	CONTROL+0
Hide row(s) containing selection	CONTROL+9	CONTROL+9
Unhide rows	CONTROL+SHIFT+(	CONTROL+SHIFT+(
Unhide columns	CONTROL+SHIFT+)	CONTROL+SHIFT+)
Select block of data containing the active cell	CONTROL+SHIFT+* (asterisk) or CONTROL+* (numeric keypad)	CONTROL+SHIFT+* (asterisk)
Select only visible cells in the current selection	ALT+;(semicolon)	COMMAND+SHIFT+Z
Select all cells with comments		COMMAND+SHIFT+O
Select entire worksheet	CONTROL+A	
Move between non-adjacent selections	CONTROL+ALT+ right or left arrow key	CONTROL+OPTION+ right or left arrow key

Entering or Editing a Formula

Begin Edit. Position the insertion point at the end of the line.	F2	
Enter the contents of the formula bar in the selected range (references are adjusted)	CONTROL+ENTER	CONTROL+ RETURN or COMMAND+ RETURN
Insert current time	CONTROL+:(colon)	CONTROL+:(colon) or COMMAND+;(semicolon)

Insert current date	CONTROL+;(semicolon)	CONTROL+;(semicolon) or COMMAND+- (hyphen)
Insert a line break	ALT+ENTER	COMMAND+OPTION+ RETURN
Toggle between values and formulas	CONTROL+~ (tilde)	CONTROL+~ or COMMAND+~ (tilde)
Display the Formula Palette (the Insert Function Step 2 dialog box) after typing a function name in a formula	CONTROL+A	CONTROL+A
Insert placeholder arguments and closing parenthesis after typing a function name in a formula	CONTROL+SHIFT+A	CONTROL+SHIFT+A
Enter data and move down one cell	ENTER	RETURN
Enter data and move up one cell	SHIFT+ENTER	SHIFT+ RETURN
Enter data and move right one cell	TAB	TAB
Enter data and move left one cell	SHIFT+TAB	SHIFT+TAB
Enter an array formula in the selected range	CONTROL+SHIFT+ ENTER	CONTROL+SHIFT+ RETURN
Toggle between relative and absolute reference	F4	COMMAND+T
Calculate and display result of selection	F9	F9 or COMMAND+=
Fill Down	CONTROL+D	CONTROL+D or COMMAND+D
Fill Right	CONTROL+R	CONTROL+R or COMMAND+R

Formatting

Apply or remove Bold formatting	CONTROL+B	CONTROL+B or COMMAND+SHIFT+B

Apply or remove Italic formatting	CONTROL+I	CONTROL+I or COMMAND+SHIFT+I
Apply or remove Underline formatting	CONTROL+U	CONTROL+U or COMMAND+SHIFT+U
Apply General number format	CONTROL+SHIFT+~ (tilde)	CONTROL+SHIFT+~ (tilde)
Apply Decimal number format (two decimal places, thousands separator, minus sign for negative numbers)	CONTROL+SHIFT+!	CONTROL+SHIFT+!
Apply Currency number format (two decimal places, negative numbers in parentheses)	CONTROL+SHIFT+$	CONTROL+SHIFT+$
Apply Percent number format (no decimal places)	CONTROL+SHIFT+%	CONTROL+SHIFT+%
Apply Scientific number format (two decimal places)	CONTROL+SHIFT+^	CONTROL+SHIFT+^
Remove all formatting		COMMAND+SHIFT+P
Apply Date number format (d-mmm-yy)	CONTROL+SHIFT+# or CONTROL+SHIFT+&	CONTROL+SHIFT+# or CONTROL+SHIFT+&
Apply Time number format (hours and minutes, AM and PM)	CONTROL+SHIFT+@	CONTROL+SHIFT+@
Apply border		COMMAND+OPTION+ arrow key

Menu Commands

New (Open a new workbook)	CONTROL+N	CONTROL+N or COMMAND+N
Display the **Open...** dialog box	CONTROL+O or CONTROL+F12	CONTROL+O or COMMAND+O or CONTROL+F12
Close	ALT+F4	COMMAND+W
Save the active workbook	CONTROL+S or F12 or ALT+F2	CONTROL+S or COMMAND+S or F12
Display the **Save** As dialog	SHIFT+F12 or	SHIFT+F12

box	ALT+SHIFT+F2	
Display the **Print...** dialog box	CONTROL+P or CONTROL+SHIFT+F12	CONTROL+P or COMMAND+P or CONTROL+SHIFT+F12
Exit/Quit		COMMAND+Q
Display the **Delete...** Rows/Columns dialog box	CONTROL+-(minus)	COMMAND+K or CONTROL+-(minus)
Display **Insert Cells...** dialog box	CONTROL+SHIFT+ +(plus)	COMMAND+I or CONTROL+SHIFT+ +(plus)
Display the **Find...** dialog box	CONTROL+F or SHIFT+F5	CONTROL+F or COMMAND+F or SHIFT+F5
Display the **Find Next** dialog box	SHIFT+F4	SHIFT+F4
Find Next		COMMAND+6 or COMMAND+SHIFT+G
Display the **Replace...** dialog box	CONTROL+H	CONTROL+H or COMMAND+H
Display the **Go To...** dialog box	CONTROL+G or F5	CONTROL+G or F5
Insert a **Cell Comment**	SHIFT+F2	SHIFT+F2 or COMMAND+SHIFT+N
Display the **Paste Function** dialog box	SHIFT+F3	SHIFT+F3
Display the **Define Name** dialog box	CONTROL+F3	CONTROL+F3 or COMMAND+L
Display the **Create Names** dialog box	CONTROL+SHIFT+F3	CONTROL+SHIFT+F3
Display the **Paste Name** dialog box.	F3	
Display **Format Cells** dialog box	CONTROL+1	COMMAND+1
Display the **Macro** dialog box.	ALT+F8	OPTION+F8
Display the **Spell Check** dialog box	F7	F7

Help	F1 or SHIFT+F1	COMMAND+/

Workbooks and Worksheets

Move to the previous workbook active window		CONTROL+SHIFT+F6
Select next sheet in workbook	CONTROL+PAGE DOWN	CONTROL+PAGE DOWN
Select previous sheet in workbook	CONTROL+PAGE UP	CONTROL+PAGE UP
Switch to next tab	CONTROL+PAGE DOWN	CONTROL+PAGE DOWN
Switch to previous tab	CONTROL+PAGE UP	CONTROL+PAGE UP
Create a new chart that uses the current selection	F11 or ALT+F1	F11
Recalculate the active worksheet	SHIFT+F9	SHIFT+F9
Recalculate all sheets in the active workbook	CONTROL+ALT+F9	
Insert a new worksheet	SHIFT+F11 or ALT+SHIFT+F1	SHIFT+F11
Toggle Window Maximize	CONTROL+F10	CONTROL+F10

Visual Basic

Halt execution of a macro	ESC	ESC or COMMAND+.(period)
Run a macro	F5	F5
Step through a macro	F8	F8
Toggle breakpoint	F9	F9
Toggle between Visual Basic Editor and Excel	ALT+F11	OPTION+F11

G

ABOUT THE CD-ROM
THAT ACCOMPANIES THIS BOOK

This appendix describes the worksheets and macros that are on the CD-ROM that accompanies this book. The workbooks are in Excel 97 format. They can be read by users with Excel 97/98 or 2000/2001. You should have no trouble using this CD with either a PC or Macintosh. If you have problems, please contact John Wiley's tech support system at (212) 850-6753.

The files on the CD-ROM are contained in the Excel for Chemists folder and are read-only. To work with a document and save the changes, you must first copy the files to your hard drive. If you are using a PC, you can run the INSTALL.EXE file on the CD an unzip the files to your hard drive. If you are a Macintosh user, copy the Excel for Chemists folder to your system.

The filenames assigned to the documents are names compatible with Excel for Windows; that is, they end with an .xls extension. If you are using a Macintosh, you can rename the documents if you wish.

Chapter 3   *Creating Advanced Worksheet Formulas*

IFdemo.xls illustrates the use of the IF function to prevent the display of error values.

MegaForm.xls illustrates the use of "megaformulas" (in this case involving text functions). Unhide columns C through F to view the separate parts of the megaformula.

Chapter 4   *Creating Array Formulas*

ArrayDem.xls illustrates a non-array-formula approach and three different array formulas for the calculation of the sum of squares of deviations.

PowerSeries.xls shows how to create and use a series of integers in array formulas.

Chapter 5   *Advanced Charting Techniques*

Methane Hydrate.xls illustrates techniques for formatting a chart and producing a smooth curve through data points.

SinCos.xls is an example of a 3-D chart. It also illustrates the use of mixed references to obtain values from a two-way table.

Creative.xls is an example of using simple plotting and formatting techniques to produce a non-standard chart type.

Chapter 6   *Using Excel's Database Features*

Database.xls is a sample database to illustrate Excel's database capabilities.

Chapter 7   *Importing Data into Excel*

NISPEC.DAT is a comma-delimited text file to be used with Excel's Data Parse or Text to Columns menu commands.

Nth.xls illustrates four different worksheet formulas to select every *N*th data point from a data table.

Chapter 8   *Adding Controls to a Spreadsheet*

DropDown.xls illustrates two drop-down list boxes that display selected names, addresses and telephone numbers from a database.

Chapter 9   *Some Mathematical Tools For Spreadsheet Calculations*

Lookup1.xls illustrates how to look up values in a one-way table.

Lookup2.xls illustrates how to look up values in a two-way table. The data table used in this example is part of the Steam Tables and is reprinted from *ASME International Steam Tables for Industrial Use*. The information in the table was provided by the National Institute for Standards and Technology (NIST) and is in the public domain.

Interpl8.xls illustrates how to perform linear interpolation.

CubicInterpl8.xls illustrates the use of the custom function CubicInterp to perform cubic interpolation.

Derivs.xls illustrates how to obtain the first and second derivatives of a data set.

NumDiff.xls illustrates how to calculate the first derivative of a function.

CurvArea.xls illustrates three worksheet formulas that can be used to obtain the area under a curve.

RK4.xls illustrates the Euler and fourth-order Runge-Kutta methods for the solution of differential equations.

Matrix.xls illustrates the tools available for matrix mathematics, with examples.

Polar.xls illustrates how to convert from polar to Cartesian coordinates.

Chapter 10   Graphical and Numerical Methods of Analysis

Goal Seek.xls illustrates the use of the Newton-Raphson method to find the roots of a polynomial. It also shows, in hidden rows 8-25, how to find a root by successive approximations.

NewtRaph.xls illustrates the use of the Newton-Raphson method to find the roots of a polynomial.

Circular.xls shows how to use an intentional circular reference with the Newton-Raphson method to find the roots of a polynomial. A "Circular Reference" error message will be displayed upon opening the workbook.

SimultEqns.xls illustrates ways to solve sets of simultaneous linear equations by using matrices.

Chapter 11   Linear Regression

CalCurv.xls is a example of linear regression to find the slope and intercept of a linear calibration curve.

Oxygen.xls illustrates the use of the LINEST function to perform multiple linear regression.

Chapter 12   Non-Linear Regression Using the Solver

NonLin.xls illustrates the use of the Solver to perform multiple non-linear regression analysis.

SolvStat.xls is a command macro that returns the standard deviations for non-linear regression analysis performed by the Solver. See "Instructions for Using SolvStat " at the end of this appendix.

Chapter 14   Programming with VBA

MsgBox.xls provides some examples of the built-in dialog box to display a message. **MsgBox** returns a value indicating which button the user pressed.

InputBox.xls provides some examples of the built-in dialog box for user input.

Chapter 15   Working with Arrays in VBA

ArrayDemos.xls contains 14 **Sub** or **Function** procedures that illustrate various features of using arrays in VBA.

Chapter 16   Creating Command Macros

ChemFormat3.xls is a **Sub** procedure that applies chemical formatting to text in a cell, in a range of cells, in a chart or in a textbox.

Labeler1.xls is the **Sub** procedure shown in Chapter 16 that adds user-specified data labels to a chart.

Labeler2.xls is a more advanced version of the Data Labeler macro. It uses a custom dialog box (not discussed in the text). See "Instructions for Using Labeler2" at the end of this appendix.

Chapter 17   Creating Custom Functions

Deming.xls is a **Function** procedure that returns the slope and intercept calculated by the method of Deming.

MolWt.xls is a **Function** procedure that returns the formula weight from text that can be interpreted as a chemical formula. Only a part of the procedure is reproduced in Chapter 17. This **Function** procedure illustrates the use of the **Optional** keyword. See "Instructions for Using MolWt" at the end of this appendix.

MolWt.xla is an **Add-In** version of the MolWt macro. Use the Add-In and keep the MolWt.xls workbook as a backup if you want to make changes at a later date.

Instructions for MolWt is a Word document that gives more details about using the custom function.

MakeArray.xls is a **Function** procedure that combines individual worksheet ranges into an array. This **Function** procedure illustrates the use of the **ParamArray** keyword.

Chapter 18   Creating Custom Menus and Menu Bars

MenuDemo1.xls is an Auto_Open **Sub** procedure that installs a new menu command in the **Tools** menu.

MenuDemo2.xls is a **Workbook_Open** event procedure that installs a new menu command in the **Tools** menu.

MenuDemo3.xls (not discussed in the text) is a **Workbook_Open** event procedure that checks to see whether the new command has already been installed before installing the menu command in the **Tools** menu.

Chapter 19   Creating Custom Toolbuttons and Toolbars

NumFmt.xls is a simple command macro that toggles between floating-point and scientific number formats. The macro can be easily assigned to a toolbutton.

FullPage.xls is a simple command macro that can be used to obtain the maximum amount of space on a page for printing a worksheet. It sets either portrait or landscape orientation, sets margins to zero and removes header and footer text. The macro can be easily assigned to a toolbutton.

Chapter 20   Analysis of Solution Equilibria

Alpha.xls is a custom function that returns α values for a polyprotic acid species.

Gran.xls illustrates the use of Gran's method to find the equivalence point of a titration.

Chapter 21   Analysis of Spectrophotometric Data

FlameCal.xls illustrates two methods for treating a curved spectrophotometric calibration curve.

3CmpSpec.xls illustrates methods for the analysis of the spectrum of a mixture of Cu^{2+}, Co^{2+} and Ni^{2+} ions.

Deconv.xls illustrates the deconvolution of a UV-visible spectrum into its component Gaussian bands.

Chapter 22   Calculation of Binding Constants

Titrat.xls is an example of the calculation of the pK_a values of a polyprotic acid from titration data, using the Solver.

NMRdata.xls is an example of the calculation of a binding constant from NMR data, using the Solver.

Chapter 23   Analysis of Kinetics Data

XII-21E.xls illustrates the use of the Solver to obtain a first-order rate constant when the final reading is unavailable.

XIII-28A.xls illustrates the use of the fourth-order Runge-Kutta method to obtain four coupled rate constants from a complex kinetic process.

Instructions for Using SolvStat

This command macro returns the standard deviations of regression coefficients obtained by using the **Solver**, plus the correlation coefficient and the RMSD; these statistical parameters are not available from the **Solver**. The array of values returned is in a format similar to that returned by LINEST.

The sheet must contain Y_{calc} values. The Y_{obsd} and Y_{calc} values must each be in either a single column or row. The regression coefficients returned by the Solver do not have to be in adjacent cells.

To use the macro, simply **Open** SolvStat.xls; it will appear on screen and then **Hide** itself. It installs a new menu command, **Solver Statistics...**, immediately under the **Solver...** command in the **Tools** menu. If the Solver Add-in has not been loaded, the **Solver Statistics...** command will be at the top of the menu. The command will remain in the menu until you exit from Excel.

Activate the document in which the **Solver** has already been used to obtain regression coefficients by minimizing the sum of squares of deviations between observed and calculated Y values. Choose **Solver Statistics...** from the menu. Follow the directions in the dialog boxes.

Instructions for Using MolWt

To install the function, simply open the Excel document MolWt. The function is available even though the macro sheet is hidden. You can examine the macro sheet by choosing **Unhide** from the **Window** menu.

Syntax: MolWt(**formula,** decimals). Formula will usually be a reference to a cell containing a text string that can be interpreted as a chemical formula, e.g., H2SO4. Decimals, an optional argument, is the number of decimal places to be displayed in the returned value.

Instructions for Using Labeler2

To use the macro, simply **Open** Labeler2.xls; it will appear on screen and then **Hide** itself. It installs a new menu command, "**Add Data Labels...**" in the **Chart** menu. To view the workbook, containing the macro and two sample charts, **Unhide** the workbook.

To add data labels to a data series, click on the chart (you can pre-select the data series or choose it later in the dialog box). Choose "**Add Data Labels...**" in the **Chart** menu. This will display the Apply Data Labels dialog box. Choose the desired options and the range of cells on the worksheet that contain the data labels, then press OK.

The custom dialog box is designed for Line plots or XY Scatter plots.

INDEX